T0140047

Inflammatory Mechanisms in Mediating Hearing Loss

Vickram Ramkumar • Leonard P. Rybak

Editors

Inflammatory Mechanisms in Mediating Hearing Loss

 Springer

Editors
Vickram Ramkumar
Department of Pharmacology
School of Medicine
Southern Illinois University
Springfield, IL, USA

Leonard P. Rybak
Division of Otolaryngology
Department of Surgery
School of Medicine
Southern Illinois University
Springfield, IL, USA

ISBN 978-3-030-06447-1   ISBN 978-3-319-92507-3   (eBook)
https://doi.org/10.1007/978-3-319-92507-3

This Springer imprint is published by the registered company Springer International Publishing AG part of Springer Nature.
The registered company address is: Gewerbestrasse 11, 6330 Cham, Switzerland

Preface

Common forms of preventable hearing loss are drug- and noise-induced hearing loss which are believed to be produced by similar mechanisms. The generation of reactive oxygen species appears to be a common mechanism mediating hearing loss produced by these different sources. As such, a number of laboratories have focused their research toward identifying the sources of ROS production in the cochlea following administration of chemotherapeutic agents or noise exposure. This led to the identification of ROS-generating enzymes, such as xanthine oxidases, nitric oxide synthase, and NADPH oxidases which are activated and/or induced during the development of hearing loss. A consequence of these findings was the implementation of antioxidants in preclinical studies for the treatment of hearing loss. These antioxidants have provided different levels of protection in animal and human studies, but none of these have been approved by the US Food and Drug Administration for the treatment of hearing loss.

More recently, it was shown that noise-induced hearing loss was associated with recruitment of inflammatory cells and mediators in the cochlea. This finding would suggest that noise could produce injury to the cochlea which stimulates local and/or circulating inflammatory cells through the release of "stress" signals. A similar finding was observed in the cochlea following administration of the anticancer drug, cisplatin, and aminoglycoside antibiotics. In addition, our laboratory and others have provided a plausible mechanism by which noise or chemotherapeutic agents could stimulate the inflammatory response. Surprisingly, this mechanism involves ROS activation of transcription factors linked to inflammatory and apoptotic processes in the cochlea. These studies have led to the use of anti-inflammatory agents for the treatment of hearing loss. Preliminary studies targeting inflammatory cytokines appear especially promising in preclinical studies.

A primary goal of this book is to describe our current understanding of the oxidant hypothesis of noise and drug-induced hearing loss and show how this relates to cochlear inflammation. Another focus of this book is to detail several different aspects of the cochlear inflammatory process, ranging from the sources of inflammatory cells to chemokines, inflammatory cytokines, and cochlea resident immune cells. Molecular pathways leading to activation of the local inflammatory process, migration of immune cells from the systemic circulation via strial capillaries into the cochlea, and a discussion of a cochlear-based corticotrophin/corticosteroid generating system will be highlighted. The relevance of certain clinically used anti-inflammatory interventions, such as trans-tympanic steroids and other treatment options, will also be discussed. Furthermore, recent clinical trials focusing on the use of anti-inflammatory agents for the treatment of drug- and noise-induced and autoimmune-mediated hearing loss will be discussed.

We believe that the ideas highlighted in the following chapters will provide novel insight into the impact of inflammatory mechanisms on hearing loss and stimulate future studies into this exciting area of research.

Springfield, IL, USA   Vickram Ramkumar
                                                                                    Leonard P. Rybak

Contents

Contributors

Esperanza Bas Department of Otolaryngology, University of Miami Ear Institute, University of Miami Miller School of Medicine, Miami, FL, USA

Christine T. Dinh Department of Otolaryngology, University of Miami Ear Institute, University of Miami Miller School of Medicine, Miami, FL, USA

Stefania Goncalves Department of Otolaryngology, University of Miami Ear Institute, University of Miami Miller School of Medicine, Miami, FL, USA

Samson Jamesdaniel Institute of Environmental Health Sciences, Wayne State University, Detroit, MI, USA

Elizabeth M. Keithley Division of Otolaryngology/Head and Neck Surgery, University of California, San Diego, La Jolla, CA, USA

Arwa Kurabi Division of Otolaryngology, Department of Surgery, University of California San Diego, La Jolla, CA, USA

Veterans Administration San Diego Healthcare System, San Diego, CA, USA

Martin L. Lesser Biostatistics Unit, Feinstein Institute for Medical Research, Manhasset, NY, USA

Department of Molecular Medicine and Population Health, Barbara and Donald Zucker School of Medicine at Hofstra/Northwell, Hempstead, NY, USA

Debashree Mukherjea Department of Surgery (Otolaryngology), Southern Illinois University School of Medicine, Springfield, IL, USA

Enrique Perez Department of Otolaryngology, University of Miami Ear Institute, University of Miami Miller School of Medicine, Miami, FL, USA

Vickram Ramkumar Department of Pharmacology, Southern Illinois University School of Medicine, Springfield, IL, USA

Michael J. Ruckenstein Department of Otorhinolaryngology-Head and Neck Surgery, University of Pennsylvania, Philadelphia, PA, USA

Allen F. Ryan Division of Otolaryngology, Department of Surgery, University of California San Diego, La Jolla, CA, USA

Department of Neurosciences, University of California San Diego, La Jolla, CA, USA

Veterans Administration San Diego Healthcare System, San Diego, CA, USA

Leonard P. Rybak Division of Otolaryngology, Department of Surgery, Southern Illinois University, School of Medicine, Springfield, IL, USA

Daniel Schaerer Division of Otolaryngology, Department of Surgery, University of California San Diego, La Jolla, CA, USA

Sandeep Sheth Department of Pharmacology, Southern Illinois University School of Medicine, Springfield, IL, USA

Xiaorui Shi Oregon Health and Science University, Portland, OR, USA

Peter S. Steyger Oregon Hearing Research Center, Oregon Health and Science University, Oregon, USA

Andrea Vambutas Department of Otolaryngology and Molecular Medicine, Barbara and Donald Zucker School of Medicine at Hofstra/Northwell, Hempstead, NY, USA

Thomas R. Van De Water Department of Otolaryngology, University of Miami Ear Institute, University of Miami Miller School of Medicine, Miami, FL, USA

Douglas E. Vetter Department of Neurobiology and Anatomical Sciences, University of Mississippi Medical Center, Jackson, MS, USA

Department of Otolaryngology and Communicative Sciences, University of Mississippi Medical Center, Jackson, MS, USA

Alanna M. Windsor Department of Otorhinolaryngology-Head and Neck Surgery, University of Pennsylvania, Philadelphia, PA, USA

Kathleen T. Yee Department of Neurobiology and Anatomical Sciences, University of Mississippi Medical Center, Jackson, MS, USA

Chapter 1
The Cochlea

Leonard P. Rybak

Abstract The mammalian cochlea is an intricately designed organ that is exquisitely sensitive to sound. It possesses unique physical and chemical properties that permit this organ to function properly. This chapter describes some of the features of the cochlea including the cells that line the fluid filled spaces of the cochlear duct and the chemical composition of the fluids that allow the tissues to produce resting and acting potentials that assist in the transduction of acoustic stimuli into electrical signals to the brain. The structure of key structures in the cochlea are illustrated with light microscopy and ultrastructural images, including transmission and scanning electron microscopy. The unusual structural and functional features of these cells allow them to function in an orderly and precise fashion to shape the special sensory function of hearing in the normal cochlea of mammals.

Keywords Cochlea · Stria vascularis · Hair cells · Spiral ligament · Spiral ganglion neurons · Organ of corti · Transduction · Tectorial membrane · Basilar membrane · Perilymph · Endolymph

The sense of hearing is controlled by the end organ called the cochlea. This structure is located in the temporal bone. It has a bony shell with a spiral shape like that of a snail. It coils around a central core called the modiolus that houses the auditory nerve. The ganglion cells are the cell bodies of the afferent auditory neurons. These cells are located in a spiral canal, Rosenthal's canal, at the periphery of the modiolus (Bohne and Harding 2008). The spiral ganglion contains two types of nerve cells—Type I and Type II. Type I ganglion cells are larger in size and more numerous than Type II neurons. They constitute around 90% of the spiral ganglion cells. They innervate only inner hair cells. Type II cells are much smaller in size than their Type I counterparts. They have thin myelination and make up only about 10% of the spiral ganglion cells. These cells innervate only outer hair cells. Efferent neurons

L. P. Rybak (✉)
Division of Otolaryngology, Department of Surgery, Southern Illinois University,
School of Medicine, Springfield, IL, USA
e-mail: lrybak@siumed.edu

© Springer International Publishing AG, part of Springer Nature 2018
V. Ramkumar, L. P. Rybak (eds.), *Inflammatory Mechanisms in Mediating Hearing Loss*, https://doi.org/10.1007/978-3-319-92507-3_1

originate in the superior olivary complex bilaterally. The number of efferent neurons is much smaller than the population of afferents. Efferent neurons supply both inner and outer hair cells (Bohne and Harding 2008).

1   Fluid Spaces in the Cochlea

If the cochlea is sectioned through its mid-modiolar plane and stained for microscopic analysis, it demonstrates three fluid filled spaces: the scala tympani, scala vestibuli and the scala media (Fig. 1.1). These fluid filled spaces have specific ion concentrations. The perilymph, which fills the scala vestibuli and scala tympani, is similar to other extracellular fluids, like the cerebrospinal fluid. These fluids contain high sodium (150 mM) and low potassium (5 mM) concentration (Sterkers et al. 1988). The scala media is located between scala vestibuli and scala tympani. The scala media contains a unique extracellular fluid called endolymph. This fluid contains high potassium (150 mM) and low sodium (2–5 mM) concentrations (Nin et al. 2016) (Table 1.1). This unusual fluid is contained within the cochlear duct. Endolymph is similar in composition to that of intracellular fluid. The scala media

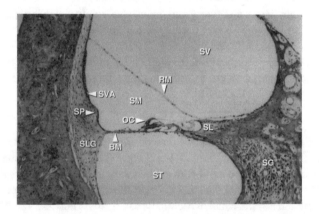

Fig. 1.1 Mid-modiolar cross section of the of the middle turn chinchilla cochlea. Magnification = 400x. Abbreviations: *BM* basilar membrane, *OC* organ of Corti, *RM* Reissner's membrane, *SLG* spiral ligament, *SL* spiral limbus, *SM* scala media, *ST* scala tympani, *SV* scala vestibuli, *SVA* stria vascularis, *SP* spiral prominence

Table 1.1 Concentrations of various ions in perilymph and endolymph

Ion	Perilymph	Endolymph
Sodium—mM	148	1.3
Potassium—mM	4.2	157
Chloride—mM	119	132
Bicarbonate—mM	21	31
Calcium—mM	1.3	0.023

Adapted from Lang et al. (2007)

has a positive potential of +80 mV with respect to that of perilymph (Von Bekesy 1952; Hibino et al. 2010).

2 Cochlear Duct

The boundaries of the endolymphatic space or cochlear duct include: Reissner's membrane superiorly, the stria vascularis laterally, and the inferior boundary. The latter includes the superior surface of the organ of Corti, Claudius cells, inner sulcus cells and epithelial cells on the upper surface of the spiral limbus (Fig. 1.1). These epithelial cells on the superior surface of the spiral limbus are known as interdental cells. The interdental cells are connected to each other and they form a comb-shaped cellular network with numerous cellular strands in the spiral limbus (Shodo et al. 2017). The medial edge of tectorial membrane is connected to interdental cells.

3 Tectorial Membrane

The tectorial membrane is an acellular sheet that covers the organ of Corti. It is composed of collagens, (including Types II, V, IX and XI), seven glycoproteins, including alpha- and beta-tectorin, carcinoembryonic antigen-related cell adhesion molecule 16 (CEACAM16), otogelin, otogelin-like, otoancorin and otolin, and two glycosaminoglycans (uronic acid and keratin sulfate) (Andrade et al. 2016). It appears that the tectorins crosslink type II collagen fibrils to connect the tectorial membrane to the spiral limbus (Andrade et al. 2016). The tips of the tallest row of the OHC stereocilia are attached to the undersurface of the tectorial membrane and form imprints in the tectorial membrane. The protein stereocilin anchors the stereocilia to the tectorial membrane (Verpy et al. 2011). The coupling of the basilar membrane vibrations to hair cell stereocilia movements are key elements of the transduction of mechanical to electrical energy by the cochlea (Andrade et al. 2016). The protein otoancorin is present at two attachment zones of the tectorial membrane, a permanent one along the top of the spiral limbus and a transient one on the surface of the developing greater epithelial ridge (Zwaenepoel et al. 2002). The exact role of the tectorial membrane has not yet been fully defined. In the past it was considered to be a rather inert structure, but it probably has a functional role in sound transduction. It may form a rigid plate that slides back and forth with vibration of the basilar membrane. It may provide a resonator that amplifies mechanical inputs to the hair bundle. It could serve as an inertial mass or a structure for propagation of the traveling wave in the cochlea. It could mediate a role in cochlear amplification in conjunction with OHCs or a regulator of electrokinetic motion (Andrade et al. 2016), finally, it may play a role in tonotopic organization of the cochlea because it changes in size from base to apex (Raphael and Altschuler

2003). Mutations of the genes encoding these proteins in the tectorial membrane cause hearing loss.

4 Basilar Membrane

The basilar membrane is a combination of cellular and acellular components. The membrane faces the scala media superiorly and the scala tympani below it. The part that faces the perilymph of the scala tympani is composed of a layer of mesothelial cells. The side of the basilar membrane that faces scala media (endolymph) has a basement membrane of the epithelium of the membranous labyrinth. The basilar membrane contains both matrix and fibers. Matrix molecules include collagens, proteoglycans, fibronectin and tenascin (Raphael and Altschuler 2003).

5 The Lateral Wall of the Cochlea

The lateral wall of the cochlea contains the spiral ligament and stria vascularis, the spiral prominence, outer sulcus cells and Claudius cells. The Claudius cells form a sheet of epithelia on the basilar membrane and go part-way up the lateral wall to the spiral prominence in the basal turn of the cochlea (Spicer and Schulte 1996; Yoo et al. 2012). These cells appear to play a role in sodium absorption by purinergic signaling (Yoo et al. 2012). Outer sulcus cells are located between the Claudius cells and the spiral prominence. Root cells are located in the outer sulcus region. Root cells underlie Claudius cells (Jagger and Forge 2013). Root cells are coupled to adjacent root cells, and to the overlying Claudius cells in the lower turns, and so form the lateral limits of the epithelial cell gap junctional network (Jagger and Forge 2013). A recent study using 3D reconstructed images of the cochlea demonstrated that the root cells were linked together to form a branched structure, similar to a tree root in the spiral ligament (Shodo et al. 2017). Superior to the spiral prominence is the stria vascularis. The spiral ligament is situated lateral to the stria vascularis (Fig. 1.1). Claudin-11 forms a barrier around the stria vascularis in the lateral wall of the human cochlea. This barrier extends inferiorly from the suprastrial region of the lateral wall to the superior epithelium of the spiral prominence (Liu et al. 2017).

6 The Spiral Ligament

The spiral ligament extends from the stria vascularis to the bony wall of the otic capsule. It helps to anchor the delicate structures of the organ of Corti and provides tension to the basilar membrane. Stress fibers that possess contractile proteins are present within tension fibroblasts (Raphael and Altschuler 2003). Five specific subtypes of spiral ligament fibrocytes have been identified (Spicer and Schulte 1996). There are gap junctions between spiral ligament fibrocytes and between these fibrocytes and the intermediate and basal cells of the stria vascularis (Forge et al. 1999; Raphael and Altschuler 2003), thus forming an electrochemical syncytium. The fibrocytes and intermediate cells which abut basal cells form the basolateral and apical surfaces for the syncytium. This configuration allows fibrocytes to be contacted by perilymph (Yoshida et al. 2016). Cochlear fibrocytes have a positive resting membrane potential (RMP). A recent in vivo study revealed that the membranes of cochlear fibrocytes appear to have a Na+ permeability that is significantly greater than that for K+ and Cl−. This Na+ permeability probably plays a key role in maintaining the unusual RMP of +5 to +12 mV. This RMP is essential for providing the K+ diffusion potential on intermediate cell membranes to set a high value for the EP (Yoshida et al. 2016). A barrier consisting of Claudin-11 observed in human cochlear specimens may protect against ion contamination of the spiral ligament (Liu et al. 2017).

7 Stria Vascularis

The marginal cells of the stria vascularis form the inner or medial layer of the lateral wall and consist of a single layer of cells connected by tight junctions at their apical surface which faces the endolymph. Marginal cells contain numerous mitochondria and are very active metabolically. These marginal cells have basolateral extensions that intertwine with apical membranes of the intermediate cells. Claudin-11 is expressed basally from the superior epithelium of the SP to the suprastrial space insulating the K+ secreting marginal cell layer in human cochlear specimens (Liu et al. 2017).

Two types of intermediate cells have been described: basal and upper subtypes. The basal subtype of intermediate cell (BIC) completely covers strial basal cells with a leaf-like horizontal process. A second, upper subtype of IC (UIC), occurs in the middle to upper strial layers and has extensive contacts with BIC's. These cells appear to play an important role in the production of the positive endocochlear potential. Intermediate cells contain melanin granules. Abnormal intermediate cells, such as those that are present in a mouse mutant (viable dominant spotting mouse) known to have a primary neural crest defect that results in an absence of melanocytes in the skin. These mice are unable to produce an endocochlear potential (EP). EP was close to zero at all ages from 6 to 20 days. Starting at about 6 days of age cells of the stria vascularis exhibited a reduced amount of inter-digitation with other cells (Steel and Barkway 1989). Ablation of intermediate cells also abolishes the EP (Kim et al. 2013). Mice with the *Mitf*-mutation lack melanocytes in the stria. These mice have extremely low EP, especially in the basal turn. This mutation results in loss of outer hair cells (Liu et al. 2016). It appears that melanocyte-like cells in the intermediate cell area are required for normal stria vascularis development and function

(Steel and Barkway 1989). However, albino animals and humans that lack melanin can hear normally (Raphael and Altschuler 2003).

Basal Cells

Basal cells are flat in appearance. They form a continuous layer of cells abutting the spiral ligament. They produce a dense network of junctional complexes with other basal cells with other adjacent cells, including the intermediate cells of the stria vascularis and fibrocytes of the spiral ligament (Yoshida et al. 2016). Basal cells have tight junctions that join them together. Their major function may consist in maintaining a diffusion barrier between the stria vascularis and the spiral ligament (Raphael and Altschuler 2003). Claudin-11 knockout mice are deaf. This indicates that tight junctions between basal cells are critical for proper function of the stria vascularis (Gow et al. 2004).

Between the three layers of cells in the stria vascularis is a small space, called the intrastrial space. This small (15 nanometer) extracellular space is enclosed in apical membranes of intermediate cells and the basolateral membranes of marginal cells (Hibino et al. 2010). It contains a low K+ concentration of about 5 mM and a high positive potential similar to that of the scala media, the EP (Yoshida et al. 2016). Study of the human cochlea revealed that the intrastrial space is separated by a continuous layer of the tight junction protein, claudin 11 (Liu et al. 2017).

A barrier system that appears to play a critical role in maintaining cochlear homeostasis is the strial cochlear vascular unit also known as the intrastrial fluid-blood barrier. This complex system includes a microvascular endothelium in close connection with numerous accessory cells (pericytes and perivascular macrophages or melanocytes) and a specific matrix of extracellular basement membrane proteins that, together, constitute a unique "cochlear vascular unit". This barrier appears to control solute and ion homeostasis in the inner ear and blocks the entry of toxic substances into the stria vascularis. If it is compromised, various clinical hearing disorders may result (Shi 2016). The enzyme, Na+/K+-ATPase alpha-1 plays a critical role in maintaining blood-labyrinth barrier integrity. Pharmacological inhibition of this transporter results in hyperphosphorylation of tight junction proteins like occluding. This results in increased permeability of the blood-labyrinth barrier. Noise trauma can break down the blood-labyrinth barrier through reduction of Na+/K+-ATPase activity leading to increase in vascular permeability (Yang et al. 2011). The ultrastructure of the stria vascularis is shown in Fig. 1.2.

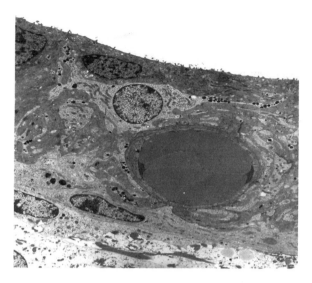

Fig. 1.2 Transmission electron micrograph of the stria vascularis from the basal turn of the adult rat cochlea. There is a single row of electron dense marginal cells facing the endolymph. A single intermediate cell with a dark staining nucleus and light staining cytoplasm is see in this section. The basal cells form a continuous layer of flat cells with a spindle shape and an elongated nucleus. The basal cells abut the spiral ligament in the lower left part of this section (Magnification = 5000×)

8   Endocochlear Potential

The endocochlear potential is a positive extracellular potential of about 80–100 mV. It is essential for hearing and is controlled by K+ transport across the lateral wall of the cochlea. The lateral wall comprises two epithelial barriers, a syncytium of multiple cell types and the marginal epithelial cells. The syncytium includes fibrocytes that are exposed to perilymph on the basolateral side. The apical part of the marginal cells form a layer adjacent to endolymph. There exists an intrastrial space (Adachi et al. 2013). This space is situated between the apical membranes of the intermediate cells and the basolateral membranes of the marginal cells. This space is electrically isolated from nearby extracellular fluids, endolymph and perilymph, and this space has a positive potential. The low K+ concentration present in the intrastrial space is maintained by Na/K/ATPases and Na, K, 2 Cl transporters present in the basolateral membranes of marginal cells (Hibino et al. 2010). The EP appears to be produced by two K+ diffusion potentials generated by gradients in K+ between intracellular and extracellular compartments in the lateral wall of the cochlea (Nin et al. 2016). A model has been proposed utilizing unidirectional K+ transport by channels and transporters in the lateral cochlear wall and combines this transport to fluxes of K+ in hair cells and simulates the current flow between endolymph and perilymph (Nin et al. 2016).

9 Organ of Corti

The organ of Corti contains the sensory epithelium that is unique to the mammalian cochlea (Nam and Fettiplace 2012). It is situated between two membranes—the basilar membrane, upon which it rests, and the tectorial membrane. The organ of Corti contains two types of mechano-sensory receptor cells known as hair cells that are held up by supporting cells. The basilar membrane is an acellular structure on which the sensory cells and the support cells are situated. The hair cells have a cuticular plate from which stereocilia project. The top of the organ of Corti forms the reticular lamina, which is made up of the head plates of inner pillar cells and the cuticular plates of the hair cells. The outer hair cells appear to amplify vibrations of the basilar membrane and they facilitate detection of these vibrations by the inner hair cells (Fettiplace and Hackney 2006). The hair cells, supporting cells and afferent and efferent nerve fibers interact to produce hearing (Hudspeth 1989). The basilar membrane movements are accelerated in a localized area. This property is called frequency tuning (Goutman et al. 2015). Recent experiments in the mouse demonstrate that outer hair cells do not amplify basilar membrane vibration directly through a local feedback. Actually they appear to actively vibrate the reticular lamina through a wide frequency range. This combined outer hair cell-reticular lamina vibration interacts with the basilar membrane traveling wave through the cochlear fluid. This interaction improves peak responses at the best frequency location in the cochlea and thereby enhances both sensitivity and frequency selectivity (Ren et al. 2016).

Both types of hair cells have apical stereocilia that can sense mechanical movements generated by sound waves (Figs. 1.3 and 1.4). Hair cells transduce mechanical energy produced by sound-induced vibrations into electrical signals (Milewski et al. 2017). The inner hair cells sense sounds and propagate acoustic information to the brain via the auditory nerve whose cell bodies are located in the spiral ganglion. The outer hair cells provide mechanical amplification to produce the exquisite tuning and

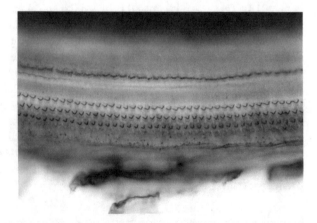

Fig. 1.3 Light microscopic view of hair cells from the basal turn cochlea of a normal chinchilla. A single row of inner hair cell stereocilia can be seen, and three rows of outer hair cells bearing v-shaped stereocilia are seen arrayed on their apical surface

Fig. 1.4 Scanning electron micrograph of the basal turn of the rat cochlea. There are three rows of outer hair cells showing rows of v-shaped stereocilia projecting from the top of the hair cells. A single row of stereocilia are seen projecting from the apical surface of the inner hair cells. Scale bar = 5 μM

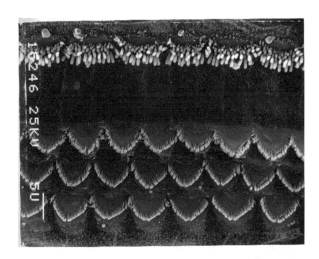

high sensitivity of the cochlea (Goutman et al. 2015). Cochlear amplification by outer hair cells appears to be achieved by somatic motility. The OHC appears to produce two types of active force: a force generated by the hair bundle and a somatic force attributed to the membrane protein prestin (Dallos et al. 2008). OHCs undergo rapid, voltage-dependent changes in cellular length. Prestin-mediated electromotility has been characterized as a two-step process. The first step involves transport of anions by an alternate access cycle. This step is followed by an anion-dependent transition generating electromotility. This electromotility is similar to piezoelectric materials that change their size depending on the voltage applied (Schaechinger et al. 2011).

Various channels or transporters have been demonstrated in the cochlea. A partial list is shown in Table 1.2.

10   Mechano-transduction

The hair cells of the organ of Corti are mechanoreceptors. Mammalian hair cells have multiple stereocilia made up of actin filaments. These filaments are cross-linked with actin-binding proteins including espin (Fettiplace 2017). They are inserted into the cuticular plate located at the apex of each hair cell. OHC stereocilia are arranged in three V or W-shaped rows (Figs. 1.3 and 1.4). The stereocilia are connected to each other by tip links. These links are required for opening MET channels. Sound waves cause deflections of stereocilia resulting in mechano-transduction. These movements result in opening of a mechano-transduction (MET) channel to generate inward currents in hair cells (Goutman et al. 2015).

Auditory hair cells contain two molecularly distinct mechano-transduction channels. One ion channel is activated by sound and is responsible for sensory transduction. This sensory transduction channel is expressed in hair cell stereocilia, and

Table 1.2 List of various channels or transporters in the cochlea

Channel/ transporter	Location	Function	Reference
MET/TMC1	Cochlear hair cells	Part of MET channel?	Fettiplace (2016)
KCNQ1 (Kv7.1)	Strial marginal cells	K+ secretion into endolymph and endolymph formation	Chang et al. (2015)
KCNQ4 (Kv7.4)	Cochlear outer hair cells	Mediates the M-like potassium current $I_{K,n}$	Kubisch et al. (1999)
KCNMA1	Cochlear inner hair cells	K+ efflux from IHCs	Molina et al. (2013)
KCNE1	Strial marginal cells	K+ secretion into endolymph and endolymph formation	Lang et al. (2007)
KCNJ10	Strial intermediate cells	Formation of EP	Lang et al. (2007)
Pendrin (SLC26A4)	Epithelial cells of inner ear	HCO_3^- secretion into endolymph	Semaan et al. (2005)
SLC22A4	Endothelial cells of stria vascularis		Ben Said et al. (2016)
BSND	Strial marginal cells	Recycling of Cl^- in strial marginal cells and endolymph formation	Lang et al. (2007)
CLCNKA/B	Strial marginal cells	Recycling of K+ in marginal cells and endolymph formation	Lang et al. (2007)
TRP	Outer hair cells	Calcium homeostasis in endolymph	Lang et al. (2007)
Aquaporins	Ubiquitous	Transport of water molecules	Chiarella et al. (2015)
KCNK5	Cochlear outer sulcus cells	K+ recycling	Cazals et al. (2015)
SLC12A7	Deiter cells	K+ efflux/influx	Lang et al. (2007)
SLC12A2	Strial marginal cells	K+ uptake into strial marginal cells and endolymph formation	Lang et al. (2007)

Table is modified from tables in Lang et al. (2007) and Mittal et al. (2017)

previous studies show that its activity is affected by mutations in the genes encoding the transmembrane proteins TMHS, TMIE, TMC1 and TMC2 (Wu et al. 2017) A protein present in stereocilia, calcium and integrin-binding protein (CIBP2) is essential for hearing. This protein binds to components of the hair cell mechano-transduction complex, TMC1 and TMC2. CIB2 appears to be essential for normal mechano-transduction in the cochlea and appears to limit the growth of stereocilia (Giese et al. 2017).

Acoustic stimuli permit K+ in endolymph to enter and excite sensory hair cells. The positive EP accelerates the K+ influx and chemically sensitizes hearing sensitivity. Subsequently, K+ effluxes from hair cells and is recirculated to the lateral wall, bringing K+ back into endolymph (Nin et al. 2016). A current and detailed review of auditory hair cell transduction has recently been published (Fettiplace

2017). The cochlea has a number of channels or transporters expressed in various cells. Some of these are listed in Table 1.2.

11 Hair Cell Synapses

Hair cells synapse with spiral ganglion axons. These highly specialized synapses are called ribbon synapses. The postsynaptic nerve endings possess AMPA-type glutamate receptors. Type Cav1.3 calcium channels at the ribbon synapse are activated by hair cell receptor potentials (Fuchs et al. 2003). This results in activation of postsynaptic glutamate receptors. This generates action potentials that are transmitted by spiral ganglion neurons to the brain (Stöver and Diensthuber 2012).

The cochlea has a highly complex molecular and biophysical structure that provides exquisite sensitivity and specificity for analyzing sounds and transmitting information to the brain for interpretation and analysis. How inflammation and immunologic responses impact hearing will be discussed in subsequent chapters.

Acknowledgement Dr. Rybak was supported by NIH grant DC02396 from NIDCD.

References

Adachi N, Yoshida T, Nin F, Ogata G, Yamaguchi S, Suzuki T, Komune S, Hisa Y, Hibino H, Kurachi Y. The mechanism underlying maintenance of the endocochlear potential by the K+ transport system in fibrocytes of the inner ear. J Physiol. 2013;591:4459–72.

Andrade LR, Salles FT, Grati M, Manor U, Kachar B. Tectorins crosslink type II collagen fibrils and connect the tectorial membrane to the spiral limbus. J Struct Biol. 2016;194:139–46.

Ben Said M, Grati M, Ishimoto T, Zou B, Chakchouk I, Ma Q, Yao Q, Hammami B, Yan D, Mittal R, Nakamichi N, Ghorbel A, Neng L, Tekin M, Shi XR, Kato Y, Masmoudi S, Lu Z, Hmani M, Liu XL. A mutation in SLC22A4 encoding an organic cation transporter expressed in the cochlea strial endothelium causes human recessive non-syndromic hearing loss DFNB60. Hum Genet. 2016;135:513–24.

Bohne BA, Harding GW. Cochlear anatomy. In: Clark WW, Ohlemiller KK, editors. Anatomy and physiology of hearing for audiologists. Clifton Park: Thomson Delmar Learning; 2008. p. 109–22.

Cazals Y, Bevengut M, Zanella S, Brocard F, Barhanin J, Gestreau C. KCNK5 channels mostly expressed in cochlear outer sulcus cells are indispensable for hearing. Nat Commun. 2015;6:8780.

Chang Q, Wang J, Li Q, Kim Y, Zhou B, Wang Y, Li H, Lin X. Virally mediated Kcnq1 gene replacement therapy in the immature scala media restores hearing in a mouse model of human Jervell and Lange-Nielsen deafness syndrome. EMBO Mol Med. 2015;7:1077–86.

Chiarella G, Petrolo C, Cassandro E. The genetics of Meniere's disease. Appl Clin Genet. 2015;8:9–17.

Dallos P, Wu X, Cheatham MA, Gao J, Zheng J, Anderson CT, Jia S, Wang X, Cheng WH, Sengupta S, He DZ, Zuo J. Prestin-based outer hair cell motility is necessary for mammalian cochlear amplification. Neuron. 2008;58:333–9.

Fettiplace R. Is TMC1 the hair cell mechanotransducer channel? Biophys J. 2016;111:3–9.

Fettiplace R. Hair cell transduction, tuning and synaptic transmission in the mammalian cochlea. Compr Physiol. 2017;7:1197–227.

Fettiplace R, Hackney CM. The sensory and motor roles of auditory hair cells. Nat Neurosci. 2006;7:19–29.

Forge A, Becker D, Casalotti S, Edwards J, Evans WH, Lench N, Souter M. Gap junctions and connexin expression in the inner ear. Novartis Found Symp. 1999;219:134–50.

Fuchs PA, Glowatzki E, Moser T. The afferent synapse of cochlear hair cells. Curr Opin Neurobiol. 2003;13:452–8.

Giese APJ, Tang Y-Q, Sinha GP, Bowl MR, Goldring AC, Parker A, Freeman MJ, Brown SDM, Riazuddin S, Fettiplace R, Schafer WR, Frolenkov GI, Ahmed Z. CiB2 interacts with TMC1 and TMC2 and is essential for mechanotransduction in auditory hair cells. Nat Commun. 2017;8(1):43.

Goutman JD, Elgoyhen AB, Gomez-Casati ME. Cochlear hair cells: the sound-sensing machines. FEBS Lett. 2015;589:3354–61.

Gow A, Davies C, Southwood CM, Frolenkov G, Chrustowski M, Ng L, Yamauchi D, Marcus DC, Kachar B. Deafness in Claudin 11-null mice reveals the critical contribution of basal cell tight junctions to stria vascularis function. J Neurosci. 2004;24:7051–62.

Hibino H, Nin F, Tsuzuki C, Kurachi Y. The specific architecture of the stria vascularis and the roles of the ion-transport apparatus. Pflugers Arch - Eur J Physiol. 2010;459:521–33.

Hudspeth AJ. How the ear's works work. Nature. 1989;341:397–404.

Jagger DJ, Forge A. The enigmatic root cell—emerging roles contributing to fluid homeostasis within the cochlear outer sulcus. Hear Res. 2013;303:1–11.

Kim HJ, Gratton MA, Lee J-H, Perez Flores MC, Wang W, Doyle KJ, Beermann F, Crognale MA, Yamoah EN. Precise toxigenic ablation of intermediate cells abolishes the "battery" of the cochlear duct. J Neurosci. 2013;33:14601–6.

Kubisch C, Schroeder BC, Friedrich T, Leutjohann B, El-Amraoui A, Marlin S, Petit C, Jentsch TJ. KCNQ4, a novel potassium channel expressed in sensory outer hair cells, is mutated in dominant deafness. Cell. 1999;96:437–46.

Lang F, Vallon V, Knipper M, Wangemann P. Functional significance of channels and transporters expressed in the inner ear and kidney. Am J Physiol Cell Physiol. 2007;293:C1187–208.

Liu H, Li Y, Chen L, Zhang Q, Pan N, Nichols DH, Zhang WJ, Fritsch B, He DZZ. Organ of Corti and stria vascularis: is there an interdependence for Survival? PLoS One. 2016;11(12):w0168953. https://doi.org/10.1371/journal.pone.0168953.

Liu W, Schrott-Fischer A, Glueckert R, Benav H, Rask-Andersen H. The human "cochlear battery" –claudin-11 barrier and ion transport proteins in the lateral wall of the cochlea. Front Mol Neurosci. 2017;10:239. https://doi.org/10.3389/fnmol.2017.00239.

Milewski AR, Maoileidigh DO, Salvi JD, Hudspeth AJ. Homeostatic enhancement of sensory transduction. Proc Natl Acad Sci U S A. 2017;114:E6794–803.

Mittal R, Aranke M, Debs LH, Nguyen D, Patel AP, Grata M, Mittal J, Yan D, Chapagain P, Eshraghi AA, Liu XZ. Indispensable role of ion channels and transporters in the auditory system. J Cell Physiol. 2017;232:743–58.

Molina L, Fasquelle L, Nouvian R, Salvetat N, Scott HS, Guipponi M, Molina F, Puel JL, Delprat B. Tmprss3 loss of function impairs cochlear inner hair cell Kcnma1 channel membrane expression. Hum Mol Genet. 2013;22:1289–99.

Nam J-H, Fettiplace R. Optimal electrical properties of outer hair cells ensure cochlear amplification. PLoS One. 2012;7(11):e50572. http://journals.plos.org/plosone/article?id=10.1371/journal.pone.0050572.

Nin F, Yoshida T, Sawamura S, Ogata G, Ota T, Higuchi T, Murakami S, Doi K, Kurachi Y, Hibino H. The unique electrical properties in an extracellular fluid of the mammalian cochlea; their functional roles, homeostatic processes, and pathological significance. Pflugers Arch - Eur J Physiol. 2016;468:1637–49.

Raphael Y, Altschuler RA. Structure and innervation of the cochlea. Brain Res Bull. 2003;60:397–422.

Ren T, He W, Kemp D. Reticular lamina and basilar membrane vibrations in living mouse cochleae. Proc Natl Acad Sci U S A. 2016;113:9910–5.

Schaechinger TJ, Gorbunov D, Halaszovich CR, Moser T, Kugler S, Fakler B, Oliver D. A synthetic prestin reveals prestin protein domains and molecular operation of outer hair cell piezoelectricity. EMBO J. 2011;30:2793–804.

Semaan MT, Alagramam KN, Megerian CA. The basic science of Meniere's disease and endolymphatic hydrops. Curr Opin Otolaryngol Head Neck Surg. 2005;13:301–7.

Shi X. Pathophysiology of the cochlear intrastrial fluid-blood barrier (review). Hear Res. 2016;338:52–63.

Shodo R, Hayatsu M, Koga D, Horii A, Ushiki T. Three-dimensional reconstruction of root cells and interdental cells in the rat inner ear by serial section scanning electron microscopy. Biomed Res (Tokyo). 2017;38(4):239–48.

Spicer SS, Schulte BA. The fine structure of spiral ligament cells relates to ion return to the stria and varies with place-frequency. Hear Res. 1996; 100(1–2):80–100.

Steel KP, Barkway C. Another role for melanocytes: their importance for normal stria vascularis development in the mammalian inner ear. Development. 1989;107:453–63.

Sterkers O, Ferrary E, Amiel C. Production of inner ear fluids. Physiol Rev. 1988;68:1083–128.

Stöver T, Diensthuber M. Molecular biology of hearing. GMS Curr Top Otorhinolaryngol Head Neck Surg. 2011;10:Doc06. https://doi.org/10.3205/cto000079. Epub 2012 Apr 26.

Verpy E, Leibovici M, Michalski N, Goodyear RJ, Houdon C, Weil D, Richardson GP, Petit C. Stereocilin connects outer hair cell stereocilia to one another and to the tectorial membrane. J Comp Neurol. 2011;519:194–210.

Von Bekesy G. Resting potentials inside the cochlear partition of the guinea pig. Nature. 1952;169:241–2.

Wu Z, Grillet N, Zhao B, Cunningham C, Harkins-Perry S, Coste B, Ranade S, Zebarjadi N, Beurg M, Fettiplace R, Patapoutian A, Mueller U. Mechanosensory hair cells express two molecularly distinct mechanotransduction channels. Nat Neurosci. 2017;20:24–33.

Yang Y, Dai M, Wilson TM, Omelchenko I, Klimek JE, Wilmarth PA, David LL, Nuttall AL, Gillespie PG, Shi X. Na$^+$/K$^+$ATPase α1 identified as an abundant protein in the blood-labyrinth barrier that plays an essential role in the barrier integrity. PLoS One. 2011;6(1):e16547. https://doi.org/10.1371/journal.pone.0016547.

Yoo JC, Kim HY, Han KH, Oh SH, Chang SO, Marcus DC, Lee JH. Na+ absorption by Claudius' cells is regulated by purinergic signaling in the cochlea. Acta Otolaryngol. 2012;132(Suppl 1):S103–8.

Yoshida T, Nin F, Murakami S, Ogata G, Uetsuka S, Choi S, Nakagawa T, Inohara H, Komune S, Kurachi Y, Hibino H. The unique ion permeability profile of cochlear fibrocytes and its contribution to establishing their positive resting membrane potential. Pflugers Arch. 2016;468:1609–19.

Zwaenepoel I, Mustapha M, Leibovici M, Verpy E, Goodyear R, Liu XZ, Nouaille S, Nance WE, Kanaan M, Avraham KB, Tekaia F, Loiselet J, Lathrop M, Richardson G, Petit C. Otoancorin, an inner ear protein restricted to the interface between the apical surface of sensory epithelia and their overlying acellular gels, is defective in autosomal recessive deafness DFNB22. Proc Natl Acad Sci U S A. 2002;99:6240–5.

Chapter 2
Oxidative Stress and Hearing Loss

Samson Jamesdaniel

Abstract Oxidative stress is considered as a central factor in acquired hearing loss. This chapter provides an introduction to the fundamental concepts of oxidative stress as well as an overview of cochlear oxidative stress pathways activated by risk factors of auditory dysfunction. It also discusses the susceptibility of the inner ear to oxidative damage, the intracellular redox sensitive mechanisms that facilitate cytotoxicity, and the cochlear targets of oxidative stress. Special focus is given to cochlear oxidative stress induced by exposure to environmental factors, such as noise, heavy metals, and organic solvents, ototoxic drugs/agents, such as aminoglycosides, cisplatin, and radiation, and aging. Potential biomarkers of oxidative stress and the utility of targeting cochlear oxidative stress to mitigate acquired hearing loss are discussed. Finally, recent developments in this field, including therapeutic compounds and strategies employed to target different steps in the oxidative stress signaling pathways as well as potential challenges to these approaches are discussed.

Keywords Oxidative stress · Free radicals · Nitrative stress · Ototoxicity · Cisplatin · Hearing loss

1 Introduction

Oxygen was discovered by Priestley and Scheele in 1774 and Fenton reported a free radical reaction as early as 1894 (Fenton 1894). Nevertheless, the concept of oxygen-related toxicity was not recognized until the late 1940s when the toxic effects of increased oxygen tension was reported (Comroe Jr et al. 1945). A few years later, Gerschman et al. (1954) suggested that both oxygen poisoning and X-irradiation occur through oxidizing free radicals. However, the discovery of superoxide dismutase by McCord and Fridovich in 1969 provided a solid foundation for subsequent

S. Jamesdaniel (✉)
Institute of Environmental Health Sciences, Wayne State University, Detroit, MI, USA
e-mail: sjamesdaniel@wayne.edu

© Springer International Publishing AG, part of Springer Nature 2018   15
V. Ramkumar, L. P. Rybak (eds.), *Inflammatory Mechanisms in Mediating Hearing Loss*, https://doi.org/10.1007/978-3-319-92507-3_2

developments in this field (McCord and Fridovich 1969), which have made a huge impact in defining the therapeutic and prophylactic strategies employed to fight diseases and promote health. At present, the critical role of oxidative stress in cancer, neurodegeneration, inflammation, cardiovascular disorders, metabolic diseases, and several other pathological conditions is well established. In addition to these diseases, a growing body of evidence from the last few decades suggests that oxidative stress also plays a causal role in acquired hearing loss.

2   Oxidative Stress

Oxidative stress refers to a state in which there is an imbalance between the generation of reactive oxygen species (ROS) or reactive nitrogen species (RNS) and the counteracting defense mechanisms. ROS and RNS are generated during both aerobic and anaerobic metabolism. Among the multiple sources that produce free radicals the most common are nicotinamide adenine dinucleotide phosphate (NADPH)-dependent enzymatic reactions during aberrant mitochondrial respiration. Under physiological conditions, these free radicals serve as regulatory messengers for maintenance of normal cellular functions. The antioxidant defense machinery, which includes endogenous antioxidants such as glutathione and thioredoxin, and free radical scavenging enzymes such as superoxide dismutase, catalase, glutathione peroxidase, and glutathione transferase, helps to maintain redox homeostasis. When the generation of free radicals exceed the detoxifying and scavenging capability of the antioxidant machinery it results in oxidative stress.

Free Radicals

ROS are oxygen-based molecules, which include free radicals that have an unpaired electron in their outer shell (Halliwell and Gutteridge 1984) and molecules that can generate free radicals. Some of the most common ROS molecules are superoxide radical (O_2^-), hydrogen peroxide (H_2O_2), and hydroxyl radical (OH^-). In addition to reactions mediated by NADPH oxidase (NOX) enzymes, xanthine oxidase and cytochrome P450 enzyme mediated reactions can also produce O_2^-. Dismutation of O_2^- by superoxide dismutase enzyme produces H_2O_2. Catalase converts H_2O_2 into water and oxygen. In the presence of metal ions such as Fe_2^-, H_2O_2 forms OH^-.

RNS are nitrogen-based molecules. The most common RNS is nitric oxide (NO). It is derived from the oxidation of L-arginine to L-citrulline by nitric oxide synthase (NOS). Another highly reactive RNS is peroxynitrite ($ONOO^-$), which is formed by the reaction between O_2^- and NO. The oxidative stress pathways that are activated in the cochlea by ototoxic agents are illustrated in the schematic (Fig. 2.1). When the defense mechanisms are overwhelmed these free radicals can trigger a cascade reaction resulting in considerable damage to the cells, tissues, or organs.

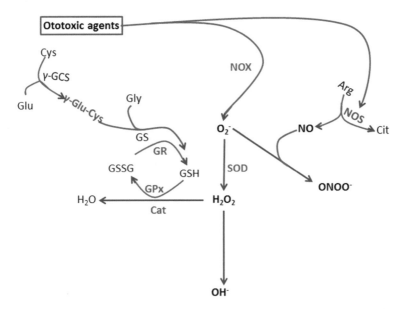

Fig. 2.1 Schematic of cochlear oxidative stress pathways. Abbreviations: *Arg* Argnine, *Cat* Catalase, *Cit* Citrulline, *Cys* Cysteine, *GCS* Glutamyl cysteine syntase, *Glu* Glutamate, *Gly* Glycine, *GPx* Glutathione peroxidase, *GR* Glutathione reductase, *GS* Glutathione syntase, *GSH* Reduced glutathione, *GSSG* Oxidized glutathione, OH⁻ Hydroxyl anion, *NO* Nitric oxide, *NOS* Nitric oxide synthase, *NOX* NADPH oxidase, O₂⁻ Superoxide anion, ONOO⁻ Peroxynitrite, *SOD* Superoxide dismutase

Oxidative Stress and Cytotoxicity

Free radicals are highly potent molecules that can directly react with proteins, lipids, and DNA bases. When the cellular levels of free radicals exceed the physiological limits they lead to oxidation/nitration of susceptible proteins resulting in the accumulation or degradation of altered proteins. These protein modifications ultimately result in loss of protein function thereby enabling cellular pathways that lead to cell death. ROS can also react with polyunsaturated fatty acids (PUFA), which are components of cell and subcellular membranes, resulting in the peroxidation of membrane lipids. This in turn compromises the function of cell membranes and can even lead to rupture of the cells. Oxidative damage to DNA causes alterations in the bases and single strand or double strand breaks in the DNA, which can inactivate key genes that regulate apoptosis (Rich et al. 2000). Alternatively, ROS/RNS also activates signaling molecules, such as mitogen activated protein kinases (MAPK), protein tyrosine phosphatases, and protein tyrosine kinases (Zhang et al. 2007; Janssen-Heininger et al. 2008). These signaling molecules can activate or inhibit the transcription of target genes resulting in cytotoxicity. For example, activation of c-Jun-N-terminal kinases (JNK) and p38 MAPK by ROS facilitates the release of cytochrome C and activation of caspase 9 and 3 leading to apoptotic cell death (Pan et al. 2009).

3 Cochlear Targets of Oxidative Stress

The oxidative stress-induced damage to the cochlea is generally focused on three discrete regions, the sensory epithelium, the lateral wall, and the modiolus. These three regions represent the sensory, vascular, and neuronal components of the cochlea, which consists of the organ of Corti, stria vascularis, and spiral ganglion neurons, respectively.

Sensory Epithelium

The inner and outer hair cells, which are the sensory receptor cells, are located in the organ of Corti along with a number of supporting cells. Stimulation of the hair cells activates the mechanoelectrical transduction channels resulting in the release of neurotransmitters at the synapses thereby initiating the transmission of auditory signals. NADPH oxidase 3 (NOX3), which is responsible for the production of O_2^- in the inner ear, is expressed in the hair cells (Banfi et al. 2004). Therefore, over-stimulation of the hair cells leads to the increased generation of O_2^-. The vulnerability of different cells in the organ of Corti to oxidative stress-induced damage varies based on the differences in the expression pattern of proteins associated with the antioxidant machinery. For, example, the outer hair cells at the base of the cochlea are more susceptible to free radical-induced damage probably due to lower levels of antioxidant proteins, such as glutathione, in the basal hair cells when compared to the apical hair cells (Sha et al. 2001a). The oxidative damage to hair cells and associated functional deficits are usually permanent because the hair cells do not have the capacity to regenerate.

Lateral Wall

The lateral wall consists of the stria vascularis, which is rich in blood vessels and has the highest metabolic rate among the different cochlear structures (Marcus et al. 1978). The integrity of vascular endothelium is critical for tissue homeostasis as the stria vascularis maintains the potassium ion concentration of the endolymph, which is essential for establishing the endocochlear potential required for signal transduction. Because vascular permeability is regulated by the release of NO, excessive production of NO, as observed in oxidative stress, can disrupt the vascular endothelium resulting in inflammation and edema of the stria vascularis (Smith et al. 1985).

Damage to the marginal cells as well as disruption of cochlear microcirculation has been attributed to increased levels of ROS in the stria vascularis (Shi and Nuttall 2003).

Modiolus

The spiral ganglion neurons, which transmit the auditory signals through the vestibulocochlear nerve, are located in this region. Similar to hair cells, the cochlear neuronal cells also express NOX3 (Banfi et al. 2004) and are vulnerable to oxidative damage because of their limited capacity to repair. Furthermore, over stimulation of the sensory cells can lead to increased release of the neurotransmitter glutamate resulting in glutamate excitotoxicity (Pujol and Puel 1999), which is facilitated by the increased calcium influx and consequent accumulation of ROS in the neuronal cells, and neuronal damage. Oxidative damage to spiral ganglion neurons has been indicated by increased levels of nitrotyrosine, a biomarker of oxidative damage to proteins, and 4-hydroxy-2-nonenal (4-HNE), a biomarker of lipid peroxidation (Jamesdaniel et al. 2012; Xiong et al. 2011).

4   Oxidative Stress in Acquired Hearing Loss

The pivotal role of ROS and RNS in initiating ototoxic cascades that result in acquired hearing loss is well documented. Hearing loss associated with aging, exposure to noise, heavy metals, organic solvents, ototoxic drugs such as aminoglycosides and cisplatin, and radiation have been attributed to oxidative stress induced activation of cochlear cell death pathways (Warchol 2010; Poirrier et al. 2010; Huth et al. 2011; Bottger and Schacht 2013; Wong and Ryan 2015). Although a multitude of other agents/drugs can induce oxidative stress in the cochlea, the discussions in this chapter are limited to some of the major risk factors of acquired hearing loss.

Environmental Exposures

Exposure to loud noise, both occupational and recreational, is one of the most common causes of acquired hearing loss as it causes permanent shift in hearing thresholds due to irreversible damage to cochlear hair cells (Liberman and Dodds 1984). Noise-induced oxidative damage has been detected not only in the hair cells but also in the supporting cells, stria vascularis, and spiral ganglion neurons. Acoustic overstimulation increases the metabolic activity of hair cells (Henderson et al. 2006) resulting in increased levels of Ca^{2+} (Fridberger et al. 1998), which in turn can stimulate mitochondrial ROS production (Peng and Jou 2010). Increased ROS leads to the generation of vasoactive lipid peroxidation products such as isoprostanes (Ohinata et al. 2000) which can lead to a decrease in cochlear blood flow due to increased vascular permeability and capillary vasoconstriction (Nuttall 1999). The consequent ischemia and reperfusion can lead to further increase in the production of ROS resulting in continued damage to the cochlear tissue even after the removal of the ototoxic stimuli. Increased levels of ROS, such as $O_2{}^-$ (Yamane et al. 1995) and OH$^-$ (Ohlemiller et al. 1999), RNS, such as NO (Heinrich et al. 2008), and biomarkers of oxidative stress, such as malondialdehyde (MDA), 4-HNE, and nitrotyrosine (Yamashita et al. 2004; Samson et al. 2008; Yuan et al. 2015) have been detected in the cochlea after noise exposure. Moreover, increase in the activity of NOX and inducible nitric oxide synthase (iNOS) enzymes in the cochlea (Vlajkovic et al. 2013; Shi and Nuttall 2003) and changes in the endogenous antioxidant machinery have been well documented after noise trauma (Ohlemiller et al. 1999; Ramkumar et al. 2004; Samson et al. 2008). Accumulation of ROS can activate the intrinsic caspase-mediated apoptotic pathway via JNK and p38 MAPK signaling resulting in cochlear cell death (Yamashita et al. 2004; Jamesdaniel et al. 2011).

Long term exposure to heavy metals such as lead, cadmium, cobalt, arsenic, and mercury has the potential to cause auditory dysfunction (Rybak 1992; Shargorodsky et al. 2011; Roth and Salvi 2016). Environmental exposure to lead and cadmium, which usually co-occur, can lead to the generation of ROS (Ercal et al. 2001; Vaziri and Khan 2007; Muthusamy et al. 2016). Exposure to lead induces degeneration of sensory receptor cells in the cochlea, disrupts the blood-cochlear labyrinth, and affects auditory nerve conduction velocity (Jones et al. 2008; Lasky et al. 1995; Liu et al. 2013; Yamamura et al. 1989). Exposure to cadmium induces an increase in the generation of ROS in the cochlea resulting in apoptosis of hair cells and an increase in hearing thresholds (Kim et al. 2008; Ozcaglar et al. 2001). Cobalt-induced ototoxicity is reported to be triggered by $O_2{}^-$ and targets the hair cells and spiral ganglion neurons (Li et al. 2015; Lee et al. 2016). Though exposure to arsenic has been associated with hearing loss and damage to cochlear tissue (Bencko and Symon 1977; Anniko and Wersall 1975) the underlying mechanism is yet to be fully understood. Exposure to methyl mercury causes auditory deficits, particularly in the higher frequencies (Wassick and Yonovitz 1985) and suppresses potassium currents of outer hair cells (Liang et al. 2003).

Occupational exposure to organic solvents such as toluene, styrene, xylene, and ethyl benzene can significantly impair auditory perception (Morata et al. 1994; Sliwinska-Kowalska et al. 2003). Styrene exposure has been reported to induce the generation of $O_2{}^-$ in the organ of Corti, spiral ganglion neurons and stria vascularis and increase the levels of 8-Isoprostane, a biomarker of lipid peroxidation, in the stria vascularis and spiral ganglion neurons (Fetoni et al. 2016). Exposure to toluene results in the loss of outer hair cells in both the mid and basal turn of the organ of Corti, particularly, in the 2–8 kHz region (Sullivan et al. 1988). Many other organic solvents also target the mid-frequency region of the cochlea (Cappaert et al. 2000; Crofton et al. 1994).

Therapeutic Drugs and Agents

Aminoglycosides, which are antibiotics used in the treatment of Gram negative infections and tuberculosis, cause hearing loss. They enter the hair cells by permeating the mechanoelectrical transduction channels (Marcotti et al. 2005) or by apical endocytosis (Hashino and Shero 1995) and cause ROS-mediated cellular damage via multiple mechanisms. The non-enzymatic formation of an aminoglycoside-iron complex can catalyze the oxidation of unsaturated fatty acids in the plasma membrane (Priuska and Schacht 1995). Enzymatic reactions catalyzed by NOX and NOS, which are activated by aminoglycosides, can lead to the generation of ROS and RNS, respectively (Jiang et al. 2006; Hong et al. 2006). Disruption of calcium homeostasis between the endoplasmic reticulum and mitochondria by aminoglycosides induces ROS in the hair cells (Esterberg et al. 2014). Aminoglycosides can also inhibit the activity of catalase (Hong et al. 2006) thereby facilitating the accumulation of ROS (Hirose et al. 1997; Choung et al. 2009), which in turn can activate downstream JNK-mediated cellular apoptosis (Wang et al. 2003). The loss of outer hair cells, particularly from the basal region (Sha et al. 2001a, b) leads to progressive loss of spiral ganglion neurons because of the lack of hair cell-derived neurotropic support (Dodson and Mohuiddin 2000).

Ototoxicity is a dose limiting side effect of cisplatin, a highly effective and widely used platinum-based anti-cancer drug. Among the cochlear cell death mechanisms reported in cisplatin ototoxicity (Dehne et al. 2001; Rybak et al. 2007; Berndtsson et al. 2007; More et al. 2010; Jamesdaniel et al. 2012; Thomas et al. 2013; Kaur et al. 2016), oxidative stress is considered to play a causal role as it activates the enzyme NOX3, and thereby increases the production of O_2^- in the inner ear (Banfi et al. 2004). Furthermore, inactivation of free radical scavenging enzymes by direct binding of cisplatin with sulfhydryl groups in these enzymes and depletion of copper and selenium that are required for these enzyme-mediated scavenging activities lead to increased ROS in the cochlea (DeWoskin and Riviere 1992). Higher levels of ROS eventually lead to apoptosis of outer hair cells in the basal region, and cells of stria vascularis and spiral ganglion neurons (Clerici et al. 1995; Rybak et al. 2007). The ROS also activates the transient receptor potential vanilloid 1 (TRPV1) resulting in apoptosis of outer hair cells and spiral ganglion neurons (Mukherjea et al. 2008). In addition to these ROS-mediated oxidative damages, the activation of iNOS pathway and the generation of NO have been detected in cisplatin ototoxicity (Li et al. 2006; Watanabe et al. 2002). This in turn can lead to an increase in the nitration of cochlear proteins in the sensory epithelium (Jamesdaniel et al. 2008). A strong correlation between cochlear protein nitration and cisplatin-induced hearing loss was observed (Jamesdaniel et al. 2012). Recent studies using a peroxynitrite decomposition catalyst (SRI110) indicated that inhibition of protein nitration prevented cisplatin-induced hearing loss (Jamesdaniel et al. 2016). Furthermore, a direct link between cisplatin-induced nitration as well as degradation of LMO4, a transcriptional regulator, and ototoxicity was detected, which suggested RNS-mediated damage to the cochlea in cisplatin-induced hearing loss (Jamesdaniel et al. 2012, 2016; Rathinam et al. 2015).

Radiotherapy, an important component of head and neck and brain tumor treatment protocols, can also cause progressive hearing loss, particularly, when the inner ear is within the radiation field (Pan et al. 2005). Free radicals are generated by the interaction of radiation with water molecules, which lead to DNA damage (Sharma et al. 2008). Radiation can also activate iNOS eventually contributing to the generation of $ONOO^-$, which damages cell membranes and DNA (Azzam et al. 2012). The vascular endothelial cells in the stria vascularis and the hair cells are highly susceptible to radiation-induced damage (Kim and Shin 1994; Winther 1969), which initially causes high frequency hearing loss and gradually affects the lower frequencies (Mujica-Mota et al. 2014).

Aging

Although age-related hearing loss is a complex multifactorial process, mitochondrial mutations and dysfunction associated with ROS appear to play a central role in mediating the cochlear pathology (Seidman et al. 2004; Someya et al. 2009). Oxidative damage accumulated over the years due to auditory insults from noise and ototoxic agents/drugs, can contribute to the hearing deficits. Therefore, the antioxidant defense mechanisms are critical for delaying age-related hearing loss. This hypothesis is consistent with the findings, which indicated that over expression of antioxidant enzyme catalase had a protective effect (Someya et al. 2009) and the lack of antioxidant enzyme superoxide dismutase enhanced age-related hearing loss (McFadden et al. 1999). Moreover, the level of endogenous antioxidant glutathione was decreased in the aging cochlea (Lautermann et al. 1995) and markers of oxidative nitrative stress were detected in the organ of Corti and spiral ganglion neurons (Jiang et al. 2007). Supplementation with antioxidants appears to retard the cochlear damage and maintain better auditory sensitivity (Seidman 2000; Seidman et al. 2002).

5   Interventional Approaches and Challenges

The key role of oxidative stress in acquired hearing loss implies that this pathway could be targeted for therapeutic interventions. The underlying mechanism could be exploited to identify the risk of hearing loss at an early stage, determine appropriate interventional strategies to minimize the pathology, and even take preventive measures before exposure to the risk factors. However, the otoprotective interventions targeting oxidative stress also has challenges as some of the ROS are key signaling molecules that regulate important cellular processes. The following section discusses the opportunities and the challenges in this field.

Biomarkers of Oxidative Stress

Although detection of oxidative stress biomarkers directly from cochlear tissue is difficult in clinical settings, other body fluids, such as plasma and urine, are reliable alternatives for assaying biomarkers of oxidative damage. The plasma levels of anti-oxidants such as glutathione and alpha-tocopherol can indicate the susceptibility to oxidative damage. The ratio of reduced to oxidized glutathione is considered to better predict the oxidative injury (Pastore et al. 2001). Other clinically relevant markers of oxidative stress include MDA, F2-isoprostane, 8-hydroxy deoxyguanosine, 3-nitrotyrosine, and protein carbonyls. MDA and F2-isoprostane, which are biomarkers of lipid peroxidation (Feng et al. 2006; Montuschi et al. 2004), can be assayed in both plasma and urine. 8-hydroxy deoxyguanosine, a DNA adduct formed during oxidative damage to nuclear and mitochondrial DNA (Marnett 1999), can be detected in urine. Oxidative stress induced post-translational modification of proteins can be detected by assaying 3-nitrotyrosine (Ceriello 2002) and protein carbonyls (Chevion et al. 2000) in plasma. In addition to these biomarkers, the plasma levels of iNOS and glutamate could indicate the severity of the oxidative damage (Chavko et al. 2008; Gopinath et al. 2000).

Otoprotection by Targeting Oxidative Stress

Direct scavenging of ROS/RNS molecules is among the commonly used approaches for otoprotection. Thiol-containing compounds, such as N-acetyl-L-cysteine (Tokgoz et al. 2011; Low et al. 2008; Feghali et al. 2001), sodium thiosulphate (Doolittle et al. 2001; Wimmer et al. 2004), diethyldithiocarbamate (Rybak et al. 1995), and D-methionine (Korver et al. 2002; Campbell et al. 1996) are used as free radical scavengers to prevent acquired hearing loss induced by oxidative stress. Administration of these thiol-containing compounds facilitates an increase in the synthesis of intracellular glutathione. Glutathione supplementation has been reported to prevent cisplatin-induced ototoxicity (Lautermann et al. 1995). In addition to thiol compounds, many other antioxidant molecules are used to scavenge ROS. Both *in-vitro* and *in-vivo* studies have demonstrated the otoprotective potential of compounds, such as vitamin E (Fetoni et al. 2004), salicylate (Sha and Schacht 1999), and herbal antioxidants that contain polyphenol flavonoids (Seidman 2000; Schmitt et al. 2009; Tian et al. 2013). Small molecule compounds that target RNS, such as SRI110, a manganese (III) bishydroxyphenyldipyrromethene-based peroxynitrite decomposition catalyst, can scavenge ONOO$^-$ and thus provide oto-protection (Jamesdaniel et al. 2016). Chelation of redox-active transition metals like iron is another approach to scavenge oxidants. Metal chelators, such as deferox-amine, have been reported to mitigate drug- and noise-induced damage to the cochlea (Song et al. 1998; Conlon et al. 1998).

Interventional strategies that specifically target intracellular antioxidant/oxidant enzymes have been highly effective in conferring otoprotection (Mukherjea et al. 2015). Both genetic and pharmacological approaches have been employed to manipulate antioxidant enzymes that detoxify ROS. Overexpression of superoxide dismutase and catalase by using adenoviral vectors prevented drug-induced ototoxicity (Kawamoto et al. 2004). Administration of superoxide dismutase linked to polyethylene glycol (Seidman et al. 1993), adenosine receptor agonists (Kaur et al. 2016), which promotes the activity of superoxide dismutase and glutathione peroxidase, and ebselen (Kil et al. 2007), which appears to act through the glutathione peroxidase like mechanism, have prevented noise- or drug-induced hearing loss. Manipulation of oxidant enzymes, for example, silencing of NOX3 by using siRNA (Rybak et al. 2012), inhibiting xanthine oxidase by using allopurinol (Seidman et al. 1993; Lynch et al. 2005), or inhibiting nitric oxide synthase by using L-N(omega)-Nitroarginine methyl ester (Watanabe et al. 2000; Ohinata et al. 2003), attenuated oxidative damage and associated hearing loss.

Small molecules that target downstream signaling pathways of ROS-mediated cell death have been successfully employed to mitigate ototoxicity. Oxidative stress-mediated damage to hair cells and neurons has been prevented by inhibition of JNK in both noise- and drug-induced ototoxicity (Pirvola et al. 2000; Eshraghi et al. 2007). Inhibition of activators of JNK also protected hair cells from oxidative damage (Bodmer et al. 2002; Battaglia et al. 2003). Moreover, inhibition of p38 MAPK prevented noise-, drug-, and radiation-induced ototoxicity (Shin et al. 2014).

Potential Interference with Cell Signaling and Drug Activity

Although targeting oxidative stress appears to be an attractive strategy for ameliorating acquired hearing loss, there are some challenges. One of the major concerns is the delivery of the antioxidants to the cochlea. Determination of appropriate dose as well as an optimal route is critical because over exuberant or inappropriate use of antioxidants can potentially interfere with ROS-mediated cellular signaling required for certain physiological processes. Furthermore, co-treatment of antioxidants for minimizing the ototoxic side effects of certain drugs may interfere with the activity of the drug itself. Nevertheless, the progress made in the past few years are promising as recent studies indicate the availability of better delivery systems and orally active compounds that could minimize oxidative stress-mediated ototoxicity without compromising the activity of life-saving drugs. Overall, the significant advancements in understanding cochlear oxidative stress are likely to lead to the discovery of effective solutions for combating acquired hearing loss.

References

Anniko M, Wersall J. Damage to the stria vascularis in the guinea pig by acute atoxyl intoxication. Acta Otolaryngol. 1975;80:167–79.

Azzam EI, Jay-Gerin JP, Pain D. Ionizing radiation-induced metabolic oxidative stress and prolonged cell injury. Cancer Lett. 2012;327:48–60.

Banfi B, Malgrange B, Knisz J, Steger K, Dubois-Dauphin M, Krause KH. NOX3, a superoxide-generating NADPH oxidase of the inner ear. J Biol Chem. 2004;279:46065–72.

Battaglia A, Pak K, Brors D, Bodmer D, Frangos JA, Ryan AF. Involvement of ras activation in toxic hair cell damage of the mammalian cochlea. Neuroscience. 2003;122:1025–35.

Bencko V, Symon K. Test of environmental exposure to arsenic and hearing changes in exposed children. Environ Health Perspect. 1977;19:95–101.

Berndtsson M, Hagg M, Panaretakis T, Havelka AM, Shoshan MC, Linder S. Acute apoptosis by cisplatin requires induction of reactive oxygen species but is not associated with damage to nuclear DNA. Int J Cancer. 2007;120:175–80.

Bodmer D, Brors D, Bodmer M, Ryan AF. [Rescue of auditory hair cells from ototoxicity by CEP-11 004, an inhibitor of the JNK signaling pathway]. Laryngorhinootologie. 2002;81:853–6.

Bottger EC, Schacht J. The mitochondrion: a perpetrator of acquired hearing loss. Hear Res. 2013;303:12–9.

Campbell KC, Rybak LP, Meech RP, Hughes L. D-methionine provides excellent protection from cisplatin ototoxicity in the rat. Hear Res. 1996;102:90–8.

Cappaert NL, Klis SF, Baretta AB, Muijser H, Smoorenburg GF. Ethyl benzene-induced ototoxicity in rats: a dose-dependent mid-frequency hearing loss. J Assoc Res Otolaryngol. 2000;1:292–9.

Ceriello A. Nitrotyrosine: new findings as a marker of postprandial oxidative stress. Int J Clin Pract Suppl. 2002;51–8.

Chavko M, Prusaczyk WK, McCarron RM. Protection against blast-induced mortality in rats by hemin. J Trauma. 2008;65:1140–5; discussion 5.

Chevion M, Berenshtein E, Stadtman ER. Human studies related to protein oxidation: protein carbonyl content as a marker of damage. Free Radic Res. 2000;33(Suppl):S99–108.

Choung YH, Taura A, Pak K, Choi SJ, Masuda M, Ryan AF. Generation of highly-reactive oxygen species is closely related to hair cell damage in rat organ of Corti treated with gentamicin. Neuroscience. 2009;161:214–26.

Clerici WJ, DiMartino DL, Prasad MR. Direct effects of reactive oxygen species on cochlear outer hair cell shape in vitro. Hear Res. 1995;84:30–40.

Comroe JH Jr, Dripps RD, Dumke PR, Deming M. Oxygen toxicity. J Am Med Assoc. 1945;128:710–7.

Conlon BJ, Perry BP, Smith DW. Attenuation of neomycin ototoxicity by iron chelation. Laryngoscope. 1998;108:284–7.

Crofton KM, Lassiter TL, Rebert CS. Solvent-induced ototoxicity in rats: an atypical selective mid-frequency hearing deficit. Hear Res. 1994;80:25–30.

Dehne N, Lautermann J, Petrat F, Rauen U, de Groot H. Cisplatin ototoxicity: involvement of iron and enhanced formation of superoxide anion radicals. Toxicol Appl Pharmacol. 2001;174:27–34.

DeWoskin RS, Riviere JE. Cisplatin-induced loss of kidney copper and nephrotoxicity is ameliorated by single dose diethyldithiocarbamate, but not mesna. Toxicol Appl Pharmacol. 1992;112:182–9.

Dodson HC, Mohuiddin A. Response of spiral ganglion neurones to cochlear hair cell destruction in the guinea pig. J Neurocytol. 2000;29:525–37.

Doolittle ND, Muldoon LL, Brummett RE, et al. Delayed sodium thiosulfate as an otoprotectant against carboplatin-induced hearing loss in patients with malignant brain tumors. Clin Cancer Res. 2001;7:493–500.

Ercal N, Gurer-Orhan H, Aykin-Burns N. Toxic metals and oxidative stress part I: mechanisms involved in metal-induced oxidative damage. Curr Top Med Chem. 2001;1:529–39.

Eshraghi AA, Wang J, Adil E, et al. Blocking c-Jun-N-terminal kinase signaling can prevent hearing loss induced by both electrode insertion trauma and neomycin ototoxicity. Hear Res. 2007;226:168–77.

Esterberg R, Hailey DW, Rubel EW, Raible DW. ER-mitochondrial calcium flow underlies vulnerability of mechanosensory hair cells to damage. J Neurosci. 2014;34:9703–19.

Feghali JG, Liu W, Van De Water TR. L-n-acetyl-cysteine protection against cisplatin-induced auditory neuronal and hair cell toxicity. Laryngoscope. 2001;111:1147–55.

Feng Z, Hu W, Marnett LJ, Tang MS. Malondialdehyde, a major endogenous lipid peroxidation product, sensitizes human cells to UV- and BPDE-induced killing and mutagenesis through inhibition of nucleotide excision repair. Mutat Res. 2006;601:125–36.

Fenton HJH. Oxidation of tartaric acid in presence of iron. J Chem Soc. 1894;65:899–910.

Fetoni AR, Sergi B, Ferraresi A, Paludetti G, Troiani D. Protective effects of alpha-tocopherol and tiopronin against cisplatin-induced ototoxicity. Acta Otolaryngol. 2004;124:421–6.

Fetoni AR, Rolesi R, Paciello F, et al. Styrene enhances the noise induced oxidative stress in the cochlea and affects differently mechanosensory and supporting cells. Free Radic Biol Med. 2016;101:211–25.

Fridberger A, Flock A, Ulfendahl M, Flock B. Acoustic overstimulation increases outer hair cell Ca2+ concentrations and causes dynamic contractions of the hearing organ. Proc Natl Acad Sci U S A. 1998;95:7127–32.

Gerschman R, Gilbert DL, Nye SW, Dwyer P, Fenn WO. Oxygen poisoning and X-irradiation: a mechanism in common. Science. 1954;119:623–6.

Gopinath SP, Valadka AB, Goodman JC, Robertson CS. Extracellular glutamate and aspartate in head injured patients. Acta Neurochir Suppl. 2000;76:437–8.

Halliwell B, Gutteridge JM. Oxygen toxicity, oxygen radicals, transition metals and disease. Biochem J. 1984;219:1–14.

Hashino E, Shero M. Endocytosis of aminoglycoside antibiotics in sensory hair cells. Brain Res. 1995;704:135–40.

Heinrich UR, Helling K, Sifferath M, et al. Gentamicin increases nitric oxide production and induces hearing loss in guinea pigs. Laryngoscope. 2008;118:1438–42.

Henderson D, Bielefeld EC, Harris KC, Hu BH. The role of oxidative stress in noise-induced hearing loss. Ear Hear. 2006;27:1–19.

Hirose K, Hockenbery DM, Rubel EW. Reactive oxygen species in chick hair cells after gentamicin exposure in vitro. Hear Res. 1997;104:1–14.

Hong SH, Park SK, Cho YS, et al. Gentamicin induced nitric oxide-related oxidative damages on vestibular afferents in the guinea pig. Hear Res. 2006;211:46–53.

Huth ME, Ricci AJ, Cheng AG. Mechanisms of aminoglycoside ototoxicity and targets of hair cell protection. Int J Otolaryngol. 2011;2011:937861.

Jamesdaniel S, Ding D, Kermany MH, et al. Proteomic analysis of the balance between survival and cell death responses in cisplatin-mediated ototoxicity. J Proteome Res. 2008;7:3516–24.

Jamesdaniel S, Hu B, Kermany MH, et al. Noise induced changes in the expression of p38/MAPK signaling proteins in the sensory epithelium of the inner ear. J Proteomics. 2011;75:410–24.

Jamesdaniel S, Coling D, Hinduja S, et al. Cisplatin-induced ototoxicity is mediated by nitroxidative modification of cochlear proteins characterized by nitration of Lmo4. J Biol Chem. 2012;287:18674–86.

Jamesdaniel S, Rathinam R, Neumann WL. Targeting nitrative stress for attenuating cisplatin-induced downregulation of cochlear LIM domain only 4 and ototoxicity. Redox Biol. 2016;10:257–65.

Janssen-Heininger YM, Mossman BT, Heintz NH, et al. Redox-based regulation of signal transduction: principles, pitfalls, and promises. Free Radic Biol Med. 2008;45:1–17.

Jiang H, Sha SH, Forge A, Schacht J. Caspase-independent pathways of hair cell death induced by kanamycin in vivo. Cell Death Differ. 2006;13:20–30.

Jiang H, Talaska AE, Schacht J, Sha SH. Oxidative imbalance in the aging inner ear. Neurobiol Aging. 2007;28:1605–12.

Jones LG, Prins J, Park S, Walton JP, Luebke AE, Lurie DI. Lead exposure during development results in increased neurofilament phosphorylation, neuritic beading, and temporal processing deficits within the murine auditory brainstem. J Comp Neurol. 2008;506:1003–17.

Kaur T, Borse V, Sheth S, et al. Adenosine A1 receptor protects against cisplatin ototoxicity by suppressing the NOX3/STAT1 inflammatory pathway in the cochlea. J Neurosci. 2016;36:3962–77.

Kawamoto K, Sha SH, Minoda R, et al. Antioxidant gene therapy can protect hearing and hair cells from ototoxicity. Mol Ther. 2004;9:173–81.

Kil J, Pierce C, Tran H, Gu R, Lynch ED. Ebselen treatment reduces noise induced hearing loss via the mimicry and induction of glutathione peroxidase. Hear Res. 2007;226:44–51.

Kim CS, Shin SO. Ultrastructural changes in the cochlea of the guinea pig after fast neutron irradiation. Otolaryngol Head Neck Surg. 1994;110:419–27.

Kim SJ, Jeong HJ, Myung NY, et al. The protective mechanism of antioxidants in cadmium-induced ototoxicity in vitro and in vivo. Environ Health Perspect. 2008;116:854–62.

Korver KD, Rybak LP, Whitworth C, Campbell KM. Round window application of D-methionine provides complete cisplatin otoprotection. Otolaryngol Head Neck Surg. 2002;126:683–9.

Lasky RE, Maier MM, Snodgrass EB, Hecox KE, Laughlin NK. The effects of lead on otoacoustic emissions and auditory evoked potentials in monkeys. Neurotoxicol Teratol. 1995;17:633–44.

Lautermann J, McLaren J, Schacht J. Glutathione protection against gentamicin ototoxicity depends on nutritional status. Hear Res. 1995;86:15–24.

Lee JN, Kim SG, Lim JY, et al. 3-Aminotriazole protects from CoCl2-induced ototoxicity by inhibiting the generation of reactive oxygen species and proinflammatory cytokines in mice. Arch Toxicol. 2016;90:781–91.

Li G, Liu W, Frenz D. Cisplatin ototoxicity to the rat inner ear: a role for HMG1 and iNOS. Neurotoxicology. 2006;27:22–30.

Li P, Ding D, Salvi R, Roth JA. Cobalt-induced ototoxicity in rat postnatal cochlear organotypic cultures. Neurotox Res. 2015;28:209–21.

Liang GH, Jarlebark L, Ulfendahl M, Moore EJ. Mercury (Hg2+) suppression of potassium currents of outer hair cells. Neurotoxicol Teratol. 2003;25:349–59.

Liberman MC, Dodds LW. Single-neuron labeling and chronic cochlear pathology. III. Stereocilia damage and alterations of threshold tuning curves. Hear Res. 1984;16:55–74.

Liu X, Zheng G, Wu Y, et al. Lead exposure results in hearing loss and disruption of the cochlear blood-labyrinth barrier and the protective role of iron supplement. Neurotoxicology. 2013;39:173–81.

Low WK, Sun L, Tan MG, Chua AW, Wang DY. L-N-Acetylcysteine protects against radiation-induced apoptosis in a cochlear cell line. Acta Otolaryngol. 2008;128:440–5.

Lynch ED, Gu R, Pierce C, Kil J. Reduction of acute cisplatin ototoxicity and nephrotoxicity in rats by oral administration of allopurinol and ebselen. Hear Res. 2005;201:81–9.

Marcotti W, van Netten SM, Kros CJ. The aminoglycoside antibiotic dihydrostreptomycin rapidly enters mouse outer hair cells through the mechano-electrical transducer channels. J Physiol. 2005;567:505–21.

Marcus DC, Thalmann R, Marcus NY. Respiratory rate and ATP content of stria vascularis of guinea pig in vitro. Laryngoscope. 1978;88:1825–35.

Marnett LJ. Lipid peroxidation-DNA damage by malondialdehyde. Mutat Res. 1999;424:83–95.

McCord JM, Fridovich I. Superoxide dismutase. An enzymic function for erythrocuprein (hemocuprein). J Biol Chem. 1969;244:6049–55.

McFadden SL, Ding D, Reaume AG, Flood DG, Salvi RJ. Age-related cochlear hair cell loss is enhanced in mice lacking copper/zinc superoxide dismutase. Neurobiol Aging. 1999;20:1–8.

Montuschi P, Barnes PJ, Roberts LJ II. Isoprostanes: markers and mediators of oxidative stress. FASEB J. 2004;18:1791–800.

Morata TC, Dunn DE, Sieber WK. Occupational exposure to noise and ototoxic organic solvents. Arch Environ Health. 1994;49:359–65.

More SS, Akil O, Ianculescu AG, Geier EG, Lustig LR, Giacomini KM. Role of the copper transporter, CTR1, in platinum-induced ototoxicity. J Neurosci. 2010;30:9500–9.

Mujica-Mota MA, Ibrahim FF, Bezdjian A, Devic S, Daniel SJ. The effect of fractionated radiotherapy in sensorineural hearing loss: an animal model. Laryngoscope. 2014;124:E418–24.

Mukherjea D, Jajoo S, Whitworth C, et al. Short interfering RNA against transient receptor potential vanilloid 1 attenuates cisplatin-induced hearing loss in the rat. J Neurosci. 2008;28:13056–65.

Mukherjea D, Ghosh S, Bhatta P, et al. Early investigational drugs for hearing loss. Expert Opin Investig Drugs. 2015;24:201–17.

Muthusamy S, Peng C, Ng JC. Effects of binary mixtures of benzo[a]pyrene, arsenic, cadmium, and lead on oxidative stress and toxicity in HepG2 cells. Chemosphere. 2016;165:41–51.

Nuttall AL. Sound-induced cochlear ischemia/hypoxia as a mechanism of hearing loss. Noise Health. 1999;2:17–32.

Ohinata Y, Miller JM, Altschuler RA, Schacht J. Intense noise induces formation of vasoactive lipid peroxidation products in the cochlea. Brain Res. 2000;878:163–73.

Ohinata Y, Miller JM, Schacht J. Protection from noise-induced lipid peroxidation and hair cell loss in the cochlea. Brain Res. 2003;966:265–73.

Ohlemiller KK, Wright JS, Dugan LL. Early elevation of cochlear reactive oxygen species following noise exposure. Audiol Neurootol. 1999;4:229–36.

Ozcaglar HU, Agirdir B, Dinc O, Turhan M, Kilincarslan S, Oner G. Effects of cadmium on the hearing system. Acta Otolaryngol. 2001;121:393–7.

Pan CC, Eisbruch A, Lee JS, Snorrason RM, Ten Haken RK, Kileny PR. Prospective study of inner ear radiation dose and hearing loss in head-and-neck cancer patients. Int J Radiat Oncol Biol Phys. 2005;61:1393–402.

Pan JS, Hong MZ, Ren JL. Reactive oxygen species: a double-edged sword in oncogenesis. World J Gastroenterol. 2009;15:1702–7.

Pastore A, Piemonte F, Locatelli M, et al. Determination of blood total, reduced, and oxidized glutathione in pediatric subjects. Clin Chem. 2001;47:1467–9.

Peng TI, Jou MJ. Oxidative stress caused by mitochondrial calcium overload. Ann N Y Acad Sci. 2010;1201:183–8.

Pirvola U, Xing-Qun L, Virkkala J, et al. Rescue of hearing, auditory hair cells, and neurons by CEP-1347/KT7515, an inhibitor of c-Jun N-terminal kinase activation. J Neurosci. 2000;20:43–50.

Poirrier AL, Pincemail J, Van Den Ackerveken P, Lefebvre PP, Malgrange B. Oxidative stress in the cochlea: an update. Curr Med Chem. 2010;17:3591–604.

Priuska EM, Schacht J. Formation of free radicals by gentamicin and iron and evidence for an iron/gentamicin complex. Biochem Pharmacol. 1995;50:1749–52.

Pujol R, Puel JL. Excitotoxicity, synaptic repair, and functional recovery in the mammalian cochlea: a review of recent findings. Ann N Y Acad Sci. 1999;884:249–54.

Ramkumar V, Whitworth CA, Pingle SC, Hughes LF, Rybak LP. Noise induces A1 adenosine receptor expression in the chinchilla cochlea. Hear Res. 2004;188:47–56.

Rathinam R, Ghosh S, Neumann WL, Jamesdaniel S. Cisplatin-induced apoptosis in auditory, renal, and neuronal cells is associated with nitration and downregulation of LMO4. Cell Death Discov. 2015;1.

Rich T, Allen RL, Wyllie AH. Defying death after DNA damage. Nature. 2000;407:777–83.

Roth JA, Salvi R. Ototoxicity of divalent metals. Neurotox Res. 2016;30:268–82.

Rybak LP. Hearing: the effects of chemicals. Otolaryngol Head Neck Surg. 1992;106:677–86.

Rybak LP, Ravi R, Somani SM. Mechanism of protection by diethyldithiocarbamate against cisplatin ototoxicity: antioxidant system. Fundam Appl Toxicol. 1995;26:293–300.

Rybak LP, Whitworth CA, Mukherjea D, Ramkumar V. Mechanisms of cisplatin-induced ototoxicity and prevention. Hear Res. 2007;226:157–67.

Rybak LP, Mukherjea D, Jajoo S, Kaur T, Ramkumar V. siRNA-mediated knock-down of NOX3: therapy for hearing loss? Cell Mol Life Sci. 2012;69:2429–34.

Samson J, Wiktorek-Smagur A, Politanski P, et al. Noise-induced time-dependent changes in oxidative stress in the mouse cochlea and attenuation by D-methionine. Neuroscience. 2008;152:146–50.

Schmitt NC, Rubel EW, Nathanson NM. Cisplatin-induced hair cell death requires STAT1 and is attenuated by epigallocatechin gallate. J Neurosci. 2009;29:3843–51.

Seidman MD. Effects of dietary restriction and antioxidants on presbyacusis. Laryngoscope. 2000;110:727–38.

Seidman MD, Shivapuja BG, Quirk WS. The protective effects of allopurinol and superoxide dismutase on noise-induced cochlear damage. Otolaryngol Head Neck Surg. 1993;109:1052–6.

Seidman MD, Khan MJ, Tang WX, Quirk WS. Influence of lecithin on mitochondrial DNA and age-related hearing loss. Otolaryngol Head Neck Surg. 2002;127:138–44.

Seidman MD, Ahmad N, Joshi D, Seidman J, Thawani S, Quirk WS. Age-related hearing loss and its association with reactive oxygen species and mitochondrial DNA damage. Acta Otolaryngol Suppl. 2004;16–24.

Sha SH, Schacht J. Salicylate attenuates gentamicin-induced ototoxicity. Lab Invest. 1999;79:807–13.

Sha SH, Taylor R, Forge A, Schacht J. Differential vulnerability of basal and apical hair cells is based on intrinsic susceptibility to free radicals. Hear Res. 2001a;155:1–8.

Sha SH, Zajic G, Epstein CJ, Schacht J. Overexpression of copper/zinc-superoxide dismutase protects from kanamycin-induced hearing loss. Audiol Neurootol. 2001b;6:117–23.

Shargorodsky J, Curhan SG, Henderson E, Eavey R, Curhan GC. Heavy metals exposure and hearing loss in US adolescents. Arch Otolaryngol Head Neck Surg. 2011;137:1183–9.

Sharma KK, Milligan JR, Bernhard WA. Multiplicity of DNA single-strand breaks produced in pUC18 exposed to the direct effects of ionizing radiation. Radiat Res. 2008;170:156–62.

Shi X, Nuttall AL. Upregulated iNOS and oxidative damage to the cochlear stria vascularis due to noise stress. Brain Res. 2003;967:1–10.

Shin YS, Hwang HS, Kang SU, Chang JW, Oh YT, Kim CH. Inhibition of p38 mitogen-activated protein kinase ameliorates radiation-induced ototoxicity in zebrafish and cochlea-derived cell lines. Neurotoxicology. 2014;40:111–22.

Sliwinska-Kowalska M, Zamyslowska-Szmytke E, Szymczak W, et al. Ototoxic effects of occupational exposure to styrene and co-exposure to styrene and noise. J Occup Environ Med. 2003;45:15–24.

Smith DI, Lawrence M, Hawkins JE Jr. Effects of noise and quinine on the vessels of the stria vascularis: an image analysis study. Am J Otolaryngol. 1985;6:280–9.

Someya S, Xu J, Kondo K, et al. Age-related hearing loss in C57BL/6J mice is mediated by Bak-dependent mitochondrial apoptosis. Proc Natl Acad Sci U S A. 2009;106:19432–7.

Song BB, Sha SH, Schacht J. Iron chelators protect from aminoglycoside-induced cochleo- and vestibulo-toxicity. Free Radic Biol Med. 1998;25:189–95.

Sullivan MJ, Rarey KE, Conolly RB. Ototoxicity of toluene in rats. Neurotoxicol Teratol. 1988;10:525–30.

Thomas AJ, Hailey DW, Stawicki TM, et al. Functional mechanotransduction is required for cisplatin-induced hair cell death in the zebrafish lateral line. J Neurosci. 2013;33:4405–14.

Tian CJ, Kim SW, Kim YJ, et al. Red ginseng protects against gentamicin-induced balance dysfunction and hearing loss in rats through antiapoptotic functions of ginsenoside Rb1. Food Chem Toxicol. 2013;60:369–76.

Tokgoz B, Ucar C, Kocyigit I, et al. Protective effect of N-acetylcysteine from drug-induced ototoxicity in uraemic patients with CAPD peritonitis. Nephrol Dial Transplant. 2011;26:4073–8.

Vaziri ND, Khan M. Interplay of reactive oxygen species and nitric oxide in the pathogenesis of experimental lead-induced hypertension. Clin Exp Pharmacol Physiol. 2007;34:920–5.

Vlajkovic SM, Lin SC, Wong AC, Wackrow B, Thorne PR. Noise-induced changes in expression levels of NADPH oxidases in the cochlea. Hear Res. 2013;304:145–52.

Wang J, Van De Water TR, Bonny C, de Ribaupierre F, Puel JL, Zine A. A peptide inhibitor of c-Jun N-terminal kinase protects against both aminoglycoside and acoustic trauma-induced auditory hair cell death and hearing loss. J Neurosci. 2003;23:8596–607.

Warchol ME. Cellular mechanisms of aminoglycoside ototoxicity. Curr Opin Otolaryngol Head Neck Surg. 2010;18:454–8.

Wassick KH, Yonovitz A. Methyl mercury ototoxicity in mice determined by auditory brainstem responses. Acta Otolaryngol. 1985;99:35–45.

Watanabe K, Hess A, Michel O, Yagi T. Nitric oxide synthase inhibitor reduces the apoptotic change in the cisplatin-treated cochlea of guinea pigs. Anticancer Drugs. 2000;11:731–5.

Watanabe K, Inai S, Jinnouchi K, et al. Nuclear-factor kappa B (NF-kappa B)-inducible nitric oxide synthase (iNOS/NOS II) pathway damages the stria vascularis in cisplatin-treated mice. Anticancer Res. 2002;22:4081–5.

Winther FO. Early degenerative changes in the inner ear sensory cells of the guinea pig following local x-ray irradiation. A preliminary report. Acta Otolaryngol. 1969;67:262–8.

Wimmer C, Mees K, Stumpf P, Welsch U, Reichel O, Suckfüll M. Round window application of D-methionine, sodium thiosulfate, brain-derived neurotrophic factor, and fibroblast growth factor-2 in cisplatin-induced ototoxicity. Otol Neurotol. 2004;25:33–40.

Wong AC, Ryan AF. Mechanisms of sensorineural cell damage, death and survival in the cochlea. Front Aging Neurosci. 2015;7:58.

Xiong M, He Q, Lai H, Wang J. Oxidative stress in spiral ganglion cells of pigmented and albino guinea pigs exposed to impulse noise. Acta Otolaryngol. 2011;131:914–20.

Yamamura K, Terayama K, Yamamoto N, Kohyama A, Kishi R. Effects of acute lead acetate exposure on adult guinea pigs: electrophysiological study of the inner ear. Fundam Appl Toxicol. 1989;13:509–15.

Yamane H, Nakai Y, Takayama M, Iguchi H, Nakagawa T, Kojima A. Appearance of free radicals in the guinea pig inner ear after noise-induced acoustic trauma. Eur Arch Otorhinolaryngol. 1995;252:504–8.

Yamashita D, Jiang HY, Schacht J, Miller JM. Delayed production of free radicals following noise exposure. Brain Res. 2004;1019:201–9.

Yuan H, Wang X, Hill K, et al. Autophagy attenuates noise-induced hearing loss by reducing oxidative stress. Antioxid Redox Signal. 2015;22:1308–24.

Zhang GX, Lu XM, Kimura S, Nishiyama A. Role of mitochondria in angiotensin II-induced reactive oxygen species and mitogen-activated protein kinase activation. Cardiovasc Res. 2007;76:204–12.

Chapter 3
Corticotropin Releasing Factor Signaling in the Mammalian Cochlea: An Integrative Niche for Cochlear Homeostatic Balance Against Noise

Douglas E. Vetter and Kathleen T. Yee

Abstract Beyond hair cells, the cochlea is composed of many cell types, and most of these cells do not directly participate in converting the acoustic signal into neural responses sent onward to the brain. Many of these "support" cells exist in niches that position them to monitor the state of the cochlea, and some are situated to signal such information to systems outside the cochlea. Others occupy positions such that they can invoke cellular responses limited to the cochlea without the need for "outside help". Inflammatory responses that occur in the inner ear are perhaps one of the best examples of this surveillance/reporting role served by the vast majority of cells in the cochlea. Understanding a complex event such as inflammation will require that we draw on many different aspects of biology. Here we will cover a wide range of topics that are likely to be of significance for understanding cochlear inflammation. These include a cochlear-based CRF signaling system that mirrors the hypothalamic-pituitary-adrenal axis, central Master clocks and peripheral clocks resident in many tissues of the body, and the molecular biology of glucocorticoids and glucocorticoid receptors. While seemingly disparate, as discussion of these topics unfolds, it will become obvious that understanding these signaling systems will be important in generating a model of cochlear inflammatory processes. Here, we seek not to cover the well-worn ground of inflammation biology. Rather, we seek to cover the signaling systems that may be involved in setting up the inflammatory state, its modulation, and its final resolution.

D. E. Vetter (✉)
Department of Neurobiology and Anatomical Sciences, University of Mississippi Medical Center, Jackson, MS, USA

Department of Otolaryngology and Communicative Sciences, University of Mississippi Medical Center, Jackson, MS, USA
e-mail: dvetter@umc.edu

K. T. Yee
Department of Neurobiology and Anatomical Sciences, University of Mississippi Medical Center, Jackson, MS, USA
e-mail: kyee@umc.edu

© Springer International Publishing AG, part of Springer Nature 2018
V. Ramkumar, L. P. Rybak (eds.), *Inflammatory Mechanisms in Mediating Hearing Loss*, https://doi.org/10.1007/978-3-319-92507-3_3

Keywords Hearing loss · Inflammation · Cochlea · CRF · Immune system · Corticosterone

1 Introduction

Inflammation may be one of the most common cellular responses to occur in the cochlea, as it has the potential to be invoked following numerous different challenges. With any type of challenge, whether from acoustic exposure to intense sounds, viral or bacterial infection, or perhaps even simple day-to-day exposures in individuals who have an enhanced metabolic response to an otherwise normal challenge, inflammation is almost surely the first response of the system. While inflammation is required to resolve many challenges, inflammation is also a long-standing suspect in many maladies, including "simple" aging. Yet, the response mounted by the cochlea to first invoke an inflammatory response, and then to modulate and finally terminate it is a relatively understudied subject. In this review, we will cover many subjects that, perhaps initially, appear to be somewhat tangential or even unrelated to inflammation in the cochlea. The goal of this chapter is not to cover the cellular/molecular process of inflammation and its consequences to hearing. Rather, we will cover the basic biology upon which inflammation is initiated and ultimately resolved, along with a relatively new signaling system resident in the cochlea that, based on similar signaling in other regions of the body, likely participates in the inflammatory processes that occur in the cochlea. It is hoped that a deeper understanding of this signaling cascade, (its basic biology and the systems it modulates) will be useful for establishing a molecular theory of how the inflammatory process occurs and is handled by the cochlea. This could also begin to reveal previously unrecognized signaling cascades that could be leveraged as new targets for drugs designed to combat hearing loss from a wide range of initiators.

The Evolution of Hypotheses Concerning Noise-Induced Hearing Loss

Previously, most research on noise-induced hearing loss (NIHL) examined damage following exposures producing permanent threshold shifts (PTS) and potential protective mechanisms against PTS. However, another form of hearing loss is characterized by temporary threshold shifts (TTS) in which hearing sensitivity naturally returns over a short period of time to pre-exposure thresholds. In a series of published papers on the phenomenon of "silent hearing loss", Kujawa and Liberman (2006, 2009, 2015) have shown that while elevated thresholds do return following TTS, suprathreshold function can be permanently compromised. While various populations of afferent fibers under inner hair cells (IHCs) have been shown to be differentially susceptible to TTS-inducing stimuli, it is not known what mechanisms are responsible for: (1) promoting the initial fiber loss; (2) allowing one population of afferent fibers (those with low threshold activation properties) to survive; and (3) most importantly, the mechanisms that limit TTS from continuing to develop into more significant, permanent loss. Less analyzed is the fact that not all TTS is the same. The same noise exposures occurring during the day that produce classic TTS instead produce greater hearing loss, much of it permanent, when delivered at night (Meltser et al. 2014; Basinou et al. 2017). Additionally, animal studies have clearly shown that pre-conditioning the auditory system to either sub-traumatic sounds (Tahera et al. 2007), or systemic stressors such as restraint (Wang and Liberman 2002) produces a protective effect against NIHL. Most individuals experience sound exposures more akin to intensities producing TTS much more often than those producing true PTS. These data begin to suggest that an underlying mechanism may exist that alters the degree of susceptibility to NIHL based on exposure levels, time of day, previous exposure history, and stress-response systems. Almost 27% of the US workforce works at night (Hamermesh and Stancanelli 2015). Data from the animal studies suggest that night and shift workers may be at risk for hearing loss beyond that encountered by their daytime co-workers. These findings may indicate circadian issues as a significant factor that needs to be assessed for hearing conservation methods going forward.

Current models of both cochlear protection against NIHL and mechanisms underlying preservation of afferent fibers following acoustic trauma do not explain the molecular basis of TTS-associated damage. Aspects of cellular metabolic insult and associated immune system signaling that individually or combined can be detrimental to hearing are very likely to be correlated, if not causative, to the observations described above. A physiologic tone (a balance of baseline activities) must exist between signaling systems critical for normal hearing (e.g. hair cell glutamatergic neurotransmission) and surveillance for and action upon metabolic insult (pro- and anti-inflammatory responses). An imbalance in this tone will lead to degradation of hearing. Examining the signaling systems that could be positioned to hold in check the TTS-like response may be an excellent way to further explore the mechanisms of cochlear tone and NIHL. The cochlear corticotropin releasing factor/hormone (CRF) signaling (Graham et al. 2011; Basappa et al. 2012) could be such a system. We have shown that a CRF signaling system exists in the mammalian cochlea that mirrors hypothalamic-pituitary-adrenal (HPA) axis signaling, and that it is critical for setting hearing thresholds and susceptibility to NIHL. Others have shown that abnormal CRF signaling is also involved in inflammation (Webster et al. 1998) and associated diseases, such as psoriasis (Tagen et al. 2007) and irritable bowel syndrome (Nozu and Okumura 2015), both diseases with an inflammatory component. These topics will be covered in other sections of this chapter. But first, what are the issues related to noise-induced damage that can lead to hearing loss?

2 Bounds Related to Current Perspectives of Noise-Induced Damage and Functional Loss: Cellular Elements of Cochlea

Most research targeting NIHL via animal experimentation focuses on hair cells, afferent synaptic contacts between IHCs and Type I spiral ganglion neurons, and/or the olivocochlear system. Additionally, the result of noise-induced damage, and thus the interpretation of whether protection has occurred following an experimental manipulation has, until very recently, focused mainly on hair cell loss, and that usually being limited to OHC loss. With the discovery of hidden hearing loss, new life has been breathed into the study of temporary threshold shifts and the associated dysfunctional state of afferent fibers and loss of specific populations of afferent synapses. These will be considered in the next sections.

Temporary Threshold Shifts (TTS) and Associated Cellular Elements at Risk for Damage

Exposure to moderate intensity sounds (typically 98–100 dB SPL, 2 h duration) produce a classically defined TTS, in which thresholds rise immediately by 20–40 dB following exposure, but over the course of 1–2 weeks, return to baseline (Liberman and Liberman 2015). Thus, TTS has historically been viewed as a relatively short-lived and fully reversible loss of hearing sensitivity without structural (i.e. hair cell or ganglion cell) loss. However, data now indicate that TTS can be accompanied by permanent cochlear injury, including supra-threshold response deficits (functional) and afferent synapse loss (structural) (Kujawa and Liberman 2006, 2009; Lin et al. 2011; Liberman and Liberman 2015). Precisely which signaling mechanisms ultimately produce threshold degradation and/or promote threshold recovery following TTS-inducing regimens are unclear, but almost certainly involve components operating in a multi-factorial or multi-modal manner. The role of low intensity inflammatory processes has never been investigated with respect to these outcomes. Whether anti-inflammatory treatments could modify the progression of TTS effects is unknown, but as will be covered in other sections, an effect could be expected.

Afferent fibers under inner hair cells (IHCs) can generally be divided into three spontaneous rate populations that also differ in ultrastructure and threshold. Low spontaneous rate fibers (in cat, ~15% of the population of fibers) typically demonstrate <0.5 spikes/second (sp/s) and high thresholds (LSR-HT fibers), while high spontaneous rate fibers (~60% of the population of fibers) maintain >18 sp/s (and can be as high as 120 sp/s), and exhibit low thresholds (HSR-LT fibers). An intermediate population (25–35% of the fiber population) exhibits spontaneous rates between 0.5 and 18 sp/s. Structural differences exist between the HSR-LT fibers and LSR-HT fibers. LSR-HT fibers are thin, mitochondrion poor (perhaps making them more prone to metabolic insult), and preferentially synapse on the modiolar side of the IHC, while the HSR-LT fibers are thicker, rich in mitochondria, and preferentially synapse with the IHCs on the pillar face (Francis et al. 2004; Taberner and Liberman 2005). These populations also differ in their susceptibility to damage. The LSR-HT fibers are prone to both age (Schmiedt and Schulte 1992) and noise-induced degeneration (Liberman and Liberman 2015), and are (together with the intermediate spontaneous rate population of afferent fibers) also the population of fibers lost following exposure to TTS-inducing noise (Furman et al. 2013). The mechanisms leading to the loss of these fibers, and conversely, salvaging of the HSR-LT fibers (these are the fibers that are initially dysfunctional in TTS, thereby producing the threshold elevations observed immediately after exposures) remains unclear beyond the simple observation that HSR-LT fibers contain greater numbers of mitochondria that may impart resistance to oxidative stress. Using CRFR1 null mice, we observed an increase in baseline ABR thresholds, and a loss of pillar-side afferent fibers on IHCs—the precise location of HSR-LT fibers (Graham and Vetter 2011). Could CRF signaling within the cochlea be responsible for their survival, and if so, what mechanism may be controlled by CRF signaling to produce these effects?

Previously Described Protective Mechanisms Against NIHL

Currently, two primary *extra-cochlear* signaling mechanisms serve as models of protection against NIHL:

1. *The Olivocochlear System*—The prevalent theory of cochlear protection against noise-induced damage is that olivocochlear (OC) system activation is protective against NIHL (Rajan 1988a, b; Rajan and Johnstone 1988). This model has not changed substantially over the past several decades. While evidence exists that OC stimulation can provide some protection against NIHL under specific conditions, long standing concerns continue to exist regarding the extent of the protective nature of the OC system, and serve to highlight some of this concept's potential shortcomings (Kirk and Smith 2003; Maison et al. 2013). Additionally, not all potentially damaging sounds appear to efficiently activate the OC system. For example, discrepancies over the intensity levels normally encountered in the environment versus those required for OC-based protection in the lab (typically greater than 100 dB) suggest that the OC system may not protect against most noise exposures outside of the laboratory setting (Kirk and Smith 2003). Slow effects from the medial OC (MOC) system normally assumed to underlie protection (Liberman and Gao 1995) are maximal only at high frequencies (Sridhar et al. 1995), and require non-physiologically intense long duration OC stimulation for its generation (Sridhar et al. 1995). Unilateral surgically de-efferented ears show no difference in threshold shifts compared to non-lesioned ears following *moderately* intense noise exposure, suggesting that the outcome is no

different in the absence or presence of the "protective" MOC (Liberman and Gao 1995; Zheng et al. 1997). Studies employing electrical or acoustic stimulation of the OC bundle often fail to provide evidence of significant protection against acoustic injury (Liberman 1991). While, in one report (Maison et al. 2013), experimental evidence has been presented that the OC system does function to protect against moderate sound-induced damage, the experimental design involves a relatively non-physiologic stimulus paradigm (1 week continuous noise exposures) that limits interpretation of the results. Further, a recent clinical report found no significant correlation between measures of efferent suppression of distortion product otoacoustic emissions (DPOAEs), indicative of OC system activation, and protection against TTS (Hannah et al. 2014). These and other reports spanning almost three decades of research underscore the lingering concerns regarding the relative efficacy and primacy of the protective nature of the OC system.

2. *The Hypothalamic-Pituitary-Adrenal (HPA) Axis*—Another model for protection against NIHL involves stress activated HPA axis signaling (Wang and Liberman 2002; Tahera et al. 2006a, 2007; Mazurek et al. 2010; Meltser and Canlon 2011), but evidence supporting this idea remains controversial and incomplete. First, exposure to noise levels typically producing TTS may not be sufficiently stressful to trigger activation of the HPA axis (Burow et al. 2005). Second, the time course of systemic responses is of concern. HPA axis activation induces release of adrenocorticotropin hormone (ACTH) into the bloodstream, but plasma ACTH levels do not peak until 2–30 min (Rivier et al. 2003) following HPA axis activation, and levels of plasma corticosterone (CORT), the effector molecule of HPA axis activity, peak 10–30 min after ACTH (De Souza and Van Loon 1982). Because it has been shown that synapse loss following TTS induction is observed immediately following the 2 h noise exposure (Liberman and Liberman 2015), this argues against a *primary* role as an early response protective system. Further, publications have linked HPA axis activity to the protective effects of pre-conditioning associated with exposures to moderate-intensity noise. However, in animals with unilateral reversible auditory meatus occlusion, pre-conditioning protection occurs only in the exposed ear (Yamasoba et al. 1999). Activation of the systemic HPA axis shoud have protected *both* ears. Additionally, pre-conditioning protection is lost when the pre-conditioning and subsequent traumatizing stimuli are separated by more than 3-octaves (Franklin et al. 1991; Miyakita et al. 1992; Subramaniam et al. 1992). These results argue *against* HPA axis activation as a primary source of protection against NIHL, but begin to hint at local (cochlear) signaling protecting against NIHL and/or a temporal "hand-off" between local immediate responses and delayed systemic responses. Finally, the only studies (to our knowledge) that sought to link HPA axis activity to protection against TTS-associated pathologies employed systemic drug delivery of RU486, a glucocorticoid receptor antagonist, and/or metyrapone, a glucocorticoid synthesis inhibitor (Tahera et al. 2006b). Data obtained from those studies, however, do not take into account that these drugs would also block the very same signaling system (cochlear HPA-like

signaling, see more in section below) now known to exist in the cochlea, confounding the results and making it impossible to assign relevance to the HPA axis as a protective system using this experimental design. A re-evaluation of systemic HPA axis and NIHL should be pursued in light of our findings of an HPA-like signaling system in the cochlea, described next.

3 Elements of Corticotropin Releasing Factor (Hormone) Signaling Systems, and Their Expression in the Mammalian Inner Ear

Broadly speaking, there are two corticotropin releasing factor receptors (CRFRs), CRFR1 and CRFR2. We have demonstrated expression of CRF (Fig. 3.1) and both CRF receptors in the cochlea, in the cochlea (Fig. 3.2) (Vetter et al. 2002; Graham et al. 2010; Graham and Vetter 2011). CRF receptors are members of the secretin-like G-protein coupled receptor (GPCR) family, also known as clan B GPCRs. Clan B receptors are distinguished by their relatively large N-termini of approximately 120 amino acids that include several cysteine residues that produce disulfide bridges. This region is important for binding ligands. Immediate downstream signaling of CRF receptors occurs via coupling with either G_s, thereby activating the adenylate cyclase pathway, and/or G_q, activating the phospholipase C/diacylglycerol pathways. Potential splice variants are numerous for CRFRs, with 12 verified and predicted splice variants for CRFR1, and 10 splice variants predicted or verified for CRFR2 (Slominski et al. 2013). In addition to the variants of the CRFRs, there are multiple structurally related peptides in addition to CRF itself that can bind the CRF receptors. Four CRF-like peptides are currently known to exist that are functional at the CRFRs: CRF, urocortin (Ucn), stresscopin (also known as urocortin III) and stresscopin-related peptide (also known as urocortin II). While CRF has a high affinity for CRFR1, its affinity is relatively low for CRFR2. Stresscopin and stresscopin-related peptide are active only at CRFR2, and exhibit no affinity for CRFR1. This is to be predicted, because both stresscopin and stresscopin-related peptide have low homology with CRF and urocortin. Urocortin (sometimes also termed Ucn1) has affinity for both CRFR1 and CRFR2, although its affinity for CRFR2 is more than 100 times greater for CRFR2 compared to CRFR1.

CRF is well recognized as the peptide released by the hypothalamus to initiate pro-opiomelanocortin (POMC) cleavage, producing among other products, ACTH that is then released into the blood stream as a true hormone. But CRF is also expressed in the brain, and is in numerous peripheral tissues. This includes expression in skin, retinal pigment epithelium, adrenal medulla, testes and ovaries, heart and blood vessels, the GI tract, pancreas, lungs, endometrium and placenta. A full accounting of this "peripheral" CRF expression is beyond the scope of this review, but information can be found in numerous reviews on the subject (Boorse and Denver 2006). However, of particular interest with respect to the topic of

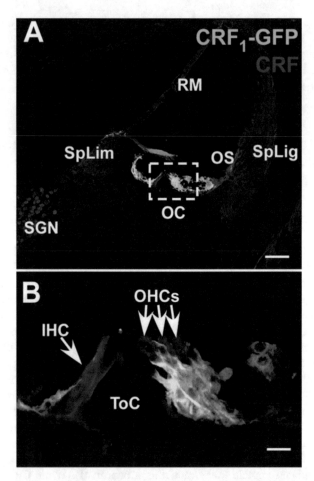

Fig. 3.1 CRFR1 expression overlaps and juxtaposes sites of CRF expression suggesting paracrine and juxtacrine signaling. (**a**) Immunofluorescent detection of GFP (green) driven by the CRFR1 promoter demonstrates expression in the inner sulcus (IS) and support cells lateral to the organ of Corti (OC). Double label with CRF (red) reveals regions of overlapping expression (inner sulcus and lateral support cells) suggesting the possibility of paracrine/autocrine signaling. Intense immunoreactivity is also observed in the organ of Corti (boxed), shown at higher magnification in (**b**). (**b**) At higher magnification, intense CRF$_1$-GFP immunofluorescence is observed in the Deiter's cells and in the border cell (BC). Overlay with CRF reveals that these CRFR1-positive cells juxtapose cells expressing CRF, including the inner hair cell (IHC) and outer hair cells (OHCs, arrows). *SGN* spiral ganglion neurons, *SpLim* spiral limbus, *OS* outer sulcus, *SpLig* spiral ligament, *RM* Reissner's membrane, *ToC* tunnel of Corti. Scale bars: (**a**) 60 μm; (**b**) 10 μm

Fig. 3.2 Non-radioactive in situ hybridization (Digoxigenin-based) with a pan-CRFR probe reveals extensive CRFR1/R2 expression in cellular populations along the basilar membrane, as well as in regions of the lateral wall. In the latter, there is heavy expression in regions known to harbor Type 1 and Type 2 fibrocytes. Boxed region in top panel indicates region of high mag visualization in bottom panel

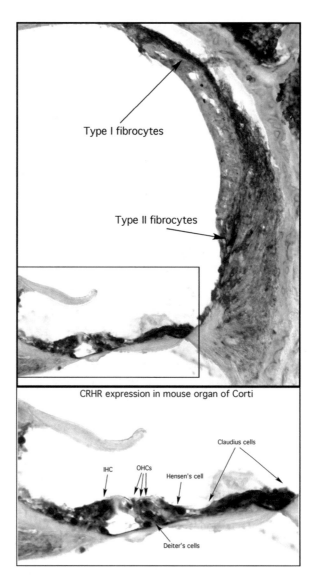

inflammation is that CRF and CRFRs are expressed by most immune cells, including lymphocytes, mast cells, and macrophages (Dermitzaki et al. 2018). This will be discussed further, below.

A Novel Theory of Local Cochlear Signaling Systems Protective Against Cellular Damage and Associated NIHL

We would argue, based on results from our lab and the longstanding controversies over current models of cochlear protection, that a paradigm shift away from current theory is required to advance our understanding of cochlear protective systems. In our ongoing efforts to identify the underlying cellular signaling involved in protection against functional and structural damage typically resulting in NIHL, we have described the expression of proteins in the mouse cochlea that collectively represent a corticotropin releasing factor (CRF)–based signaling system. This system includes local expression of all the major stress-response signaling molecules (POMC, ACTH, CRFR1 and CRFR2) commonly associated with HPA axis-mediated signaling (Fig. 3.3) (Graham et al. 2011; Graham and Vetter 2011). Many cells within the organ of Corti express CRF, the ligand for the CRF receptors. Unlike the hierarchical arrangement of information flow along the systemic HPA axis, however, the cochlear system is arranged in a manner more conducive to paracrine signaling. In this model, activation of these receptors initiates an HPA-like cochlear stress axis response. One may envision a spreading involvement of cells releasing CRF that expands with increasing intensity of sound and consequent organ of Corti displacement. Thus, the summed magnitude of CRF release offers functional feature selectivity in the model, and may encode the magnitude of local cellular stress correlated with sound exposures. Two possible mechanisms of homeostatic maintenance in the face of sound over-exposure follow from the expression pattern of the receptors. First, shearing forces that occur during exposures to moderate sound intensities may activate CRF signaling to either induce a pre-conditioning protective effect, or when faced with moderately intense sounds, protect against, or limit, significant structural and excitotoxicity damage characteristic of TTS. Second, with increasing sound exposures, CRF signaling in lateral support cells (as one example) may modulate ion recycling/homeostasis, thereby maintaining proper ionic balance and protecting hair cells against damage or death. Since this "cochlear stress axis" is completely contained within the cochlea, no delays would be incurred between acoustic over-exposure and activation of protective mechanisms, in contrast to the current models of cochlear protection based on extra-cochlear feedback-based mechanisms.

Importantly, we have shown that global ablation of CRFR1 (thus leaving only CRFR2 as the signalling system for CRF release) results in significant elevation of auditory thresholds even when supplemental corticosterone is provided (Graham and Vetter 2011), while ablation of CRFR2 (thus leaving only CRFR1 as the signal system for CRF release) results not only in significantly greater hearing sensitivity, but also a greater degree of hearing loss following exposure to loud (100 dB) sound,

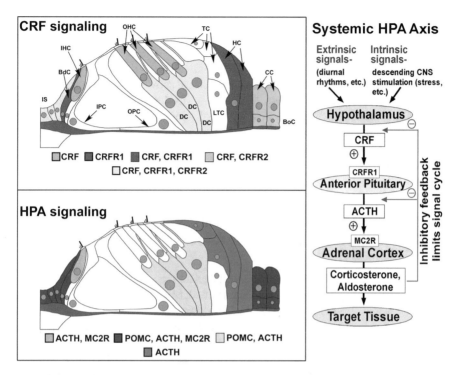

Fig. 3.3 Expression of CRF signaling and HPA-related molecules in organ of Corti. Hair cells express CRF, but not CRF receptors. CRF receptors are expressed widely throughout the cochlear support cell populations. Unlike the hierarchical arrangement of the systemic HPA axis pathway illustrated on the right, the cochlear HPA-equivalent system is assembled to allow paracrine signaling between local cell populations. Ultimate signal output of the systemic HPA axis is release of steroid hormones from the adrenal glands following ACTH stimulation of the MC2R receptor. In the cochlea, MC2R is expressed exclusively by support cell populations flanking the organ of Corti. *IS* inner sulcus, *BdC* Border cell, *IHC* inner hair cell, *IPC* inner phalangeal cell, *OPC* outer phalangeal cell, *OHC* outer hair cell, *DC* Deiter's cell, *TC* tectal cell, *LTC* lower tectal cell, *HC* Hensen's cell, *CC* Claudius cells, *BoC* Boetcher's cells

and even low/moderate intensity (50 dB) noise not usually injurious to wild type mice (Graham et al. 2010). Our data indicate that this occurs at least partly by abnormally high GluR expression in cochleae of CRFR2 null mice and loss of normal Akt signaling, known to function as a protective signaling cascade against glutamate excitotoxicity in the hippocampus (Kim et al. 2002). These data highlight the need to balance CRFR1 and CRFR2 activity within the cochlea.

Finally, we have recently provided preliminary evidence that activation of the cochlear HPA-equivalent system triggers the local release of steroid hormones (Vetter and Yee, ARO abstract 370, 2017). These data suggest a significant protective role against hearing loss for local CRF-induced steroid signaling originating within the cochlea. Despite data highlighting shortcomings of previous cochlear protection models, it is critical to recognize that extant data also suggest that *each of the current major models described above could explain cochlear protection under certain circumstances.* It is possible, perhaps even likely, that a continuum exists in the cochlea's response to sound that includes aspects of the major models previously proposed, plus cochlear stress axis signaling. We envision that in this hybrid model, there is an initial immediate local activation of the cochlear stress axis by moderate intensity sounds, extending to involvement of the systemic HPA axis with higher intensity exposures, and finally including the classic OC system effects under the most intense sound exposures when modulation of basilar membrane mechanics is required to minimize damage. This model predicts cross talk and synergistic activity between protective signaling systems.

4   Inflammatory/Immune Responses in the Cochlea Are Driven by Acoustic Over-Exposure

The cochlea has two major responses to acoustic over-exposure. The first is the well-recognized physical response to sounds that, at a minimum, can result in loss of synapses and afferent fibers under the IHCs (known as silent hearing loss following TTS), and in the extreme, results in rupture of the reticular lamina, destruction of OHCs, etc. The second major response is an immune response that can lead to metabolic damage (lipid peroxidation, as one example) if allowed to proceed unchecked. The immune response primarily involves migration of immune system cells into the cochlea, along with biochemical responses of the fibrocyte populations and stria cells that results in production and release of pro-inflammatory cytokines (Fujioka et al. 2006). Marrow-derived cells carrying molecular markers for macrophages invade the cochlea after acoustic trauma (Hirose et al. 2005; Tornabene et al. 2006). In addition, it has been shown that the vast majority of genes up-regulated following acoustic trauma are involved in immune defense functions (Yang et al. 2016).

Most models of systems involved in protection against NIHL typically suggest a mechanism by which protection occurs following activation of the system under study. In the case of olivocochlear protection, it is clear that a mechanical effect on basilar membrane motion ensues following olivocochlear activity, and that resultant dampening of mechanical shearing forces produces some degree of protection. Damage associated with metabolic insult includes an up-regulation of molecular stressors such as oxidative damage via free radicals, excitotoxicity related to exuberant glutamatergic neurotransmission, and inflammation. Models of protection against metabolic insult then describe mechanisms that involve alterations of metabolic states, scavenging free radicals, etc. If the cochlear CRF signaling system is involved in protection against damage following noise exposure, we must uncover unique mechanisms by which it could produce protection. One often overlooked pathway to damage involves an interaction between the immune system, pro- and anti-inflammation pathways, the cells of the cochlea, and signaling systems that initiate and terminate these molecular/cellular responses.

In general, protection against metabolic damage, assumed to be the product of inflammatory responses, has invoked HPA axis activity. Given that the end-product of HPA axis activation is, among other things, the release of steroid hormones such as corticosterone (cortisol in humans), the logical assumption has been that the HPA axis is important for limiting inflammation in the inner ear. Certainly, much evidence exists that corticosterone dampens inflammation throughout the body, in addition to numerous other actions (gluconeogenesis, as an example). But absent from the coverage of this topic up to now has been a cogent discussion of what the induction mechanism is that sets into action the inflammation reaction presumed to follow an acoustic challenge. The inflammatory response includes an up-regulation of pro-inflammatory mediators that include altered signaling via cytokines and chemokines. Thus, while cytokines must enter into such an equation, what is the initiation signal for cytokine release that then leads to the full inflammatory response?

CRF Signaling Is Involved with the Inflammatory Process

Inflammation is a well-defined process that occurs in the inner ear. As described above, acoustic trauma induces both immune cell migration into the cochlea (Hirose et al. 2005), and local inflammatory and immune signaling (Fujioka et al. 2014); the majority of genes up-regulated following acoustic trauma serve an immune defense function (Yang et al. 2016); and many immune system-related genes associated with Toll-like receptor activity are expressed in the cochlea, predominantly in the supporting cell population (Cai et al. 2014). Yet, relatively little is understood concerning protection against metabolic (immune-related) damage, and what is known is most often couched in responses following high–intensity exposure, not responses to moderate sound intensity associated damage. Given that CRF signaling occurs in the inner ear, a dichotomy therefore exists between a local CRF signaling system expressed in the cochlea and a systemic HPA axis. Both seem to function similarly, and therefore one is left with trying to unravel the advantage of having two systems that could flood the inner ear with steroid hormones to battle metabolic disruption. However, distilling CRF signaling down to a simple mechanism useful only in release of steroid hormones is ignoring a complex and burgeoning field of cell:cell signaling involving CRF.

CRF signaling is intimately involved in immune signaling. It is expressed in immune system organs such as the thymus and spleen, while also being expressed by various immune cells, including lymphocytes and monocytes (Karalis et al. 1997). Paradoxically, CRF is both a pro-inflammatory and an anti-inflammatory molecule. This apparent contradiction can be resolved by recognizing that differential expression patterns of CRF can explain varied modes of action. Local (peripheral) CRF plays a *direct* role as an autocrine/paracrine pro-inflammatory cytokine (e.g. Karalis et al. (1991)), while systemic CRF (via its actions on the HPA axis) act *indirectly* as an anti-inflammatory molecule, via its well-known stimulatory role in glucocorticoid (corticosterone/cortisol) release.

5   Local (Cochlear) CRF Signaling Versus Systemic HPA Axis: A Tale of Timing, Balance of Mechanistic Actions, and Direct Versus Indirect Signaling

CRF signaling acts at two different levels. One is via its actions at the systemic level, and this occurs in an indirect manner via its actions on steroid hormone release from the adrenal glands. While this is certainly important for body-wide integration of a stress response, one may argue that the mechanism by which CRF functions, in a local (direct) manner, is equally important, but less well studied compared to the systemic (indirect) responses of CRF. Direct CRF signaling occurs without intermediaries, and involves neurons, immune cells, and cells within the cochlea, as a few examples.

The Indirect CRF Signaling System

CRF signaling is perhaps most well-known for its role in orchestrating anti-inflammatory signaling. This is produced, indirectly, by suppression of a large number of cytokines (Kunicka et al. 1993) due to the actions of the glucocorticoids released by the adrenal glands in response to CRF signals. Because factors that initiate CRF signaling between the hypothalamus and the pituitary is commonly considered to be any manner of "stress", CRF signaling is typically considered to be an arm of classic paracrine/endocrine signaling mediating responses emanating from the intersection of neuronal encoding of stress (cellular and systemic) and immune system function.

The Direct CRF Signaling System, and Its Potential Role in Cochlear Function and Protection from Noise-Induced Damage

Our previously published work has shown significant expression of CRFR1 in the lateral wall and support cells of the cochlea (Vetter et al. 2002). The lateral wall accumulates increased numbers of various immune cells following noise exposure, as described above. We have continued to explore the expression dynamics of CRF in the cochlea, and our preliminary data demonstrate that not only is CRF present in regions that respond to acoustic trauma with an immune response, but that there is a significant up-regulation of CRF expression in these areas in response to noise exposure (manuscript in preparation).

The significance of CRF and CRFRs in these regions of the cochlea can seem perplexing if one considers CRF signaling simply as the initiator of a cascade of events that ultimately end in steroid hormone release, i.e. only working as described for the systemic, indirect mechanism. Following this argument, CRF signaling should ultimately result in anti-inflammatory cascades. Assuming only this mechanism of action, one is left with an impression of redundancy for cochlear CRF signaling; that peripheral CRF signaling in, for example the cochlea, will act identically to the systemic activities induced by hypothalamic CRF signaling to the pituitary. But it has been known for decades that numerous immune system cells such as lymphocytes and macrophages express CRFRs and release POMC cleavage products such as ACTH and beta-endorphin (Kavelaars et al. 1989, 1990). This supports the idea that CRF signaling can *directly* modulate immune responses to challenges (Webster et al. 1997a, b). By considering that CRF signaling effects are not limited to classic adrenal responses that simply produce and release steroid hormones, CRF signaling can be modeled as a fast, direct pro-inflammatory signal, effectively acting as an autocrine or paracrine inflammatory cytokine, that only later, following systemic HPA axis activity, and local glucocorticoid release, reverts to an anti-inflammatory response. Although on the surface of the argument, this is apparently complex, this combined pro- and anti-inflammatory role should not be unexpected. The immune response is a balance between the need to immediately address a problem (cellular damage, invading pathogen, etc.), followed by shutting down the initial response as part of a final resolution step. Without resolution, the initial response of inflammation, which includes recruitment of numerous cellular debris scavengers, as well as signaling systems that are potent inducers of cell death, can begin to damage the surrounding normal tissue. This "chronic inflammatory state" can lead to disease states that in other tissues/organs include such pathologies as chronic pain and tissue destruction that could even lead to autoimmune events. Examples of disease states with a suggested etiology of chronic inflammation include rheumatoid arthritis (Crofford et al. 1992), colitis (La et al. 2008), and psoriasis (Reich 2012; Rivas Bejarano and Valdecantos 2013). As the panoply of molecular signals modulated by noise increasingly begins to highlight CRF-mediated cochlear responses, a novel set of potential therapeutics that hold the promise of treatment for such issues as idiopathic or noise-related hearing loss begins to emerge. These therapeutics include CRFR antagonists. Experiments designed to investigate the utility of such drugs have included investigations of CRFR antagonists as anti-inflammatory drugs for the GI tract (Wlk et al. 2002). However, more work needs to be done examining the role of CRF signaling and the potential for using CRFR antagonists in a therapeutic manner. Various studies have shown both a protective effect of CRF against neurodegenerative states (Pedersen et al. 2001; Hanstein et al. 2008), and protection from ensuing neurodegeneration (Carroll et al. 2011), and a limitation of infarct size following, traumatic brain injury (Roe et al. 1998) with use of CRFR antagonists.

6 Of Glucocorticoids, Clocks, Entrainment, and a Potential Interaction Between Systemic and Peripheral CRF Signaling Systems Important for Inflammatory Responses in the Cochlea

What has been missing in research into mechanisms underlying cochlear protection from noise-induced damage, and responses to viral and bacterial infection, etc., is a hypothesis-driven approach that views the cochlea as an organ first and foremost, despite its specialized activity, and as such, is subject to interactions with the wider biology of the body. Viewing the cochlea first as an organ also helps to bring into focus the potential role(s) of the supporting cells, and allows one to move away from a "neuro/hair cell only" approach to understanding hearing loss and protective systems. The idea that an insult can elicit a body-wide response that can impact the cochlea has more often been undervalued, misinterpreted as a trauma affecting the cochlea alone (divorced from systemic responses), or even ignored. In the same way, it must be recognized that while generalized (non-auditory) trauma can affect the cochlea, basal systemic physiology can also underlie normal cochlear processing and the ability of the cochlea to respond to more directed (auditory) challenges. In this section, we will begin describing examples of these phenomena as they may pertain to inflammatory responses of the cochlea.

Glucocorticoid Function

One is hard pressed to find a cell that does not respond to glucocorticoids. While the actions of glucocorticoids are diverse, the most well studied effects of glucocorticoids are on carbohydrate metabolism and immune system function. With respect to carbohydrate metabolism, glucocorticoids maintain normal concentrations of blood glucose. This is accomplished by the most well-known mechanisms of action of glucocorticoids: stimulation of gluconeogenesis in the liver, mobilizing amino acids from tissues other than the liver for use as substrates for gluconeogenesis, and the breakdown of fat from adipose tissue to be used as secondary energy sources, while the glycerol can also be used as substrate for gluconeogenesis. With respect to immune system function, glucocorticoids have effects on both the innate and adaptive immune systems. With respect to the innate immune system, glucocorticoids decrease the complete blood count of eosinophils, monocytes, and lymphocytes, while raising the total count of neutrophils. At a cellular level, leukocytes have a reduced ability to adhere to the vascular endothelium (resulting from a loss of adhesion molecule expression by the leukocytes and endothelial cells), thus preventing diapedesis (and hence, the rise in neutrophilia). This directly impairs the initial inflammatory response at the site of injury. Monocytes and macrophages exhibit a diminished production of cytokines. Lymphocytes, especially T-cells, are redistributed to the spleen, lymph nodes, and bone marrow, keeping them out of the circulation. Mast cell degranulation is inhibited by glucocorticoids, while inhibition of cytokine production in Mast cells occurs via transcriptional repression of these genes. Concerning the adaptive immune system, glucocorticoids induce a reduction in circulating dendritic cells via apoptotic mechanisms on dendritic cells, and on their precursor cells. Finally, glucocorticoids can also (at high doses) significantly deplete T-cells via numerous mechanisms that include impaired release from lymphoid tissue, induction of apoptosis, and loss of interleukin-2, which is a major growth factor for T-cells, and associated downstream IL-2 signaling. Expanded coverage of these subjects can be found in numerous reviews (Barnes 2010; Coutinho and Chapman 2011).

We have presented preliminary data showing that the cochlea produces and releases its own glucocorticoids and mineralocorticoids (specifically, corticosterone and aldosterone; Vetter and Yee, ARO abstract 377, 2017, manuscript in preparation). Because the data have yet to be peer reviewed, it is inappropriate to comment extensively on that data, but assuming for the sake of argument that the cochlea proves to be an extra-adrenal steroidogenic tissue (joining the retina, skin, GI tract, and other tissues in this category), one must ask the question "What advantage would the cochlea enjoy by fully recapitulating the systemic HPA axis?" We endeavor to answer this question below.

Glucocorticoid Receptors

Twenty-eight years ago, the first (Rarey and Luttge 1989) of a series of studies (ten Cate et al. 1992, 1993; Rarey et al. 1993) was published demonstrating the expression of glucocorticoid receptors in the mammalian inner ear. This group went on to show that acoustic stress (Rarey et al. 1995) and then generalized stress (induced via restraint) (Curtis and Rarey 1995) alters the level of glucocorticoid receptor expression in the cochlea, and that this in turn also alters *de novo* protein synthesis occurring in the inner ear (Yao et al. 1995). Since plasma levels of corticosterone also increase during stress events, one naturally assumed that the source of glucocorticoids impacting the cochlea was derived as adrenal output. Complexity of the story of glucocorticoid receptors increases, however, when regulation of the glucocorticoid receptor is taken into account.

Responses to glucocorticoids are determined by the expression level of the glucocorticoid receptor, and its ability to associate with transcriptional regulatory binding sites on the DNA of genes susceptible to glucocorticoid regulation (those genes harboring a glucocorticoid response element, GRE). One of the main mechanisms that glucocorticoid receptor levels are altered is by the action of glucocorticoids themselves (Svec and Rudis 1981; Svec 1985), with the presence of glucocorticoids down-regulating the expression level of glucocorticoid receptors. In addition to glucocorticoid-induced negative feedback on expression level changes, the Master Clock gene, Clock (circadian mechanisms are discussed in the next section), has been implicated in directly altering the ability of glucocorticoid receptors to bind to

DNA. This occurs through acetylation of the hinge region of the glucocorticoid receptor (Charmandari et al. 2011), presumably altering affinity for the glucocorticoid responsive element on a gene domain. Acetylation of glucocorticoid receptors was found to be highest in the morning in peripheral blood mononuclear cells, in agreement with levels of expression of various glucocorticoid responsive genes. Given the findings that time of day can impact basic physiology, what are the underlying mechanisms, and are they important for the basic physiology of the cochlea, and more importantly, to susceptibility to inflammation and other mechanisms leading to noise-induced hearing loss?

Physiological Oscillations

All terrestrial organisms are exposed to variations in the environment, from the diurnal light/dark cycle each day, to seasonal variations, and cycles existing with periods in between. Most organisms have adapted to these variations by evolving an inner timekeeping mechanism that allows the body to "predict" the coming variations in physiological needs, and alter relevant processes accordingly. A major goal of chronobiological research is to understand how these processes are communicated throughout the body, and to begin unraveling the consequences that follow dysfunction of the internal timekeeping system. Surprisingly, these dysfunctions can quickly develop into life threatening issues, but more often will result in damage to the tissue/organ under study, or at least compromise its function. Inflammation is one of the more well-known processes that is tied to rhythmic activity of the body. A discussion follows below on the basics of biological oscillations and their potential role in hearing. This will not cover the molecular biology of the timekeeping mechanisms, however. The reader will find many references covering these topics elsewhere. Rather, we seek here to cover broad topic issues that may impact auditory processes.

The Master Clock of the Suprachiasmatic Nucleus

Hypothalamic-pituitary-adrenal (HPA) axis activity is the main initiation signal for glucocorticoid release from the adrenals (while there is also a neural component, we will ignore this aspect for the sake of clarity/brevity of our arguments here, but the reader is invited to explore the role of autonomic (sympathetic) nervous system regulation of adrenal outflow for a more complete understanding of events). It is well understood that basal release of glucocorticoids occurs in an oscillatory manner, more correctly termed a circadian cycle, with an approximately (more on this later) 24-h peak-to-peak oscillation. There is a peak of release just at the time of initiating the animal's active period (in nocturnal rodents, i.e. those found in a standard lab setting, this is coincident with the time of "lights off" in the vivarium), and a trough at the start of the rest phase (in nocturnal rodents, this is coincident with "lights on"). Since the circadian response is also synchronized to the day/night cycle, the system is also often referred to as diurnal.

The HPA axis is a well-studied circadian response system that serves as a stereotypical circadian model. The amplitude of basal activity, defined as release of corticosterone (or cortisol in humans), varies by time of day. Such physiological variation follows from the organism's activities, which revolve around rhythmic events, the most obvious being the activity state (sleep/wake cycles) that is driven by the environment's normal light/dark cycles. Very briefly, cell autonomous pacemaker signals from the suprachiasmatic nucleus (SCN, considered the Master Clock) are synchronized (entrained) by detection of light via direct photic input from the retina along the retinohypothalamic pathway. This pathway starts with the melanopsin retinal ganglion cells (mRGCs), which convey non-vision forming information to regions of the brain known to be critical for circadian rhythms, pupil regulation, melatonin expression, cognition, and sleep, and have been implicated in various disease states such as Alzheimer's (Feng et al. 2016). The SCN, in turn, imparts rhythmicity of the basic physiological state of the organism, and includes control over such basal functions as thermoregulation, glucose and fatty acid metabolism, regulation of cholesterol, osmoregulation, and blood pressure (Chaix et al. 2016; Panda 2016). However, in addition to basic physiology, circadian rhythms also exert effects over complex behaviors such as the anticipatory drive to seek food, general cognition and learning and memory, social interactions, etc. (Benca et al. 2009). Finally, abnormalities in circadian physiology have been linked to various disease states and/or treatment outcomes (Bechtold et al. 2010). Of particular interest with respect to the subject of this volume is the fact that there is now accumulating evidence of circadian influences over inflammation. Patients with rheumatoid arthritis experience more severe symptoms early in the day (Straub and Cutolo 2007; Cutolo and Straub 2008), while some patients experience their greatest asthma symptoms in the evening, apparently following normal daily fluctuations in lung physiology that include narrowing of the airways and bronchial responsiveness (Ferraz et al. 2006). Disease states can also feedback and modulate the circadian clock (Takahashi et al. 2001; Okada et al. 2008), which then feeds forward and participates in physiologic dysfunction, setting up a downward spiral for homeostasis.

In addition to the basal circadian characteristics of glucocorticoid release via HPA axis activity, it has also been shown that a circadian cycle exists for glucocorticoid response to stressors. This implicates the Master clock in setting both the susceptibility and overall response amplitude to stress. In human studies, it has been shown that stress responses are lowest in the morning, a time when basal glucocorticoid levels are highest in humans (morning being the time of day for onset of activity), and stress responses are highest in the evening hours as basal circulating (circadian oscillation) glucocorticoids are at the lowest (as the rest phase begins). In these studies, CRF challenge at any time of day produced a similar ACTH response (i.e. signal to the adrenals), and therefore, the differential adrenal response must have been due to either a change in sensitivity to ACTH, or a sensitivity to glucocorticoid synthesis (Kudielka et al. 2004).

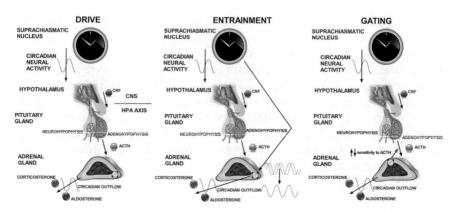

Fig. 3.4 Schematic of the HPA axis and various hypothesized interactions between the Master clock residing in the SCN and the adrenal glands. Under the Drive model, rhythms of the HPA axis drive the circadian output of the adrenal glands directly. The Entrainment model suggests that the peripheral clock in the adrenal glands and the Master clock, located in the SCN, are different, and that the peripheral clock (red waveform) floats out of phase with the absolute time-keeper of the Master clock (blue waveform). The Master clock re-registers the peripheral clock, bringing it into phase with that of the Master clock. This resets the circadian output of the adrenal glands to keep it timed properly for the rest of the body. The Gating principle suggests that the adrenal clock alters the adrenal gland's sensitivity to HPA axis activity, and therefore that adrenal output is the product of the coincidence between the adrenal clock and the phase of the Master clock

Peripheral Clocks

In addition to the central Master Clock represented by the SCN, it is now well understood that other regions of the body express clock genes and can "keep their own time", even in the absence of SCN input (e.g. under *in vitro* conditions). Other portions of the brain that express rhythmicity independent of the SCN, known as "central oscillators", include the pineal gland, arcuate nucleus, and hippocampus, among others. In addition to these central (but not Master) clocks, numerous "peripheral clocks" exist in tissues outside of the brain. These include (but are not limited to) the liver, muscle, adipose tissue, lung, ovary, testes, and heart (Oster et al. 2017). Additionally, many immune cells, including macrophages and lympho-cytes, also express their own circadian clock (Boivin et al. 2003; Bollinger et al. 2011; Borniger et al. 2017). These peripheral clocks are also termed Slave Clocks, given their sensitivity to the activity of the Master Clock of the SCN.

The adrenal glands also express their own peripheral clock, which can serve to help describe circadian outflow of glucocorticoids as well as act as a general model for Master and Slave clock interactions. The interaction between the adrenal periph-eral clock and the Master clock of the SCN is now thought to control sensitivity to the ACTH signal generated by pituitary activation (Oster et al. 2006b). Systemic circadian regulation over glucocorticoid secretion from the adrenals probably occurs via an interaction between the clocks (Fig. 3.4) Currently, such interactions are thought to take one of three forms. In the Drive model, the Master (SCN) clock drives the circadian release of glucocorticoids from the adrenal glands. In the Entrainment model, the Master clock is thought to modulate the peripheral clock, which alters the adrenal output of glucocorticoids (slaving to the Master clock circadian cycle). Finally, in the Gating model, the peripheral (adrenal) clock modulates sensitivity to the HPA signal. In reality, the exact mechanism is likely to be a hybrid between these models. For example, it is well known that peripheral clocks can begin to drift out of phase without the Master clock resetting the peripheral clock, yet without the master clock, glucocorticoids are still released.

Many cells expressing these peripheral clocks also express glucocorticoid receptors. The cells expressing peripheral clocks are therefore sensitive to the circadian release of glucocorticoids, and thus are indirectly controlled by the Master Clock (note that the SCN is one of the very few examples of a tissue that does not express glucocorticoid receptors, and therefore cannot be influenced by the circadian output of the adrenals). While some peripheral clocks maintain their rhythmicity in the absence of circadian glucocorticoid release (for example, following adrenalectomy), others slowly desynchronize (the circadian phase begins to float away from the previous cycles) without the circadian input of glucocorticoid signals (the importance of this idea will be covered below). With exposure to stressors, the normal rhythmicity of the circadian glucocorticoid output can be overridden, producing a transient up-regulation of glucocorticoids. This transient up-regulation then reconfigures (re-initiates) the peripheral clocks in tissues sensitive to circulating glucocorticoids via transcriptional activity of clock genes.

Since many peripheral clocks have been described, it is not unexpected that the cochlea also expresses clock genes. Further, it has been demonstrated that this may have functional ramifications. Thus, it has been shown that the time of noise challenge affects the final outcome of that challenge with respect to the severity of noise-induced threshold shifts (Meltser et al. 2014). When noise exposure designed to create a temporary threshold shift (TTS) was delivered during the daylight hours (in nocturnal animals such as mice, this coincides with increasing levels of circulating glucocorticoids that peak at the "lights-off" time just at the beginning of the active period), thresholds returned to normal. But when the same challenge was experienced at night (when circulating glucocorticoids are decreasing toward their nadir as the animal progresses toward the rest phase), thresholds remained elevated 2 weeks after insult. While these results have been attributed to circadian oscillations of BDNF, to date no studies have examined the role of circadian-based sensitivity to inflammation (expected based on the circadian rhythmicity of systemic glucocorticoid release/availability) and how this may also play a major role in the degree of damage following noise exposures.

Because glucocorticoids have such a wide range of physiological actions, and because glucocorticoids are under the control of circadian time signals emanating from the Master Clock, one may begin to refer to the effects of glucocorticoids as being "chronophysiological", thereby implicating that any expected actions of glucocorticoids must take into consideration the time along the 24-h circadian cycle. This comes about in large part not only from the oscillating presence of systemic glucocorticoids, but also due to the circadian influence over the expression of the glucocorticoid receptor, as touched upon above.

In light of the existence of the Master clock and various peripheral "Slave" clocks, the nature of interactions between the Master and peripheral clocks may be critical to the pathophysiological response of the cochlea to intense sounds. If one presumes that not only basal circadian oscillations of circulating glucocorticoids can impact peripheral clock tissues such as the cochlea, but also that sound challenges will elicit independent HPA axis and inner ear responses, one must consider what the outcome of this complex interaction could be. A hint at a possible scenario can be found in the brain, where a central clock residing in the oval nucleus of the Bed Nucleus of the Stria Terminalis (BNSTovl) expresses its own circadian rhythm (Amir et al. 2004). Expression in the BNSTovl of per2, a gene critical for producing circadian rhythms, is significantly down-regulated in adrenalectomized mice. This strongly suggests that the systemic circadian release of glucocorticoids (under the control of the Master clock in the SCN) is critical for maintenance of this central (not Master) clock. In general, similar to clock-based modulation of HPA axis/adrenal glucocorticoid release, there are four models of interactions between the Master clock driven circadian glucocorticoid rhythms and central and peripheral clocks (Fig. 3.5). Here we will model potential interactions that could take place in the cochlea. In the simplest model, termed the Drive model, the Master clock induces systemic circadian glucocorticoid modulation of cochlear physiology (given the expression of glucocorticoid receptors in the cochlea (Rarey and Luttge 1989; Rarey et al. 1993)), which could include release of local glucocorticoids. A closely related model, termed the Entrainment model, suggests that the systemic circadian glucocorticoid release acts on the peripheral clocks to reset the peripheral clocks, which carry out their normal physiology via transcriptome modulation as an example. A third model, the Gating model, incorporates the findings that circulating glucocorticoids can alter glucocorticoid receptor availability, thereby modulating sensitivity to Master clock signals. Further evidence of circadian influences on the transcriptome that directly impact a cell's ability to respond to circadian cues includes the adrenal gland itself, which undergoes circadian oscillations of the ACTH receptor and numerous components of the downstream signaling pathway (Oster et al. 2006a, b). Thus, in the Gating model, the peripheral clock affects the ability, in a circadian fashion, of the glucocorticoid receptor to bind to its target DNA (see above). In addition, previous glucocorticoid exposures may also down-regulate the availability of glucocorticoid receptors. Finally, a Permissive model suggests that Master clock derived systemic glucocorticoids are required for the peripheral clock to function in a rhythmic manner, effectively producing the ability of the peripheral clock to oscillate. This model does not necessarily suggest that potentially different clocks expressed by various cells in the cochlea are fully entrained and therefore slaved to the Master clock, but that some level of glucocorticoid signal is required to allow each clock to operate, perhaps with its own cycle.

It is entirely possible that different models (or combinations of models) may explain different tissues. Regardless of the model most relevant to the peripheral tissue examined, one must recognize that the overall biology under consideration is

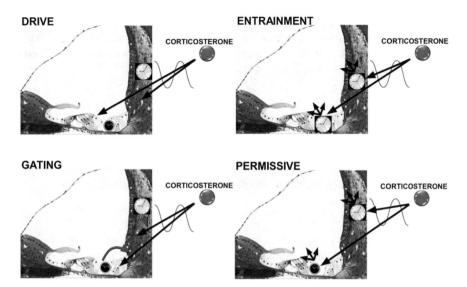

Fig. 3.5 Just as the adrenal clock plays a role in shaping glucocorticoid outflow, the peripheral clock of the cochlea could play a similar role(s) in shaping the local response to physiological challenges. It is possible that different cell populations of the cochlea could run under their own clock (signified by the different clock faces). Interactions between the cochlear clock(s) and the circadian output of the adrenals (signified by the wave function) can be modeled in one of four possible scenarios. The Drive hypothesis suggests that circadian oscillations of glucocorticoids directly impact and produce the circadian physiology of the cochlea, which would include circadian sensitivity to NIHL, as an example. The Entrainment suggests that the circadian release of glucocorticoids synchronizes the various peripheral clocks of the cochlea to be in phase with the Master (absolute) clock. This would then establish a strict diurnal rhythm to cochlear physiology. In the Gating model, much like the Gating model of the adrenal glands, sensitivity to circadian glucocorticoid levels are modulated by the peripheral clocks. This model can result in complex interplays between local cochlear, and extra-cochlear signals if different clocks exist in the cochlea. This model could also explain the differential susceptibility of certain cells to noise-induced damage (for example, the root cells in the lateral wall being the first to suffer morphological damage/ loss following exposures to intense sounds). Finally, the Permissive model indicates that the circadian release of glucocorticoids is required for the cochlear clocks to establish their own rhythmic physiology

complex, and simple questions may be posited—what biological advantage is attained by this complexity, and what pressures may be in place to maintain it? While the second question may be difficult to answer in any but a philosophical way, it is clear that the entire system of multiple clocks may be the simplest way of maintaining rhythmicity throughout the body. The Master clock is not perturbed by stressors of almost any kind, owing to the lack of steroid hormone receptors expressed by the SCN clock. Thus, the central clock acts as a standard clock that slaves the peripheral clocks into a daily reset of their oscillatory activity, which is important to meet the predictable stressors of the day that the particular tissue is sensitive to. When considering the complex signaling underlying inflammation, it is easy to see that multiple points of modulation exist that, when the organism is functioning properly, will limit the degree of the inflammatory response. The exact balance of the response will depend on the interactions between the Master clock and the various slave clocks, and in the cochlea, that will include the support cells and the resident and recruited immune system cells.

7 A Confluence of Biological Disciplines That Includes Circadian Biology, Immunology, and Cochlear Molecular Neurobiology Converge to Suggest a Role for Cochlear CRF Signaling in Modulating Inflammatory Responses in the Cochlea

We have covered many lines of independent analyses from various biological disciplines. These include studies on both Master and Slave clocks and modulation of their associated genes and their impacts on circadian physiology, regulation and function of immune system cells, and the molecular biology of cochlear responses to sound. These begin to converge and tell a story of which signals shape the cochlear response to acoustic challenges. These data strongly suggest that to understand the mechanisms giving rise to an integrated cochlear response to acoustic challenges, a multifaceted (multidisciplinary) theory will be required. Inclusion of data from fields outside the realm of cochlear neurobiology will be critical in advancing not only our basic understanding of these processes, but also in developing new therapies useful in combating hearing loss. This demands a treatment of the cochlea as an integrative organ within which various biological signaling, ranging from immune system signaling, to signals involved in steroid synthesis, to signals involved in classic cochlear physiology, converge. Thus, experiments designed to probe the cochlea's responses to challenges must include influences stemming from the larger system in general.

A Theory of Cochlear Inflammatory Response to Traumatic Noise Would Include the Following Aspects of Signaling

1. As previously described, the time of day that noise exposure occurs is important for setting up the response to that challenge. Response to the exposure will immediately be shaped by the circadian "place" at which the exposure occurs along the 24-h oscillation. The systemic circadian rhythm will dictate the basal level of glucocorticoids currently in circulation, and how much glucocorticoid may be released in response to the insult. The time along the circadian oscillation will also dictate how sensitive various tissues/cells are to the glucocorticoids released into the systemic circulation due to the rhythmic nature of glucocorticoid receptor expression.

2. Our preliminary data indicate that the cochlea responds to intense sounds immediately by upregulating the level of CRF expressed in the lateral wall. Because CRFRs are expressed by many immune system cells, and because CRF activates immune cells via CRFR1, CRF is recognized as a cytokine. Thus, the first steps of inflammation in the inner ear are likely to be produced in response to CRF release. Because the cochlea has a small number of resident immune cells, these will likely be activated and then amplify the response via further release of cytokines and chemokines, resulting in recruitment of more immune cells to the region. The exact cue that initiates CRF release is yet to be determined, but is likely to be exaggerated motion (absolute displacement and/or velocity) of the basilar membrane.

3. Because the cochlea contains cells that can be classified as peripheral clocks, the circulating basal level of glucocorticoids following circadian oscillatory behavior of the general HPA axis most likely resets the peripheral clock of the inner ear. This is actually an important step in preparedness for potential future stressors. Such resetting of peripheral clocks is required because the circadian cycle is approximately (not exactly) the length of one day. Without a daily reset, the peripheral clocks will slowly fall out of phase both with the absolute time of day and with each other. But as important, stressors that produce transient glucocorticoid peaks associated with HPA axis activity can reset the peripheral clocks and prepare the tissue for reactions to the initiating trauma.

4. Not all acoustic stressors will activate the systemic HPA axis, however. Additionally, as described above, the time that it takes for the HPA axis to be activated and produce a glucocorticoid response can be significantly delayed from the onset of the stressor, perhaps leaving the inner ear vulnerable during the delay. Thus, local CRF signaling can exert its direct initial pro-inflammatory role as described above. But because CRF is also expressed in support cells of the cochlea, which themselves also express the biosynthetic enzymes required to produce glucocorticoids, CRF can induce a local anti-inflammatory response by eliciting the local release of glucocorticoids. This should be recognized as a critical step in ultimately balancing the severity of the initial inflammatory response, and also in finally resolving the inflammation.

8 Conclusions

It becomes increasingly clear that a general understanding of immune system function and its regulation by glucocorticoids, along with a general understanding of chronobiology, is critically important to understanding how the cochlea responds to acoustic challenges. This is especially true when considering a complex phenomenon such as inflammation, which by definition will demand a multifaceted approach to understanding how it forms, how it is resolved, and what the consequences are when resolution does not occur in a timely manner. These discussions in combination with the finding of CRF-based signaling in the cochlea lead one to several important and novel conclusions: (1) the cochlea expresses a CRF-based signaling system seemingly mirroring the classic HPA axis; (2) the cochlea is an extra-adrenal steroidogenic tissue that produces its own glucocorticoids and expresses glucocorticoid receptors; (3) the cochlea expresses its own clock genes, and thus must be considered a peripheral clock tissue; (4) the systemic HPA axis, which itself is impacted by the central master clock in the SCN, delivers glucocorticoids throughout the body via the blood supply giving rise to the diurnal cyclic oscillation of glucocorticoid release. The manner of interaction between glucocorticoids produced by systemic HPA axis and the peripheral, cochlear, clock likely includes an entrainment of the cochlear clock by systemic glucocorticoids, a gating effect of the cochlear clock that establishes sensitivity of the cochlea to systemic glucocorticoid levels, or a permissive effect of the systemic glucocorticoids on the activity of the cochlear clock and its glucocorticoid output. There is no reason that multiple models cannot be true, especially with respect to acoustic challenges.

Acknowledgements The work from the Vetter lab described in this chapter was funded by the NIH (R01DC006258, R21DC015124), and grants from The Richard and Susan Smith Family Foundation, and the Russo Family Award.

References

Amir S, Lamont EW, Robinson B, Stewart J. A circadian rhythm in the expression of PERIOD2 protein reveals a novel SCN-controlled oscillator in the oval nucleus of the bed nucleus of the stria terminalis. J Neurosci. 2004;24:781–90.

Barnes PJ. Mechanisms and resistance in glucocorticoid control of inflammation. J Steroid Biochem Mol Biol. 2010;120:76–85.

Basappa J, Graham CE, Turcan S, Vetter DE. The cochlea as an independent neuroendocrine organ: expression and possible roles of a local hypothalamic-pituitary-adrenal axis-equivalent signaling system. Hear Res. 2012;288:3–18.

Basinou V, Park J-S, Cederroth CR, Canlon B. Circadian regulation of auditory function. Hear Res. 2017;347:47–55.

Bechtold DA, Gibbs JE, Loudon AS. Circadian dysfunction in disease. Trends Pharmacol Sci. 2010;31:191–8.

Benca R, Duncan MJ, Frank E, McClung C, Nelson RJ, Vicentic A. Biological rhythms, higher brain function, and behavior: gaps, opportunities, and challenges. Brain Res Rev. 2009;62:57–70.

Boivin DB, James FO, Wu A, Cho-Park PF, Xiong H, Sun ZS. Circadian clock genes oscillate in human peripheral blood mononuclear cells. Blood. 2003;102:4143–5.

Bollinger T, Leutz A, Leliavski A, Skrum L, Kovac J, Bonacina L, Benedict C, Lange T, Westermann J, Oster H, Solbach W. Circadian clocks in mouse and human CD4+ T cells. PLoS One. 2011;6:e29801.

Boorse GC, Denver RJ. Widespread tissue distribution and diverse functions of corticotropin-releasing factor and related peptides. Gen Comp Endocrinol. 2006;146:9–18.

Borniger JC, Walker Ii WH, Gaudier-Diaz MM, Stegman CJ, Zhang N, Hollyfield JL, Nelson RJ, DeVries AC. Time-of-day dictates transcriptional inflammatory responses to cytotoxic chemotherapy. Sci Rep. 2017;7:41220.

Burow A, Day HE, Campeau S. A detailed characterization of loud noise stress: intensity analysis of hypothalamo-pituitary-adrenocortical axis and brain activation. Brain Res. 2005;1062:63–73.

Cai Q, Vethanayagam RR, Yang S, Bard J, Jamison J, Cartwright D, Dong Y, Hu BH. Molecular profile of cochlear immunity in the resident cells of the organ of Corti. J Neuroinflammation. 2014;11:173.

Carroll JC, Iba M, Bangasser DA, Valentino RJ, James MJ, Brunden KR, Lee VM, Trojanowski JQ. Chronic stress exacerbates tau pathology, neurodegeneration, and cognitive performance through a corticotropin-releasing factor receptor-dependent mechanism in a transgenic mouse model of tauopathy. J Neurosci. 2011;31:14436–49.

Chaix A, Zarrinpar A, Panda S. The circadian coordination of cell biology. J Cell Biol. 2016;215:15–25.

Charmandari E, Chrousos GP, Lambrou GI, Pavlaki A, Koide H, Ng SS, Kino T. Peripheral CLOCK regulates target-tissue glucocorticoid receptor transcriptional activity in a circadian fashion in man. PLoS One. 2011;6:e25612.

Coutinho AE, Chapman KE. The anti-inflammatory and immunosuppressive effects of glucocorticoids, recent developments and mechanistic insights. Mol Cell Endocrinol. 2011;335:2–13.

Crofford LJ, Sano H, Karalis K, Webster EL, Goldmuntz EA, Chrousos GP, Wilder RL. Local secretion of corticotropin-releasing hormone in the joints of Lewis rats with inflammatory arthritis. J Clin Invest. 1992;90:2555–64.

Curtis LM, Rarey KE. Effect of stress on cochlear glucocorticoid protein. II. Restraint. Hear Res. 1995;92:120–5.

Cutolo M, Straub RH. Circadian rhythms in arthritis: hormonal effects on the immune/inflammatory reaction. Autoimmun Rev. 2008;7:223–8.

De Souza EB, Van Loon GR. Stress-induced inhibition of the plasma corticosterone response to a subsequent stress in rats: a nonadrenocorticotropin-mediated mechanism. Endocrinology. 1982;110:23–33.

Dermitzaki E, Venihaki M, Tsatsanis C, Gravanis A, Avgoustinaki PD, Liapakis G, Margioris AN. The Multi-faceted Profile of Corticotropin-releasing Factor (CRF) Family of Neuropeptides and of their Receptors on the Paracrine/Local Regulation of the Inflammatory Response. Curr Mol Pharmacol. 2018;11(1):39–50

Feng R, Li L, Yu H, Liu M, Zhao W. Melanopsin retinal ganglion cell loss and circadian dysfunction in Alzheimer's disease (Review). Mol Med Rep. 2016;13:3397–400.

Ferraz E, Borges MC, Terra-Filho J, Martinez JA, Vianna EO. Comparison of 4 AM and 4 PM bronchial responsiveness to hypertonic saline in asthma. Lung. 2006;184:341–6.

Francis HW, Rivas A, Lehar M, Ryugo DK. Two types of afferent terminals innervate cochlear inner hair cells in C57BL/6J mice. Brain Res. 2004;1016:182–94.

Franklin DJ, Lonsbury-Martin BL, Stagner BB, Martin GK. Altered susceptibility of 2f1-f2 acoustic-distortion products to the effects of repeated noise exposure in rabbits. Hear Res. 1991;53:185–208.

Fujioka M, Kanzaki S, Okano HJ, Masuda M, Ogawa K, Okano H. Proinflammatory cytokines expression in noise-induced damaged cochlea. J Neurosci Res. 2006;83:575–83.

Fujioka M, Okano H, Ogawa K. Inflammatory and immune responses in the cochlea: potential therapeutic targets for sensorineural hearing loss. Front Pharmacol. 2014;5:287.

Furman AC, Kujawa SG, Liberman MC. Noise-induced cochlear neuropathy is selective for fibers with low spontaneous rates. J Neurophysiol. 2013;110:577–86.

Graham CE, Vetter DE. The mouse cochlea expresses a local hypothalamic-pituitary-adrenal equivalent signaling system and requires corticotropin-releasing factor receptor 1 to establish normal hair cell innervation and cochlear sensitivity. J Neurosci. 2011;31:1267–78.

Graham CE, Basappa J, Vetter DE. A corticotropin-releasing factor system expressed in the cochlea modulates hearing sensitivity and protects against noise-induced hearing loss. Neurobiol Dis. 2010;38:246–58.

Graham CE, Basappa J, Turcan S, Vetter DE. The cochlear CRF signaling systems and their mechanisms of action in modulating cochlear sensitivity and protection against trauma. Mol Neurobiol. 2011;44:383–406.

Hamermesh DS, Stancanelli E. Long workweeks and strange hours. ILR Rev. 2015;68:1007–18.

Hannah K, Ingeborg D, Leen M, Annelies B, Birgit P, Freya S, Bart V. Evaluation of the olivo-cochlear efferent reflex strength in the susceptibility to temporary hearing deterioration after music exposure in young adults. Noise Health. 2014;16:108–15.

Hanstein R, Lu A, Wurst W, Holsboer F, Deussing JM, Clement AB, Behl C. Transgenic overex-pression of corticotropin releasing hormone provides partial protection against neurodegenera-tion in an in vivo model of acute excitotoxic stress. Neuroscience. 2008;156:712–21.

Hirose K, Discolo CM, Keasler JR, Ransohoff R. Mononuclear phagocytes migrate into the murine cochlea after acoustic trauma. J Comp Neurol. 2005;489:180–94.

Karalis K, Sano H, Redwine J, Listwak S, Wilder RL, Chrousos GP. Autocrine or paracrine inflammatory actions of corticotropin-releasing hormone in vivo. Science (New York, NY). 1991;254:421–3.

Karalis K, Muglia LJ, Bae D, Hilderbrand H, Majzoub JA. CRH and the immune system. J Neuroimmunol. 1997;72:131–6.

Kavelaars A, Ballieux RE, Heijnen CJ. The role of IL-1 in the corticotropin-releasing factor and arginine- vasopressin-induced secretion of immunoreactive beta-endorphin by human periph-eral blood mononuclear cells. J Immunol. 1989;142:2338–42.

Kavelaars A, Berkenbosch F, Croiset G, Ballieux RE, Heijnen CJ. Induction of beta-endorphin secretion by lymphocytes after subcutaneous administration of corticotropin-releasing factor. Endocrinology. 1990;126:759–64.

Kim AH, Yano H, Cho H, Meyer D, Monks B, Margolis B, Birnbaum MJ, Chao MV. Akt1 regu-lates a JNK scaffold during excitotoxic apoptosis. Neuron. 2002;35:697–709.

Kirk EC, Smith DW. Protection from acoustic trauma is not a primary function of the medial olivo-cochlear efferent system. J Assoc Res Otolaryngol. 2003;4:445–65.

Kudielka BM, Schommer NC, Hellhammer DH, Kirschbaum C. Acute HPA axis responses, heart rate, and mood changes to psychosocial stress (TSST) in humans at different times of day. Psychoneuroendocrinology. 2004;29:983–92.

Kujawa SG, Liberman MC. Acceleration of age-related hearing loss by early noise exposure: evi-dence of a misspent youth. J Neurosci. 2006;26:2115–23.

Kujawa SG, Liberman MC. Adding insult to injury: cochlear nerve degeneration after "temporary" noise-induced hearing loss. J Neurosci. 2009;29:14077–85.

Kujawa SG, Liberman MC. Synaptopathy in the noise-exposed and aging cochlea: primary neural degeneration in acquired sensorineural hearing loss. Hear Res. 2015;330(Pt B):191–9.

Kunicka JE, Talle MA, Denhardt GH, Brown M, Prince LA, Goldstein G. Immunosuppression by glucocorticoids: inhibition of production of multiple lymphokines by in vivo administration of dexamethasone. Cell Immunol. 1993;149:39–49.

La JH, Sung TS, Kim HJ, Kim TW, Kang TM, Yang IS. Peripheral corticotropin releasing hor-mone mediates post-inflammatory visceral hypersensitivity in rats. World J Gastroenterol. 2008;14:731–6.

Liberman MC. The olivocochlear efferent bundle and susceptibility of the inner ear to acoustic injury. J Neurophysiol. 1991;65:123–32.

Liberman MC, Gao WY. Chronic cochlear de-efferentation and susceptibility to permanent acous-tic injury. Hear Res. 1995;90:158–68.

Liberman LD, Liberman MC. Dynamics of cochlear synaptopathy after acoustic overexposure. J Assoc Res Otolaryngol. 2015;16:205–19.

Lin HW, Furman AC, Kujawa SG, Liberman MC. Primary neural degeneration in the Guinea pig cochlea after reversible noise-induced threshold shift. J Assoc Res Otolaryngol. 2011;12:605–16.

Maison SF, Usubuchi H, Liberman MC. Efferent feedback minimizes cochlear neuropathy from moderate noise exposure. J Neurosci. 2013;33:5542–52.

Mazurek B, Haupt H, Joachim R, Klapp BF, Stover T, Szczepek AJ. Stress induces transient audi-tory hypersensitivity in rats. Hear Res. 2010;259:55–63.

Meltser I, Canlon B. Protecting the auditory system with glucocorticoids. Hear Res. 2011;281:47–55.

Meltser I, Cederroth CR, Basinou V, Savelyev S, Lundkvist GS, Canlon B. TrkB-mediated protection against circadian sensitivity to noise trauma in the murine cochlea. Curr Biol. 2014;24:658–63.

Miyakita T, Hellstrom PA, Frimanson E, Axelsson A. Effect of low level acoustic stimulation on temporary threshold shift in young humans. Hear Res. 1992;60:149–55.

Nozu T, Okumura T. Corticotropin-releasing factor receptor type 1 and type 2 interaction in irritable bowel syndrome. J Gastroenterol. 2015;50:819–30.

Okada K, Yano M, Doki Y, Azama T, Iwanaga H, Miki H, Nakayama M, Miyata H, Takiguchi S, Fujiwara Y, Yasuda T, Ishida N, Monden M. Injection of LPS causes transient suppression of biological clock genes in rats. J Surg Res. 2008;145:5–12.

Oster H, Damerow S, Hut RA, Eichele G. Transcriptional profiling in the adrenal gland reveals circadian regulation of hormone biosynthesis genes and nucleosome assembly genes. J Biol Rhythms. 2006a;21:350–61.

Oster H, Damerow S, Kiessling S, Jakubcakova V, Abraham D, Tian J, Hoffmann MW, Eichele G. The circadian rhythm of glucocorticoids is regulated by a gating mechanism residing in the adrenal cortical clock. Cell Metab. 2006b;4:163–73.

Oster H, Challet E, Ott V, Arvat E, Ronald de Kloet E, Dijk DJ, Lightman S, Vgontzas A, Van Cauter E. The functional and clinical significance of the 24-hour rhythm of circulating glucocorticoids. Endocr Rev. 2017;38:3–45.

Panda S. Circadian physiology of metabolism. Science (New York, NY). 2016;354:1008–15.

Pedersen WA, McCullers D, Culmsee C, Haughey NJ, Herman JP, Mattson MP. Corticotropin-releasing hormone protects neurons against insults relevant to the pathogenesis of Alzheimer's disease. Neurobiol Dis. 2001;8:492–503.

Rajan R. Effect of electrical stimulation of the crossed olivocochlear bundle on temporary threshold shifts in auditory sensitivity. I. Dependence on electrical stimulation parameters. J Neurophysiol. 1988a;60:549–68.

Rajan R. Effect of electrical stimulation of the crossed olivocochlear bundle on temporary threshold shifts in auditory sensitivity. II. Dependence on the level of temporary threshold shifts. J Neurophysiol. 1988b;60:569–79.

Rajan R, Johnstone BM. Binaural acoustic stimulation exercises protective effects at the cochlea that mimic the effects of electrical stimulation of an auditory efferent pathway. Brain Res. 1988;459:241–55.

Rarey KE, Luttge WG. Presence of type I and type II/IB receptors for adrenocorticosteroid hormones in the inner ear. Hear Res. 1989;41:217–21.

Rarey KE, Curtis LM, ten Cate WJ. Tissue specific levels of glucocorticoid receptor within the rat inner ear. Hear Res. 1993;64:205–10.

Rarey KE, Gerhardt KJ, Curtis LM, ten Cate WJ. Effect of stress on cochlear glucocorticoid protein: acoustic stress. Hear Res. 1995;82:135–8.

Reich K. The concept of psoriasis as a systemic inflammation: implications for disease management. J Eur Acad Dermatol Venereol. 2012;26(Suppl 2):3–11.

Rivas Bejarano JJ, Valdecantos WC. Psoriasis as autoinflammatory disease. Dermatol Clin. 2013;31:445–60.

Rivier CL, Grigoriadis DE, Rivier JE. Role of corticotropin-releasing factor receptors type 1 and 2 in modulating the rat adrenocorticotropin response to stressors. Endocrinology. 2003;144:2396–403.

Roe SY, McGowan EM, Rothwell NJ. Evidence for the involvement of corticotrophin-releasing hormone in the pathogenesis of traumatic brain injury. Eur J Neurosci. 1998;10:553–9.

Schmiedt RA, Schulte BA. Physiologic and histopathologic changes in quiet- and noise-aged gerbil cochleas. In: Dancer AL, Henderson D, Salvi RJ, Hammernik RP, editors. Noise-induced hearing loss. St. Louis: Mosby; 1992. p. 246–58.

Slominski AT, Zmijewski MA, Zbytek B, Tobin DJ, Theoharides TC, Rivier J. Key role of CRF in the skin stress response system. Endocr Rev. 2013;34:827–84.

Sridhar TS, Liberman MC, Brown MC, Sewell WF. A novel cholinergic "slow effect" of efferent stimulation on cochlear potentials in the guinea pig. J Neurosci. 1995;15:3667–78.

Straub RH, Cutolo M. Circadian rhythms in rheumatoid arthritis: implications for pathophysiology and therapeutic management. Arthritis Rheum. 2007;56:399–408.

Subramaniam M, Henderson D, Campo P, Spongr V. The effect of 'conditioning' on hearing loss from a high frequency traumatic exposure. Hear Res. 1992;58:57–62.

Svec F. Glucocorticoid receptor regulation. Life Sci. 1985;36:2359–66.

Svec F, Rudis M. Glucocorticoids regulate the glucocorticoid receptor in the AtT-20 cell. J Biol Chem. 1981;256:5984–7.

ten Cate WJ, Curtis LM, Rarey KE. Immunochemical detection of glucocorticoid receptors within rat cochlear and vestibular tissues. Hear Res. 1992;60:199–204.

ten Cate WJ, Curtis LM, Small GM, Rarey KE. Localization of glucocorticoid receptors and glucocorticoid receptor mRNAs in the rat cochlea. Laryngoscope. 1993;103:865–71.

Taberner AM, Liberman MC. Response properties of single auditory nerve fibers in the mouse. J Neurophysiol. 2005;93:557–69.

Tagen M, Stiles L, Kalogeromitros D, Gregoriou S, Kempuraj D, Makris M, Donelan J, Vasiadi M, Staurianeas NG, Theoharides TC. Skin corticotropin-releasing hormone receptor expression in psoriasis. J Invest Dermatol. 2007;127:1789–91.

Tahera Y, Meltser I, Johansson P, Hansson AC, Canlon B. Glucocorticoid receptor and nuclear factor-kappa B interactions in restraint stress-mediated protection against acoustic trauma. Endocrinology. 2006a;147:4430–7.

Tahera Y, Meltser I, Johansson P, Bian Z, Stierna P, Hansson A, Canlon B. NF-kappaB mediated glucocorticoid response in the inner ear after acoustic trauma. J Neurosci Res. 2006b;83:1066–76.

Tahera Y, Meltser I, Johansson P, Salman H, Canlon B. Sound conditioning protects hearing by activating the hypothalamic-pituitary-adrenal axis. Neurobiol Dis. 2007;25:189–97.

Takahashi S, Yokota S, Hara R, Kobayashi T, Akiyama M, Moriya T, Shibata S. Physical and inflammatory stressors elevate circadian clock gene mPer1 mRNA levels in the paraventricular nucleus of the mouse. Endocrinology. 2001;142:4910–7.

Tornabene SV, Sato K, Pham L, Billings P, Keithley EM. Immune cell recruitment following acoustic trauma. Hear Res. 2006;222:115–24.

Vetter DE, Li C, Zhao L, Contarino A, Liberman MC, Smith GW, Marchuk Y, Koob GF, Heinemann SF, Vale W, Lee K-F. Urocortin-deficient mice show hearing impairment and increased anxiety-like behavior. Nat Genet. 2002;31:363–9.

Wang Y, Liberman MC. Restraint stress and protection from acoustic injury in mice. Hear Res. 2002;165:96–102.

Webster EL, Elenkov IJ, Chrousos GP. Corticotropin-releasing hormone acts on immune cells to elicit pro-inflammatory responses. Mol Psychiatry. 1997a;2:345–6.

Webster EL, Elenkov IJ, Chrousos GP. The role of corticotropin-releasing hormone in neuroendocrine-immune interactions. Mol Psychiatry. 1997b;2:368–72.

Webster EL, Torpy DJ, Elenkov IJ, Chrousos GP. Corticotropin-releasing hormone and inflammation. Ann N Y Acad Sci. 1998;840:21–32.

Wlk M, Wang CC, Venihaki M, Liu J, Zhao D, Anton PM, Mykoniatis A, Pan A, Zacks J, Karalis K, Pothoulakis C. Corticotropin-releasing hormone antagonists possess anti-inflammatory effects in the mouse ileum. Gastroenterology. 2002;123:505–15.

Yamasoba T, Dolan DF, Miller JM. Acquired resistance to acoustic trauma by sound conditioning is primarily mediated by changes restricted to the cochlea, not by systemic responses. Hear Res. 1999;127:31–40.

Yang S, Cai Q, Vethanayagam RR, Wang J, Yang W, Hu BH. Immune defense is the primary function associated with the differentially expressed genes in the cochlea following acoustic trauma. Hear Res. 2016;333:283–94.

Yao X, Buhi WC, Alvarez IM, Curtis LM, Rarey KE. De novo synthesis of glucocorticoid hormone regulated inner ear proteins in rats. Hear Res. 1995;86:183–8.

Zheng XY, Henderson D, McFadden SL, Hu BH. The role of the cochlear efferent system in acquired resistance to noise-induced hearing loss. Hear Res. 1997;104:191–203.

Chapter 4
Cochlear Vascular Pathology and Hearing Loss

Xiaorui Shi

Abstract Normal vascular function is essential for hearing. Abnormal blood flow to the cochlea is an etiologic factor contributing to various hearing disorders and vestibular dysfunctions, including noise-induced hearing loss, sudden deafness, presbyacusis, genetically-linked hearing loss, and endolymphatic hydrops such as Meniere's disease. Progression in blood flow pathology can parallel progression in hair cell loss and hearing impairment. To sustain hearing acuity, a healthy blood flow must be maintained. The blood supply not only provides oxygen and glucose to the hearing organ, it is also responsible for transporting hormones and neurotrophic growth factors to the tissue critical for organ health. Study of the vascular system in the inner ear has a long and rich history. There is a large body of evidence demonstrating a relationship between disturbances in cochlear microcirculatory homeostasis and decreased auditory sensitivity. This chapter focuses on recent discoveries relating the physiopathology of the microvasculature in the cochlear lateral wall to hearing function.

Keywords Cochlear blood flow · Aging · Noise · Ototoxic drug · Hearing loss

1 Introduction

The volume of cochlear blood flow is extremely small, on the order of 1/1,000,000 of the total cardiac output in a human (Axelsson 1968; Nakashima et al. 2003), but this blood flow is critical for maintaining the cochlear homeostasis essential for hearing. Blood flow to the stria vascularis is particularly important for generating the endocochlear potential (EP), on which transduction of sound in hair cells depends. Perturbations in the microcirculation can lead to significant cochlear and vestibular dysfunction (Nakashima 1999; Seidman et al. 1999), including sound-induced hearing loss (*endothelial injury*), age-related hearing loss (*lost vascular density*), genetic hearing loss (*Norrie disease—strial avascularization*), and

X. Shi (✉)
Oregon Health & Science University, Portland, OR, USA
e-mail: shix@ohsu.edu

© Springer International Publishing AG, part of Springer Nature 2018
V. Ramkumar, L. P. Rybak (eds.), *Inflammatory Mechanisms in Mediating Hearing Loss*, https://doi.org/10.1007/978-3-319-92507-3_4

autoimmune inner ear disease (*hydrops, IgG deposit on the vessel wall concurrent with ionic, osmotic, or metabolic imbalance*) (Ding et al. 2002; Gratton et al. 1997; Kellerhals 1972; Kurata et al. 2016; Neng et al. 2015; Prazma et al. 1990; Rehm et al. 2002; Schulte and Schmiedt 1992; Yang et al. 2011). Recent studies have shown primary strial vascular dysfunction, such as vascular degenerative changes in the stria, lead to progressive reduction of the EP and hearing loss (Chen et al. 2014; Ingham et al. 2016). Chronic hypo-perfusion of the blood flow often produces permanent hypoxia in the cochlea, which can significantly accelerate deterioration of organ function. Sensory cells are particularly vulnerable to hypoxia (Nuttall 1987).

Under normal conditions, the inner ear is a remarkably stable homeostatic system for maintaining the stability of inner ear fluids and nutrients (Juhn et al. 2001). The stable homeostasis is sustained by a variety of regulatory mechanisms, including autoregulation of blood flow (Brown and Nuttall 1994) and control of ion transport in and out of the inner ear through the blood-labyrinth barrier (Juhn et al. 2001; Juhn and Rybak 1981). This chapter concludes by underscoring the importance of understanding cochlear function from a microvascular perspective, as this may build the foundation for diagnosis, prevention, and treatment of many vascular related hearing disorders.

2   Blood Supply to the Lateral Wall

The cochlear blood supply is principally from the inner ear artery (labyrinthine artery), which is a branch of the anterior inferior cerebellar artery, (AICA). Blood is supplied directly to the cochlea by the spiral modiolar artery (SMA), a branch of the AICA (Axelsson 1968; Hawkins 1976; Nakai et al. 1992; Penha et al. 1999). The spiral modiolar arterioles centrifugally radiate over the scala vestibuli and form into different microvascular networks in the cochlea.

Two major microvascular networks in the cochlear lateral wall support the larger portion of cochlear blood flow. In experiments on rabbits using microspheres, over 80% of cochlear blood flow was shown to distribute to the lateral wall region, 9% to the middle region of the organ of Corti, and 9% to the modiolar region (Angelborg et al. 1984). In rats, the distribution is 57, 19, and 24% (Nakashima et al. 2001). These findings show the predominant portion of blood supplies the lateral wall region (Gyo 2013).

The two major capillary networks in the cochlea lateral wall are the capillaries of the stria vascularis and capillaries of the spiral ligament. The two networks are anatomically distant (>100 μm) from sensory hair cells in the organ of Corti, an arrangement which minimizes the effect of perturbations in blood flow on hearing (Axelsson 1968). The capillaries of the spiral ligament are generally divided into three major sectors: the pre-capillaries (red in Fig. 4.1), capillaries (purple in Fig. 4.1), and post-capillaries (blue in Fig. 4.1). The microvessels in the spiral ligament are also characterized as "arteriovenous anastomosing" vessels, passing directly across the ligament from arteriole to venule (Axelsson 1968). Microvessels of the spiral

Fig. 4.1 Schematic view of CoBF supply in the cochlear lateral wall (Illustration from the Shi. Lab). The two major microvessel networks in the cochlear lateral wall are the microvessels of the spiral ligament and the microvessels of the stria vascularis. A rich capillary network is also present in the region of the spiral prominence (green)

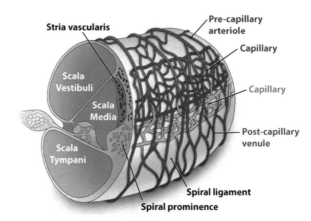

ligament play a critical role in regulating the blood flow in the cochlear lateral wall (Wangemann and Liu 1996). In contrast, the capillaries of the stria vascularis are highly specialized vascular epithelia forming into polygonal loops and constituting a unique intra-strial fluid–blood barrier, critical for ion transport and fluid balance in the inner ear, particularly for maintaining the EP, the essential driving force for sensory hair cell transduction (Nuttall and Lawrence 1980; Spiess et al. 2002; Wangemann 2002).

3   Features of the Microvascular Networks in the Cochlear Lateral Wall

Pericytes

The two microvascular networks in the cochlear lateral wall are richly populated by pericytes (Shi et al. 2008). A particularly high density of pericytes (~1220/stria vascularis) is distributed on the capillaries of the stria vascularis in mouse cochlea (Fig. 4.2b) (Neng et al. 2015). The ratio of pericytes to endothelial cells in the stria vascularis of guinea pig cochlea is between 1:1 and 1:2 (Shi et al. 2008), similar to that in retina (1:1), but higher than in brain (1:5), lung (1:10), and skeletal muscle (1:100) (Frank et al. 1987; Shepro and Morel 1993).

Pericytes are specialized mural cells located on the abluminal surface of microvessels (Shepro and Morel 1993). Extensively branched, the pericytes tightly embrace vessel walls (Fig. 4.3a–d). Pericyte interaction with the endothelium is vital for vascular development, regulation of blood flow, vascular integrity, angiogenesis, and tissue fibrogenesis (Dore-Duffy et al. 2006; Greenhalgh et al. 2013; Hall et al. 2014; Peppiatt et al. 2006; Quaegebeur et al. 2010). Pericyte pathology leads to vascular dysfunction, which is also seen in brain stroke, heart infarct, retinal disease (Greenhalgh et al. 2015; Greif and Eichmann 2014; Liu et al. 2012b; O'Farrell and Attwell 2014), and diabetic retinopathy (Kim et al. 2016; Pfister et al. 2008).

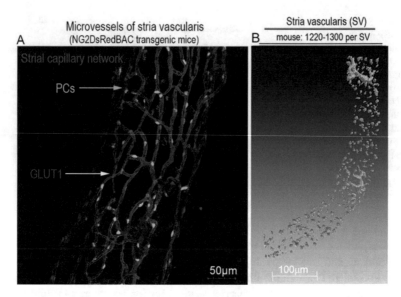

Fig. 4.2 Pericyte (PC) distribution on microvessels of the cochlear stria vascularis (SV) in a NG2 transgenic mouse model at age 6 weeks. (**a**) The confocal projection images show pericytes (green) have a characteristic "bump on a log" shape on the strial vessel wall (labeled by antibody for glucose transporter I, Glut1, red). (**b**) A super-resolution image shows the high density of pericytes (labeled with NG2, gray) in the mouse SV (Shi 2016)

Pericytes are morphologically heterogeneous from organ to organ and also heterogeneous within organs (Sims 2000). Consistent with the heterogeneity seen in other organs, pericytes in the cochlear lateral wall also show morphological heterogeneity (Shi et al. 2008). The majority of pericytes on true capillaries in the lateral wall have a polygonal-shaped cell body and long, slender processes, while pericytes in precapillary areas near the radiating arterioles have a prominent soma and display band-like processes which completely encircle the vessel. Most pericytes in postcapillary venule areas have flattened cell bodies and circumferential band-like, encircling processes (see Fig. 4.3).

Morphology and numbers of pericytes are thought to reflect specific functional features of the microvessels, particularly where blood flow is tightly coupled to metabolic demand (Sims 2000). Cochlear pericytes also show differences in function depending on location. For example, pericytes on vessels of the spiral ligament express contractile proteins such as α-SMA, desmin, F-actin, and tropomyosin (Franz et al. 2004; Shi et al. 2008), as well as exhibit vasocontractility (Dai et al. 2009). In contrast, pericytes on vessels of the stria vascularis, which lack expression of α-SMA or tropomyosin, instead richly express structural proteins such as desmin. The data support the view that pericytes on the vessels of the spiral ligament play a role in control of local blood flow, whereas it is hypothesized pericytes on the vessels of the stria vascularis primarily serve to maintain the structural strength of the microvessel wall (Shi et al. 2008). Recent studies also demonstrate the role of PCs in the stria vascularis in regulating the expression of tight junctions (TJs)

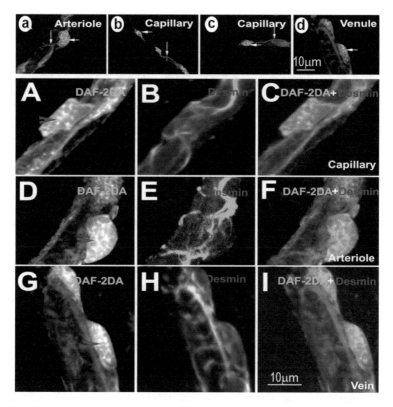

Fig. 4.3 Different shapes of pericytes on cochlear microvessels in guinea pig. The pericytes were double-labeled with the pericyte marker protein, desmin (red), combined with fluorescent DAF-2DA for marking intracellular nitric oxide (green). (**a–c**) shows the morphology of a pericyte on a capillary. The pericyte has a polygonal-shaped cell body with relatively few long longitudinal processes and shorter fine circumferential processes. (**c**) is a merged image of (**a**) and (**b**). (**d–f**) show the morphology of a pericyte on a precapillary. The pericyte has "bump shaped" soma and relatively large processes that encircle the capillary (**e**). (**f**) is a merged image of (**d**) and (**e**). (**g–i**) show the morphology of pericytes on a postcapillary. They have a flattened cell body (**g**) and short processes encircling the vessel (**h**). (**i**) is a merged image of (**g**) and (**h**) (Shi et al. 2008)

between endothelial cells and maintaining the functional stability of the blood-tissue barrier in the stria vascularis. The pericytes may be playing a structural role related to microvessel-wall integrity (Neng et al. 2013b; Shi et al. 2008).

Resident Macrophages

A population of perivascular resident macrophages (PVMs) are also in direct contact with the capillary network in the normal cochlear lateral wall (O'Malley et al. 2016; Shi 2016). Approximately 500–600 PVMs are found in the microvessel

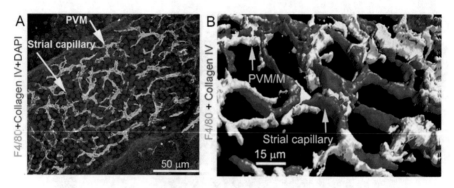

Fig. 4.4 (**a**) PVMs are labeled with an antibody for F4/80 (green), capillaries with antibody for colleagen IV (red). Cell nuclei are labeled with DAPI (blue). (**b**) A 3D rendering of the confocal stacks show the ramified processes of PVMs interfacing with the endothelial tube. Capillaries are labeled with antibody for collagen IV (Zhang et al. 2012)

network of the stria vascularis in normal adult mouse cochlea (Neng et al. 2013b). The resident macrophages are highly invested on the abluminal surface of the capillaries through multiple thin membrane protrusions (shown in Fig. 4.4), closely associated with microvessels, and structurally intertwined with endothelial cells and pericytes (see Fig. 4.4a, b).

PVMs in the stria vascularis are a hybrid cell type not fitting the classical phenotype of a tissue resident macrophage. They display characteristics of both macrophage and melanocyte (Neng et al. 2013a; Zhang et al. 2012). For example, they are positive for several macrophage surface markers, including F4/80, CD68, and CD11b. They constitutively express scavenger receptor classes A_1 and B_1 and accumulate blood-borne proteins such as horseradish peroxidase and acetylated low-density lipoprotein (Shi 2010). They also exhibit melanocyte characteristics, showing significant amounts of melanin and expressing the melanocyte marker proteins, glutathione S-transferase alpha 4 (Gstα4) and Kir 4.1, the latter the fiduciary marker of intermediate cells (Zhang et al. 2012). In early studies Cable and Steel (1991) and Spicer and Schulte (2005) mention two subclasses of intermediate cells in the stria vascularis. Others later discovered some of these intermediate cells interact with strial capillaries through gap junctions (Spicer and Schulte 2005). PVMs are considered the equivalent of this latter subclass of intermediate cell. Phylogenetic origin of PVMs is not clear, although it is generally accepted that cochlear melanocytes derive from the neural crest and migrate to the stria vascularis during development (Freyer et al. 2011; Steel and Barkway 1989; Steel et al. 1992; Wakaoka et al. 2013). A recent study has shown that the majority of PVMs in the stria vascularis are capable of self-renewal and turn over within a 10 month time frame from circulating blood cells (Shi 2010).

Tissue resident macrophages, in general, exist in many tissues, including in brain and retina (Cuadros and Navascués 1998; Hess et al. 2004). In each organ, resident macrophages have a role in immunological defense and repair (Cui et al. 2009;

Ekdahl et al. 2009), ingesting and degrading dead cells, debris, and foreign material, as well as orchestrating inflammatory processes by producing superoxide anions, nitric oxide, and inflammatory cytokines (Block and Hong 2005; Block et al. 2007; Chéret et al. 2008; Hanisch and Kettenmann 2007; Varol et al. 2015). PVMs may play a similar role in immunological defense against pathological agents as a homeostatic "safeguard" of the tissue in the cochlear lateral wall.

PVMs, as melanocytes, act as biosensors responsive to biological and physico-chemical signals in the local environment by producing melanin pigment in response to noxious factors (Slominski et al. 2012; Sulaimon and Kitchell 2003). Earlier studies document the melanin has a role in buffering calcium, scavenging heavy metals, and promoting antioxidant activity (Bush and Simon 2007; Dräger 1985; Murillo-Cuesta et al. 2010; Ohlemiller et al. 2009; Plonka et al. 2009; Slominski 2009; Slominski et al. 2012).

Recent studies show that PVMs in the stria vascularis, similar to tissue resident macrophages in other organs such as the glial cells in brain and retina (Abbott et al. 2006; Adams 2009; Prat et al. 2001), have a role in regulation of barrier integrity in the stria vascularis through upregulation of TJ proteins between endothelial cells (Zhang et al. 2012). However, we are only beginning to understand the dynamic role of PVMs in cochlear function, and much work remains to be done.

4 Cochlear-Vascular-Unit

'Pericyte-Fibrocyte Coupling' Regulates Blood Flow

The cochlea is a high energy demand organ which transduces acoustic input to electrical signals within a time scale of microseconds. Sound stimulation of the inner ear imposes a peak of energy demand that requires both efficient delivery of oxygen and nutrients and rapid removal of metabolic waste (Nuttall 1999). Adequate cochlear blood supply is crucial for auditory function (Shi 2011; Wangemann 2002). Regulation of cochlear blood flow, under the prevailing model, is hypothesized to include both local auto-regulation and central control via neuronal pathways. In particular, cochlear blood flow is thought in the main to be regulated in the end arterial system of the cochlea, specifically in the spiral modiolar artery and its branching arterioles (Jiang et al. 2007; Wangemann 2002). The model incorporates neural- and autocrine/paracrine-based regulation of vasoconstriction and dilation at the level of the artery and arterioles (Miller and Dengerink 1988; Wangemann 2002). Capillary-mediated local control of perfusion has been less studied but was first reported on by (Wangemann and Liu 1996). Recent findings which show cochlear capillaries in the spiral ligament densely populated by contractile protein expressing pericytes (Shi et al. 2008) and exhibiting vasocontractility (Dai et al. 2009) reopen the question on the role of capillary-based local blood-flow control. A recent study has demonstrated contractility of pericytes to affect flow resistance in

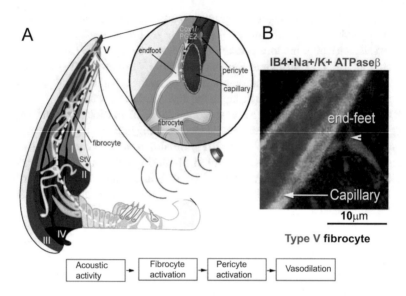

Fig. 4.5 Fibro-vascular coupled control of regional blood perfusion in the cochlear lateral wall. (**a**) The schematic diagram illustrates cochlear blood flow is coupled to sound activity through type V fibrocyte-pericyte-vascular coupling mechanisms. (**b**) A representative confocal projection image demonstrates morphological evidence of a Type V foot process in contact with a capillary in the spiral ligament (Dai and Shi 2011)

the vascular network and alter overall blood flow (Dai et al. 2009). Further study has shown a local control mechanism to regulate inner ear blood flow involving fibrocyte signaling to vascular cells, including pericytes [see Fig. 4.5; Dai et al. (2011)].

Fibrocytes are known to facilitate generation of the endocochlear potential, recycling K^+ from hair cell transduction through gap junctions to strial intermediate cells and marginal cells to the endolymph (Adams et al. 2009; Doherty and Linthicum Jr 2004; Moon et al. 2006; Nakashima et al. 2003; Qu et al. 2007; Spicer and Schulte 1991, 2002; Trowe et al. 2008; Wangemann 2002; Wu and Marcus 2003). Fibrocytes in the cochlear lateral wall are classified as types I to V based on morphological appearance, staining pattern, and general location (Kikuchi et al. 2000; Spicer and Schulte 1991, 1996). Type V fibrocytes are found to be morphologically associated with pre-capillaries in the spiral ligament through "end-feet" structures [Dai and Shi (2011), as shown in Fig. 4.6], analogous to the "neurovascular units" (NVUs) of astrocyte/pericyte junctions in brain. Fibrocyte to vascular cell coupled signaling mediates sound stimulated increase in cochlear blood flow. Local metabolic substances, such as NO, ATP, Cyclooxygenase-1 (COX-1), and K^+, have a vasoactive effect on microvessel diameter via pericyte contraction and dilation (Dai et al. 2009; Dai and Shi 2011), significantly affecting the flow resistance of the vascular network and profoundly impacting overall blood flow.

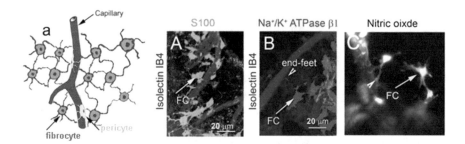

Fig. 4.6 Confocal images show fibro-vascular coupled morphology. The working model illustrates how cochlear blood flow is locally regulated to meet metabolic demand (**a**). (**A**) Type V fibrocytes positive for S100 (green) abut capillary walls labeled by isolectin IB4 (blue). (**B**) The type V fibrocytes are positive for Na⁺/K⁺ ATPase β1 (red). (**C**) Type V fibrocytes also contain high levels of NO, as detected with the intracellular NO indicator, DAF-2DA (gray) (Dai and Shi 2011)

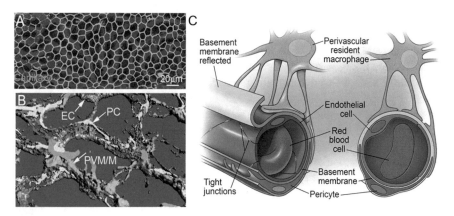

Fig. 4.7 (**a**) Strial microvessels (blue) lie beneath the marginal cell layer (green). (**b**) Isosurface renderings give a perspective on the interface between the ramified processes of PVMs, ECs, and PCs in the intra-strial fluid–blood barrier. (**c**) The illustration of a strial microvessel in cross-section shows the major components of the barrier. The vessel lumen is comprised of ECs connected by TJs. ECs are ensheathed by a dense basement membrane shared with pericytes. PVM end-feet cover a large portion of the capillary surface. *PVM/M* perivascular resident macrophage-type melanocyte, *PC* pericyte, *EC* endothelial cell (Neng et al. 2013a)

'Pericyte-Endothelial Cell-Resident Macrophage Coupling' Controls the Integrity of the Fluid-Blood-Barrier in the Stria Vascularis

The strial microvessel network lies beneath the marginal cell layer, as shown in Fig. 4.7a. The coupling of strial pericytes, endothelial cells, and PVMs in the stria vascularis forms a unique intra-strial fluid-blood barrier [Fig. 4.7b, c (Shi 2011, 2016)]. This barrier is essential for sustaining the active metabolism required for secretion of endolymph, production of the endolymphatic potential (EP), and prevention of toxic substances from entering the cochlea (Hibino et al. 2010; Ohlemiller et al. 2008; Salt et al. 1987; Spicer and Schulte 1996, 2005; Thomopoulos et al. 1997; Wangemann 2002). In particular, the integrity of the barrier is crucial for maintaining low K^+ levels (~5 mmol/L) in the intrastrial space between marginal and basal cell layers, preventing an intrastrial electric shunt, and setting the conditions for K^+ to be secreted into endolymph by marginal cells to generate the EP.

This specialized barrier in the cochlear lateral wall is also a system rich in proteins for molecular transport. Using a mass-spectrometry, shotgun-proteomics approach, in combination with a novel method for "sandwich-dissociation" of the strial capillary, our lab showed more than 42% of total spectral counts are transporters and 19% are related to metabolic processes (Yang et al. 2011). The ion transporter ATP1A1 was the most abundant protein in the intra-strial fluid–blood barrier. Metabolic enzymes are also highly expressed in the intra-strial fluid–blood barrier, including glutathione S-transferase (GST), prosaposin, leukotriene A4 hydrolase, and glutamate oxaloacetate transaminase (Yang et al. 2011).

At the cellular level, the intra-strial fluid-blood barrier comprises cochlear microvascular endothelial cells connected to each other by TJs, an underlying basement membrane, and a second line of support consisting of cochlear pericytes and PVMs (as illustrated in Fig. 4.7c). Physical interactions between the endothelial cells, pericytes, and PVMs, as well as signaling between the cells, are critical for controlling vascular permeability (Zhang et al. 2012). In particular, strial pericytes have a significant role in the regulation of TJ protein expression between endothelial cells in the capillaries of the stria vascularis (Neng et al. 2013a, b). Normal PVM function is also essential for stabilizing the integrity of the barrier. Equally important is the production of pigment epithelium derived factor (PEDF), a 50-kDa secreted glycoprotein of the noninhibitory serpin family, first identified in retinal pigment epithelium cells (Liu et al. 2012a). Secretion of PEDF by cochlear PVMs has direct and broad effects on the expression of several TJ-associated proteins, including occludin, ZO-1, and ve-cadherin (Zhang et al. 2012). Most recently, PEDF was also shown to be the most potent endogenous inhibitor of vasopermeability in other organs (Liu et al. 2004; Ueda et al. 2010). Studies have shown that PEDF binding to its receptor counteracts vascular endothelial growth factor (VEGF)-induced vascular permeability (Liu et al. 2004; Ueda et al. 2010).

5   Blood Flow Dysfunction in Hearing Disorders

A reduction in blood flow to the ear can rapidly lead to a shortage of nutrients and oxygen in the tissue, creating a "toxic" environment with accumulation of harmful metabolites (Nuttall 1987). Many experimental studies have shown reduced blood flow and alterations in the blood barrier to characterize a wide range of conditions. These include aging-related hearing loss (Gratton et al. 1996; Gratton et al. 1997; Ohlemiller et al. 2009; Neng et al. 2015), noise-induced hearing loss (Shi and Nuttall 2003; Shi 2009), autoimmunodiseases (Lin and Trune 1997; Ruckenstein et al. 1999), genetic hearing disorders (Chen et al. 2014; Ingham et al. 2016; Jabba et al. 2006), and hearing loss caused by ototoxic drugs (Campbell et al. 1999; Cardinaal et al. 2000; Kohn et al. 1991; Meech et al. 1998).

Ageing-Related Hearing Loss

Age-related hearing loss (ARHL, also known as presbycusis) is one of the most prevalent chronic degenerative conditions, characterized by a gradual decline in auditory function (Tavanai and Mohammadkhani 2017). Growing evidence suggests this form of hearing loss is associated with both reduced cognitive functioning and incidental dementia (Mudar and Husain 2016). Loss of sensory hair cells, spiral ganglion (SG) neurons, and cochlear vascular cells are typically seen in afflicted individuals (Gates and Mills 2005).

Loss of Vessels and Blood Flow Perfusion

Cochlear vascular regression and degeneration of the stria vascularis (density loss) has long been observed in both animal models and humans (Carraro and Harrison 2016; Carraro et al. 2016; Gratton et al. 1996, 1997; Neng et al. 2013a; Ohlemiller et al. 2008, 2009; Prazma et al. 1990; Schulte and Schmiedt 1992). For example, substantial strial capillary density loss is shown in both aged C57 BL/J (Neng et al. 2015) and genetically deficient NOD.NON-H2nb1 mice (Ohlemiller et al. 2008). In human temporal bone studies, presbycusis patients often show atrophy of the stria vascularis (Kurata et al. 2016; Sprinzl and Riechelmann 2010). An earlier functional study reported reduction in blood supply to the cochlea in old gerbils compared to young animals. Using a microsphere technique to quantify blood flow, they found diminished flow in morphologically normal-appearing basal turn capillaries (Prazma et al. 1990). Changes in whole blood viscosity and red cell rigidity have also been correlated with high-frequency hearing loss in elderly human subjects (Gatehouse and Lowe 1991). Furthermore, a series of *in vivo* intravital microscopy experiments on the cochlear microvasculature demonstrated age-dependent, statistically significant reductions in mean red blood cell velocity accompanied by increases in capillary permeability (Seidman et al. 1999). Brown et al. (1995) found old mice less reactive to topical application of nitroprusside, a vasodilating agent. Suzuki et al. (1998) demonstrated that autoregulation was significantly reduced in an aged group.

Thickening of the Vascular Basement Membrane

A thickened vascular basement membrane and increased immunoglobulin and laminin deposits are found in aged strial capillaries (Gratton et al. 1996; Sakaguchi et al. 1997a, b). Basement membrane (BM) thickness was increased 65–85% in strial capillaries of gerbils aged 33 months or older, and the thickness was often observed several-fold that in young controls (Thomopoulos et al. 1997). Increased immunoglobulin and laminin deposits accompany the thickened basement membranes in aged strial capillaries (Sakaguchi et al. 1997a, b). In humans, a gradual loss of capillaries in the spiral ligament of the scala vestibuli and stria vascularis was observed (Kurata et al. 2016; Sprinzl and Riechelmann 2010). Significant thickening of vessel walls in the modiolar artery and strial vessels was also noticed (Kurata et al. 2016).

Pericyte and PVM Abnormality

Aged animals exhibit a significant decrease in pericyte and PVM number, marked by morphological changes that are seen in all regions of the stria vascularis (Neng et al. 2015). Pericytes in young C57/6BJ animals (<3 months) display with a flat and slender morphology tightly associated with endothelial cells (pericyte density of 21 ± 2/mm capillary). Pericytes in older animals (>6 months) are less abundant and have a prominent round body in less physical contact with endothelia, a morphology previously described as a sign of pericyte migration (Pfister et al. 2008). The 21-month-old animals had a density of 13 ± 1/mm capillary. At the ultrastructural level, pericytes from aged animals show a loss of cytoplasmic organelles, presenting with a vacuolated appearance detached from endothelial cells (Fig. 4.8), (Neng et al. 2015). The abundant pericyte coverage of strial capillaries suggests pericytes have an important role in contributing to vessel stability and regulation of BLB function. An abnormally low number of pericytes in the aged ear may be contributing to vascular instability and malfunction.

PVM morphology also shows dramatic differences between young and aged animals. In younger animals most PVMs exhibit a branched morphology (see Fig. 4.8). The cells are arranged in a self-avoidance pattern. At 6, 9, 12, and 21 months the animals display with smaller PVMs and shorter processes. In some regions, the PVMs are flat and amoeboid-shaped (Fig. 4.8) and in less physical contact with capillaries. The biochemistry of the PVMs also undergoes changes. A terminal galactopyranosyl group is now exposed on its membrane surface, as detected by binding to the lectin GS-IB4. This binding is the hallmark of macrophage activation (Maddox et al. 1982; Neng et al. 2015) (see Fig. 4.8).

Noise-Induced Hearing Loss (NIHL)

Noise-induced hearing loss (NIHL) is another common sensorineural hearing deficit. Loud sound damages auditory sensory cells and also has effects on the cochlear lateral wall which include impairment of the microvasculature in the stria vascularis (Canlon 1987, 1988; Yoshida et al. 1999; Wang et al. 2002; Kujawa and Liberman 2015; Liberman et al. 2015; Ohlemiller and Gagnon 2007; Kamogashira et al. 2015; Hultcrantz and Nuttall 1987; Shi and Nuttall 2007; Shi 2009).

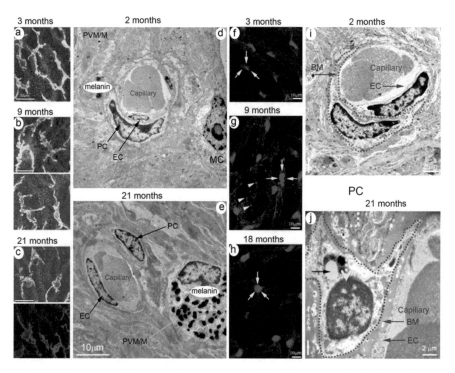

Fig. 4.8 Perivascular resident macrophage activation and the "dropping out" of pericytes in aged animals. (**a**) The confocal image shows PVM morphology and distribution on strial capillaries in a young animal (PVMs are labeled with antibody for F4/80, green, and strial capillaries with GS-IB4, red). PVMs exhibit a long branched morphology. (**b**) PVMs in middle aged animals show reduced branching and withdrawal of ramifications. (**c**) PVMs in the old animals show much shorter processes in less physical contact with strial capillaries. The PVMs are also positive for GS-IB4, an indication of activation. (**d**) Transmission electron micrographs of cochlear PVMs at 2 months show a flat cell body and cells which contain a modest amount of melanin. (**e**) PVMs at 21 months appear dark, owing to the abundance of melanin in the cytoplasm. (**f**) PCs exhibit a flat and slender cell body in a young animal (transgenic mice with fluorescent labeled NG2, red). (**g**) and (**h**) PC morphology changes in middle and old aged animals. Apical PCs display a "prominent round" cell body in less physical contact with strial capillaries. (**i**) In transmission electron micrographs of cochlear pericytes at <3 months PCs appear as long and slender polymorphic cells located on the abluminal side of the ECs (yellow arrow). (**j**) The irregular shaped PCs of a 21 month old mouse are sparse in caveolae. *PVM/M* perivascular resident macrophage-type melanocyte, *PC* pericyte, *EC* endothelial cell (Neng et al. 2015)

Hypoxia

The stria vascularis is a region of high metabolic demand. Dr. Nuttall has documented the mechanical and metabolic vulnerability of strial capillaries to extremely loud sounds (Nuttall 1987, 1999). Numerous experimental studies have shown that loud sound causes cochlear hypoxia in its immediate aftermath, with prolonged effects after the sound is terminated (Lamm et al., 2000). For example, Misrahy et al. (1958) observed a striking decrease in endocochlear oxygen tension during

130–135 dB SPL sound exposure. Induction of HIF-1α translocation is also detected as early as 30 min after the sound exposure (Shi 2009).

Generation of the EP in the stria vascularis requires efficient production of ATP (Marcus et al. 1981). A shortfall in blood flow results in a lack of oxygen (hypoxia) to cell mitochondria and causes an imbalance in ion and hemostasis resulting in excess production of reactive oxygen species ROS (Chance et al. 1979). The increased ROS significantly affects the function of many transporters/pumps, these including the Na^+/K^+ pump and Na^+-K^+-$2Cl^-$ co-transporter in marginal cells. These transporters are essential for secretion of K^+ to the scala media (Komune et al. 1993).

The mechanisms of noise-induced hypoxia (hypo-perfusion) is complicated. However, loud sound-induced vasoconstriction and endothelial impairment are respectively attributed to the early and late stages of cochlear hypoxia (Shi and Nuttall 2007). Loss and regression of capillaries (vascular degenerative change) after extensive noise-induced trauma has been attributed to the cochlear hypoxia (Axelsson and Dengerink 1987; Miller et al. 2003; Yamane et al. 1991).

Endothelial Dysfunction

Loud sound trauma can cause destructive changes in the cochlear endothelia, including vessel shutdown and "intra-vascular strand formation" (Axelsson and Dengerink 1987; Axelsson and Vertes 1982; Dai and Gan 2010; Hawkins 1973; Kellerhals 1972; Shaddock et al. 1984; Yamane et al. 1991). In an earlier study, Hawkins (1973, 1971) reported that when animals were exposed to wideband noise at 118–120 dB SPL for 8 h, red blood cells are observed trapped by swollen capillary endothelial cells and avascular channels replace capillaries. These mechanisms in the endothelium are not fully understood. However, vascular damage mechanisms involving mechanical destruction and intense metabolic (hypoxic/toxic) activity have been proposed. Significantly increased ROS (nitric oxide, a natural by-product of aerobic metabolism) (Le Prell et al. 2007a, b; Shi and Nuttall 2002) and increased inflammatory cytokines, including interleukin-6 (IL-6), interleukin-1β (IL-1β), NF-kappaB, ICAM, and VEGF are frequently seen in loud sound exposed animals (Goldwyn and Quirk 1997; Hillerdal et al. 1987; Hultcrantz and Nuttall 1987; Lamm and Arnold 1999; Reif et al. 2013; Scheibe et al. 1993; Seidman et al. 1999; Shi and Nuttall 2002, 2007; Suzuki et al. 2002; Yamamoto et al. 2009), particularly in supporting cells and ligament fibrocytes (Adams et al. 2009; Fujioka et al. 2006; Hirose et al. 2005; Jamesdaniel et al. 2011; Le Prell et al. 2003; Miyao et al. 2008; Nakamoto et al. 2012; Sato et al. 2008; Shi and Nuttall 2007; Tornabene et al. 2006; Yamamoto et al. 2009). These events closely parallel the pathogenesis of noise-induced hearing loss (Goldwyn and Quirk 1997; Hultcrantz and Nuttall 1987; Nuttall 1987; Shi and Nuttall 2002). Therapeutic approaches that utilize free radical scavengers and anti-inflammatories have been shown to ameliorate noise-induced vascular pathology (Fujioka et al. 2014; Honkura et al. 2016; Lamm and Arnold 1999; Le Prell et al. 2007a, b).

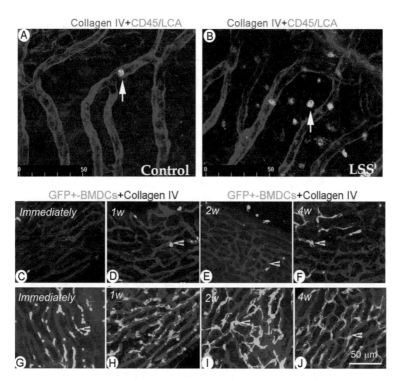

Fig. 4.9 Leukocytes and bone marrow derived cells infiltrate the cochlear lateral wall under noise conditions. Double labeling of leukocytes and vessel walls with a marker for CD45/LCA (green) and collagen IV (Red) in control and LSS animals. Panels (**a** and **b**) show the distribution of vessels of the spiral ligament in a control and LSS mouse. No emigrated leukocytes were found under control conditions. In contrast, some emigrated leukocytes are identified on vessels of the spiral ligament after LSS. Noise initiates robust GFP+-BMDC infiltration into the damaged cochlear blood barrier in the stria vascularis. (**c–f**) show GFP+-BMDC infiltration (green and white arrowheads) from capillaries of the stria vascularis (blue, labeled with antibody for collagen type IV) in a control mouse. (**g–j**) show GFP+-BMDC infiltration from capillaries of the stria vascularis in a noise-exposed mouse at week 1, week 2, and week 4. Images shown are projections of confocal z-stacks taken at the tested time points (Shi 2016; Shi and Nuttall 2007)

Leukocyte and Bone Marrow Derived Cells Infiltration

'Sterile' inflammation is seen in noise exposure conditions. Early studies from our lab (Shi and Nuttall 2003) found that loud-sound stress activates the expression of adhesion molecular proteins in the cochlear lateral wall. In particular, increased expression of inflammatory adhesive molecules such as P-selectin and PECAM-1 are found, predominantly in the vessels of the spiral ligament, and correspondingly, increased populations of migrated leukocytes are also observed in the area of the spiral ligament [as shown in Fig. 4.9, (Shi and Nuttall 2007)]. This was also reported by other scientists (Hirose et al. 2005). In support with previous findings using a constitutional mouse model, we found that GFP-labeled circulating bone marrow derived cells (GFP+-BMDC) adhere to endothelial cells immediately after noise exposure and (Fig. 4.9g) transmigrate through the vessel wall about 1 week after noise exposure (Fig. 4.9h). The migrated blood cell undergoes morphological changes. At an early stage (approximately 1 week after noise exposure), infiltrated GFP$^+$-BMDCs are frequently spherical or nodular shaped (possibly caught in the act of transmigration, Fig. 4.9h). Approximately 2 weeks after noise exposure, most infiltrated GFP$^+$-BMDCs develop ramified processes, appear dendriform in shape, and are irregularly distributed on the capillaries of the stria vascularis (Fig. 4.9i). Approximately 4 weeks after noise exposure, the majority of infiltrated BMDCs are elongated and display an orientation—that is, their long processes parallel the vessels of the stria vascularis (Fig. 4.9j). Some of these infiltrated BMDCs can be identified as macrophages. An intrinsic signaling of inducible nitric oxide synthase (iNOS) is identified to mediate GFP(+)-BMDC infiltration.

Vascular Leakage, Pericyte Migration, and PVM Activation

Noise destabilizes the intra-strial fluid-blood barrier in the stria vascularis. An early study showed increased accumulation of high molecular weight horseradish peroxidase in the stria vascularis following intense sound exposure (Hukee and Duvall III 1985). Recent studies have further revealed the finer details of structural and molecular changes in the intra-strial fluid-blood barrier after acoustic trauma, including decreased expression of tight- and adherens-junction proteins, loosened TJs between ECs, and increased vascular permeability [Fig. 4.10b, d, (Yang et al. 2011; Zhang et al. 2013)]. PCs are particularly vulnerable. Upon exposure to loud sound, some PCs show irregularities in their processes and migration from their normal locations attached to endothelial cells, resulting in destabilization of the intra-strial fluid-blood barrier [Fig. 4.10g, h (Shi 2009)]. Acoustic trauma also causes some PVMs to activate, as shown in Fig. 4.11c, d. The traumatized PVMs produce less PEDF, leading to down-regulation of TJ-associated proteins and subsequent vascular leakage (Zhang et al. 2013). PEDF produced by normal PVMs is essential for stabilizing the intra-strial fluid–blood barrier, as the PEDF regulates expression of TJ-associated proteins such as ZO-1 and VE-cadherin (Zhang et al. 2012). The signaling which causes cochlear pericytes to migrate and PVMs to activate is not yet clear. The mechanism of the permeability change in the cochlea is also complicated and remains to be fully understood.

Meniere's Disease and Autoimmune Hearing Loss

Ménière's disease is an inner ear disorder characterized by vertigo attacks, fluctuating and progressive hearing loss, tinnitus, and aural fullness in the affected ear (Oberman et al. 2017). While the pathophysiology of Ménière's disease remains elusive, it has long been thought to be caused by hydrops in the inner ear (Pender

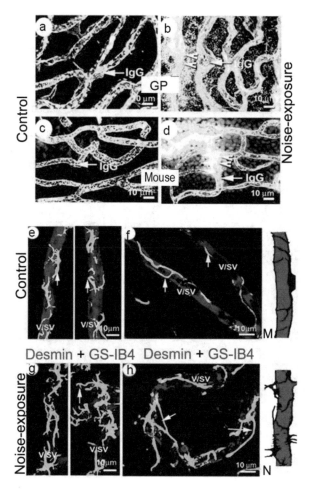

Fig. 4.10 Noise induces pericyte detachment, and blood barrier breakdown. (**a**) and (**b**) Serum protein IgG is confined to blood plasma (IgG/arrow) in vessels of the stria vascularis in normal mice (**a**) and guinea pigs (**b**). (**c**) and (**d**), Serum protein IgG leaks from vessels (arrow/IgG) in noise-exposed mice (**c**) and guinea pigs (**d**). Arrowheads indicate sites of vascular leakage. Pericytes containing desmin filaments are evenly distributed on the vessel walls of the stria vascularis in normal guinea pigs (**e**) and mice (**f**). Pericytes are labeled with an antibody for desmin (green), and vessels with an antibody for isolectin IB4 (red). (**g**) and (**h**): Confocal fluorescent images from noise-exposed guinea pigs and mice show abnormal pericyte morphology. Arrows point to irregular pericyte foot processes turning away from vessel walls (**K**) and detached from them (**L**). **M** and **N**: Drawings illustrate the pattern of pericyte distribution on vessel walls in normal and noised-exposed animals. *V/SV* vessel of the stria vascularis, *NE* noise exposure, *GP* guinea pig, *MS* mouse, exposure, *GP* guinea pig, *MS* mouse (Shi 2009)

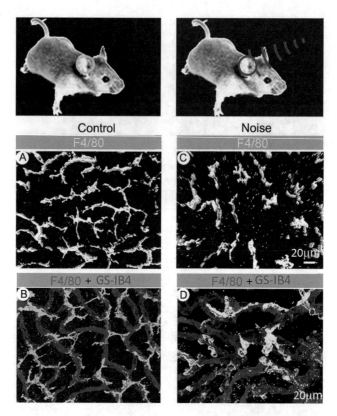

Fig. 4.11 Noise induces PVM activation, pericyte detachment, and blood barrier breakdown. (**a** and **b**) Confocal images show the morphology of PVMs on strial capillaries labeled with GS-IB4 (red) in a control animal. (**c** and **d**) Activated PVMs in noise-exposed animals show reduced branching and withdrawal of ramifications, displaying less physical contact with capillaries. images from noise-exposed guinea pigs and mice show abnormal pericyte morphology (Zhang et al. 2013)

2014). Previous reports have shown dysfunctional autoregulation of cochlear blood flow and impairment of the blood barrier (Fujioka et al. 2014; Goodall 2015; Greco et al. 2012; Hughes et al. 1983; Kim et al. 2014; Sara et al. 2014; Tagaya et al. 2011). Recent progress in understanding blood-labyrinth-barrier pathophysiology in animal models and human postmortem specimens highlight the importance of blood barrier integrity to ion homeostasis, prompting the question whether dysfunction of cochlear vascular function is key to understanding Ménière's disease. A recent study involving examination of human specimens reveals ultrastructural changes in the microvasculature of the stria vascularis. The changes include pericyte detachment and disruption of the basement membrane surrounding the endothelium, severe vacuolization or frank necrosis of vascular endothelial cells, and loss of subcellular organelles in the endothelial cells. Results have shown severe degeneration of the blood labyrinth barrier associated with a high degree of basement membrane thickening and edematous changes in the vestibular stroma in the macula utricle of Meniere's disease patients (Ishiyama et al. 2017). An earlier study also revealed increased gadoteridol (Gd) intensity in the endolymphatic compartment of patients with Ménière's disease examined with 3 T magnetic resonance imaging (MRI) (Tagaya et al. 2011).

Autoimmune disease in the inner ear often leads to progressive sensorineural hearing loss such as sudden deafness or vestibular symptoms (i.e., Meniere's disease) (Fujioka et al. 2014; Goodall 2015; Greco et al. 2012; Hughes et al. 1983; Kim et al. 2014; Sara et al. 2014). While the pathophysiology of autoimmune related hearing loss remains largely unknown, accumulating evidence is showing that stria capillaries are one of the targets of autoimmune disease (Ågrup and Luxon 2006; Goodall 2015; Lin and Trune 1997; Takahashi and Harris 1988). Substantial evidence implicates deposition of immune-complexes and direct attack by auto-antibodies on the endothelium as common features in these hearing and vestibular disorders (Ågrup and Luxon 2006; Goodall 2015; Lin and Trune 1997). Research using a C3H/lpr autoimmune mouse model has demonstrated the primary defect is breakdown of strial blood barrier integrity, circulating IgG deposition within strial capillaries, and thickening of the basement membrane (Lin and Trune 1997; Trune et al. 1998; Wong et al. 1992; Young et al. 1988). Clinically, blood drawn from patients with autoimmune hearing disorder often show high levels of anti-endothelial and anti-phospholipid antibodies, including anti-choline transporter-like protein 2 (CTL2) and anti-heat shock protein (HSP70) (Cadoni et al. 2002; Mijovic et al. 2013; Mouadeb and Ruckenstein 2005; Nair et al. 2004; Ottaviani et al. 1999; Toubi et al. 2004; Yehudai et al. 2006). Targeting strial vascular function may be effective in treating autoimmune hearing loss.

Inflammation

Inflammatory factors (viral and bacterial infections) disrupt vascular integrity in the stria vascularis, causing breakdown of the blood-labyrinth barrier and imbalance in endolymph ion homeostasis (Hilger 1952; Trune and Nguyen-Huynh 2012). For example, Zhang et al. 2015 recently showed lipopolysaccharide-induced middle ear inflammation to disrupt the cochlear intra-strial fluid-blood barrier by down-regulating TJ protein expression (Zhang et al. 2015). Correspondingly, Hirose et al. (2014) demonstrated that lipopolysaccharide increases entry of serum fluorescein into the perilymph via the blood barrier (Hirose et al. 2014). A study by Quintanilla-Dieck et al. (2013) demonstrated that lipopolysaccharide-induced inflammation increases cytokine levels in the murine cochlea. The cytokines may be one of the causes for the increased permeability of the intra-strial fluid–blood barrier. Previous studies showing viral and bacterial infection to induce anti-endothelial (anti-phospholipid) antibody attack of glycocalyx components in the barrier (Blank et al. 2007) are a strong indication systemic and local inflammatory events perturb the normal function of the blood barrier, resulting in homeostatic imbalance and hearing loss. Cytomegalovirus (CMV) infection is one of the common causes of congenital hearing loss in children (Carraro and Harrison 2016; Carraro et al. 2016). Studies reveal that CMV not only damages cochlear sensory cells, but also cochlear vasculature. CMV causes a primary lesion in the stria vascularis and adjacent spiral limbus capillary network (Carraro et al. 2017). Capillary beds of the spiral ligament are generally less affected. The initial vascular damage is in the mid-apical turn and appears to progress to more basal cochlear regions. Results suggest the initial auditory threshold losses caused by the strial vascular dysfunction are due to poor development or maintenance of the endocochlear potential (Carraro et al. 2017).

Ototoxicity

Drug ototoxicity is a known cause of hearing loss (Rybak and Ramkumar 2007; Rybak et al. 2007; Yorgason et al. 2011). A variety anticancer agents (e.g., cisplatin, carboplatin, nedaplatin, and oxaliplatincisplatin), aminoglycoside antibiotics (e.g., gentamicin, streptomycin), as well as loop diuretics such as furosemide and salicylate family (aspirin-like drugs), cause either irreversible or reversible hearing loss dependent on the dose and duration of exposure (Ding et al. 2012; Kamogashira et al. 2015; Karasawa and Steyger 2011; Kaur et al. 2011; Oishi et al. 2012; Rybak et al. 2007; Schacht et al. 2012).

The cochlear vascular system is actively involved in drug-induced ototoxicity. Cochlear vessels, particularly in the stria vascularis, provide access for the drugs to enter the cochlea. The microvessel network in the stria vascularis is richly articulated and the velocity of blood flow is extremely slow, giving sufficient time for the drugs to pass through channels. The intra-strial fluid–blood barrier is the main port of entry for ototoxic drugs from the blood into cochlear fluids (Dai and Steyger 2008; Laurell et al. 2000; Wang and Steyger 2009). Enhanced drug uptake and significantly increased hearing damage is seen when the barrier is disrupted by diuretics or noise exposure (Ding et al. 2007; Li et al. 2015). Vasoactive peptides also modulate cochlear uptake of ototoxic drugs such as gentamicin (Aksoy et al. 2015; Koo et al. 2011). The glycocalyx lining the endothelium of cochlear vessels in the stria vascularis is negatively charged by proteoglycans such as heparin sulfate and chondroitin sulfate. The negative charged molecules may facilitate movement of positively charged cation drugs into the cochlea. Recent studies have shown that transient receptor potential cation channel V4 (TRPV4), present in strial capillaries, especially facilitate the entry of gentamicin into the inner ear (Ishibashi et al. 2009; Karasawa et al. 2008).

The mechanism underlying ototoxic drug induced hearing loss is complicated, but a few studies have shown vascular pathology is a contributor. For example, an early study by Miettinen et al. (1997) showed that cisplatin at high dosage reduces cochlear blood flow by 20–30%. Dr. Rybak et al., in a review article, point out that cisplatin causes a temporary reduction of the EP and associated strial edema (Rybak et al. 2007). Recent studies have also shown that inflammatory cells from the blood infiltrate to the cochlear lateral wall when animals receive ototoxic drugs in combination with LPS treatment (Hirose et al. 2014; Wood and Zuo 2017). Circulating inflammatory factors may play a critical role as a causal agent in ototoxic induced hearing loss. Upregulation of pro-inflammatory cytokines, such as TNF-α, IL-1β, and IL-6, are found in the cochlea of cisplatin plus LPS treated mice (Oh et al. 2011). Ototoxic drug studies have increased our understanding of the mechanisms of drug-induced hearing loss, and in particular have informed us of the relationship between inflammation and infiltration in the cochlear vascular system. The insights may be clinically useful for preventing drug-induced hearing loss by targeting inflammation and preventing recruitment of inflammatory cells from the blood circulation.

Summary

The etiology of otologic disorders such as sudden sensorineural hearing loss, presbyacusis, noise-induced hearing loss, and certain vestibulopathies involves disturbance of cochlear blood flow. Reduction in blood supply to the cochlea leads to tissue hypoperfusion and hypoxia/ischemia, contributing to reduced auditory sensitivity. Recently, the use of transgenic animals and more sophisticated imaging systems has enabled researchers to identify the unique characteristics of pericyte-fibrocyte coupling in the spiral ligament and specialized blood-tissue barrier in the stria vascularis, as well as other properties of the microcirculation in the cochlear lateral wall. These studies have demonstrated that cochlear vascular coupling systems are highly regulated and tightly controlled to provide efficient delivery of ions, oxygen, and nutrients to the cochlea and rapid removal of metabolic waste from cochea. Yet we do not have the full picture of blood flow pathophysiology in hearing. A primary constraint is the inaccessibility of the hearing organ and the difficulties in imaging cochlear blood flow with current research tools. Nevertheless, progress is rapidly being made and new findings are revealing the importance of inner ear microcirculation in hearing. Therapeutic targeting of the cochlear microcirculation in parallel with a strategy for restoring hair cells, neurons, and other cochlear cells may offer opportunities to facilitate recovery of hearing.

Acknowledgments Most of the data presented in this review reflects the efforts of my colleagues and students at the Oregon Hearing Research Center. In particular, the author is deeply indebted to Dr. Alfred Nuttall for stimulating discussion and advice. The author also thanks Mr. Allan Kachelmeier and Ms Janice Moore for editorial assistance, and Christine Casabar for assistance with the references.

    This work was supported by National Institutes of Health grants R03 DC008888, DC008888S1, R01 DC010844 (X. Shi), R21 DC1239801 (X. Shi.); P30-DC005983 (Peter Barr-Gillespie); R01 DC000105 (Alfred L. Nuttall); R21 DC016157 (X. Shi.) and R01 DC015781 (X. Shi).

References

Abbott NJ, Rönnbäck L, Hansson E. Astrocyte–endothelial interactions at the blood–brain barrier. Nat Rev Neurosci. 2006;7:41–53.

Adams J. Immunocytochemical traits of type IV fibrocytes and their possible relations to cochlear function and pathology. J Assoc Res Otolaryngol. 2009;10:369–82.

Adams JC, Seed B, Lu N, Landry A, Xavier RJ. Selective activation of nuclear factor kappa B in the cochlea by sensory and inflammatory stress. Neuroscience. 2009;160:530–9.

Ågrup C, Luxon LM. Immune-mediated inner-ear disorders in neuro-otology. Curr Opin Neurol. 2006;19:26–32.

Aksoy F, Dogan R, Ozturan O, Yildirim YS, Veyseller B, Yenigun A, Ozturk B. Betahistine exacerbates amikacin ototoxicity. Ann Otol Rhinol Laryngol. 2015;124:280–7.

Angelborg C, Axelsson A, Larsen H-C. Regional blood flow in the rabbit cochlea. Arch Otolaryngol. 1984;110:297–300.

Axelsson A. The vascular anatomy of the cochlea in the guinea pig and in man. Acta Otolaryngol. 1968;Suppl 243:3+.

Axelsson A, Dengerink H. The effects of noise on histological measures of the cochlear vasculature and red blood cells: a review. Hear Res. 1987;31:183–91.

Axelsson A, Vertes D. Histological findings in cochlear vessels after noise, new perspectives on noise-induced hearing loss. New York: Raven Press; 1982. p. 49–68.

Blank M, Barzilai O, Shoenfeld Y. Molecular mimicry and auto-immunity. Clin Rev Allergy Immunol. 2007;32:111–8.

Block ML, Hong J-S. Microglia and inflammation-mediated neurodegeneration: multiple triggers with a common mechanism. Prog Neurobiol. 2005;76:77–98.

Block ML, Zecca L, Hong J-S. Microglia-mediated neurotoxicity: uncovering the molecular mechanisms. Nat Rev Neurosci. 2007;8:57–69.

Brown JN, Nuttall AL. Autoregulation of cochlear blood flow in guinea pigs. Am J Physiol Heart Circ Physiol. 1994;266:H458–67.

Brown JN, Miller JM, Nuttall AL. Age-related changes in cochlear vascular conductance in mice. Hear Res. 1995;86:189–94.

Bush WD, Simon JD. Quantification of Ca2+ binding to melanin supports the hypothesis that melanosomes serve a functional role in regulating calcium homeostasis. Pigment Cell Res. 2007;20:134–9.

Cable J, Steel KP. Identification of two types of melanocyte within the stria vascularis of the mouse inner ear. Pigment Cell Res. 1991;4:87–101.

Cadoni G, Fetoni AR, Agostino S, Santis AD, Manna R, Ottaviani F, Paludetti G. Autoimmunity in sudden sensorineural hearing loss: possible role of anti-endothelial cell autoantibodies. Acta Otolaryngol. 2002;122:30–3.

Campbell KC, Meech RP, Rybak LP, Hughes LF. D-Methionine protects against cisplatin damage to the stria vascularis. Hear Res. 1999;138:13–28.

Canlon B. Acoustic overstimulation alters the morphology of the tectorial membrane. Hear Res. 1987;30:127–34.

Canlon B. The effect of acoustic trauma on the tectorial membrane, stereocilia, and hearing sensitivity: possible mechanisms underlying damage, recovery, and protection. Scand Audiol Suppl. 1988;27:1–45.

Chance B, Sies H, Boveris A. Hydroperoxide metabolism in mammalian organs. Physiol Rev. 1979;59:527–605.

Cardinaal RM, de Groot JC, Huizing EH, Veldman JE, Smoorenburg GF. Dose-dependent effect of 8-day cisplatin administration upon the morphology of the albino guinea pig cochlea. Hear Res. 2000;144:135–46.

Carraro M, Harrison RV. Degeneration of stria vascularis in age-related hearing loss; a corrosion cast study in a mouse model. Acta Otolaryngol. 2016;136:385–90.

Carraro M, Park AH, Harrison RV. Partial corrosion casting to assess cochlear vasculature in mouse models of presbycusis and CMV infection. Hear Res. 2016;332:95–103.

Carraro M, Almishaal A, Hillas E, Firpo M, Park A, Harrison RV. Cytomegalovirus (CMV) infection causes degeneration of cochlear vasculature and hearing loss in a mouse model. J Assoc Res Otolaryngol. 2017;18:263–73.

Chance B, Sies H, Boveris A. Hydroperoxide metabolism in mammalian organs. Physiological reviews 1979;59:527–605.

Chen J, Ingham N, Kelly J, Jadeja S, Goulding D, Pass J, Mahajan VB, Tsang SH, Nijnik A, Jackson IJ. Spinster homolog 2 (spns2) deficiency causes early onset progressive hearing loss. PLoS Genet. 2014;10(10):e1004688.

Chéret C, Gervais A, Lelli A, Colin C, Amar L, Ravassard P, Mallet J, Cumano A, Krause K-H, Mallat M. Neurotoxic activation of microglia is promoted by a nox1-dependent NADPH oxidase. J Neurosci. 2008;28:12039–51.

Cuadros MA, Navascués J. The origin and differentiation of microglial cells during development. Prog Neurobiol. 1998;56:173–89.

Cui Q, Yin Y, Benowitz L. The role of macrophages in optic nerve regeneration. Neuroscience. 2009;158:1039–48.

Dai C, Gan RZ. Change in cochlear response in an animal model of otitis media with effusion. Audiol Neurootol. 2010;15:155–67.

Dai M, Shi X. Fibro-vascular coupling in the control of cochlear blood flow. PLoS One. 2011;6:e20652.

Dai M, Shi X. Fibro-vascular coupling in the control of cochlear blood flow. PloS One. 2011;6(6):e20652.

Dai CF, Steyger PS. A systemic gentamicin pathway across the stria vascularis. Hear Res. 2008;235:114–24.

Dai M, Nuttall A, Yang Y, Shi X. Visualization and contractile activity of cochlear pericytes in the capillaries of the spiral ligament. Hear Res. 2009;254:100–7.

Ding D, McFadden SL, Woo JM, Salvi RJ. Ethacrynic acid rapidly and selectively abolishes blood flow in vessels supplying the lateral wall of the cochlea. Hear Res. 2002;173:1–9.

Ding D, Jiang H, Wang P, Salvi R. Cell death after co-administration of cisplatin and ethacrynic acid. Hear Res. 2007;226:129–39.

Ding D, Allman BL, Salvi R. Review: ototoxic characteristics of platinum antitumor drugs. Anat Rec. 2012;295:1851–67.

Doherty JK, Linthicum FH Jr. Spiral ligament and stria vascularis changes in cochlear otosclerosis: effect on hearing level. Otol Neurotol. 2004;25:457–64.

Dore-Duffy P, Katychev A, Wang X, Van Buren E. CNS microvascular pericytes exhibit multipotential stem cell activity. J Cereb Blood Flow Metab. 2006;26:613–24.

Dräger U. Calcium binding in pigmented and albino eyes. Proc Natl Acad Sci U S A. 1985;82:6716–20.

Ekdahl C, Kokaia Z, Lindvall O. Brain inflammation and adult neurogenesis: the dual role of microglia. Neuroscience. 2009;158:1021–9.

Frank RN, Dutta S, Mancini MA. Pericyte coverage is greater in the retinal than in the cerebral capillaries of the rat. Invest Ophthalmol Vis Sci. 1987;28:1086–91.

Franz P, Helmreich M, Stach M, Franz-Italon C, Böck P. Distribution of actin and myosin in the cochlear microvascular bed. Acta Otolaryngol. 2004;124:481–5.

Freyer L, Aggarwal V, Morrow BE. Dual embryonic origin of the mammalian otic vesicle forming the inner ear. Development. 2011;138:5403–14.

Fujioka M, Kanzaki S, Okano HJ, Masuda M, Ogawa K, Okano H. Proinflammatory cytokines expression in noise-induced damaged cochlea. J Neurosci Res. 2006;83:575–83.

Fujioka M, Okano H, Ogawa K. Inflammatory and immune responses in the cochlea: potential therapeutic targets for sensorineural hearing loss. Front Pharmacol. 2014;5:287.

Gatehouse S, Lowe G. Whole blood viscosity and red cell filterability as factors in sensorineural hearing impairment in the elderly. Acta Otolaryngol. 1991;111:37–43.

Gates GA, Mills JH. Presbycusis. Lancet. 2005;366:1111–20.

Goldwyn BG, Quirk WS. Calcium channel blockade reduces noise-induced vascular permeability in cochlear stria vascularis. Laryngoscope. 1997;107:1112–6.

Goodall AF. Current understanding of the pathogenesis of autoimmune inner ear disease: a review. Clin Otolaryngol. 2015;40(5):412–9.

Gratton MA, Schmiedt RA, Schulte BA. Age-related decreases in endocochlear potential are associated with vascular abnormalities in the stria vascularis [corrected and republished article originallly printed in Hear Res 1996 May;94(1–2):116–24]. Hear Res. 1996;102:181–90.

Gratton MA, Schulte BA, Smythe NM. Quantification of the stria vascularis and strial capillary areas in quiet-reared young and aged gerbils. Hear Res. 1997;114:1–9.

Greco A, Gallo A, Fusconi M, Marinelli C, Macri G, De Vincentiis M. Meniere's disease might be an autoimmune condition? Autoimmun Rev. 2012;11:731–8.

Greenhalgh SN, Iredale JP, Henderson NC. Origins of fibrosis: pericytes take centre stage. F1000Prime Rep. 2013;5:37.

Greenhalgh SN, Conroy KP, Henderson NC. Healing scars: targeting pericytes to treat fibrosis. QJM. 2015;108:3–7.

Greif DM, Eichmann A. Vascular biology: brain vessels squeezed to death. Nature. 2014;508:50–1.

Gyo K. Experimental study of transient cochlear ischemia as a cause of sudden deafness. World J Otorhinolaryngol. 2013;3:1–15.

Hall CN, Reynell C, Gesslein B, Hamilton NB, Mishra A, Sutherland BA, O'Farrell FM, Buchan AM, Lauritzen M, Attwell D. Capillary pericytes regulate cerebral blood flow in health and disease. Nature. 2014;508:55–60.

Hanisch U-K, Kettenmann H. Microglia: active sensor and versatile effector cells in the normal and pathologic brain. Nat Neurosci. 2007;10:1387–94.

Hawkins JE Jr. The role of vasoconstriction in noise-induced hearing loss. Ann Otol Rhinol Laryngol. 1971;80:903–13.

Hawkins J. Comparative otopathology: aging, noise, and ototoxic drugs, otophysiology. Karger Publishers; 1973. p. 125–41.

Hawkins J. Microcirculation in the labyrinth. Eur Arch Otorhinolaryngol. 1976;212:241–51.

Hess DC, Abe T, Hill WD, Studdard AM, Carothers J, Masuya M, Fleming PA, Drake CJ, Ogawa M. Hematopoietic origin of microglial and perivascular cells in brain. Exp Neurol. 2004;186:134–44.

Hibino H, Nin F, Tsuzuki C, Kurachi Y. How is the highly positive endocochlear potential formed? The specific architecture of the stria vascularis and the roles of the ion-transport apparatus. Pflugers Arch. 2010;459:521–33.

Hilger JA. The common ground of allergy, autonomic dysfunction and endocrine imbalance. Trans Am Acad Ophthalmol Otolaryngol. 1952;57:443–6.

Hillerdal M, Sperber G, Bill A. The microsphere method for measuring low blood flows: theory and computer simulations applied to findings in the rat cochlea. Acta Physiol. 1987;130:229–35.

Hirose K, Discolo CM, Keasler J, Ransohoff R. Mononuclear phagocytes migrate into the murine cochlea after acoustic trauma. J Comp Neurol. 2005;489(2):180–94.

Hirose K, Hartsock JJ, Johnson S, Santi P, Salt AN. Systemic lipopolysaccharide compromises the blood-labyrinth barrier and increases entry of serum fluorescein into the perilymph. J Assoc Res Otolaryngol. 2014;15:707–19.

Honkura Y, Matsuo H, Murakami S, Sakiyama M, Mizutari K, Shiotani A, Yamamoto M, Morita I, Shinomiya N, Kawase T. Nrf2 is a key target for prevention of noise-induced hearing loss by reducing oxidative damage of cochlea. Sci Rep. 2016;6:19329.

Hughes G, Kinney S, Barna B, Calabrese L. Autoimmune reactivity in Meniere's disease: a preliminary report. Laryngoscope. 1983;93:410–7.

Hukee MJ, Duvall AJ III. Cochlear vessel permeability to horseradish peroxidase in the normal and acoustically traumatized chinchilla: a reevaluation. Ann Otol Rhinol Laryngol. 1985;94:297–303.

Hultcrantz E, Nuttall AL. Effect of hemodilution on cochlear blood flow measured by laser-Doppler flowmetry. Am J Otolaryngol. 1987;8:16–22.

Ingham NJ, Carlisle F, Pearson S, Lewis MA, Buniello A, Chen J, Isaacson RL, Pass J, White JK, Dawson SJ. S1PR2 variants associated with auditory function in humans and endocochlear potential decline in mouse. Sci Rep. 2016;6:28964.

Ishibashi T, Takumida M, Akagi N, Hirakawa K, Anniko M. Changes in transient receptor potential vanilloid (TRPV) 1, 2, 3 and 4 expression in mouse inner ear following gentamicin challenge. Acta Otolaryngol. 2009;129:116–26.

Ishiyama G, Lopez IA, Ishiyama P, Vinters HV, Ishiyama A. The blood labyrinthine barrier in the human normal and Meniere's disease macula utricle. Sci Rep. 2017;7:253.

Jabba SV, Oelke A, Singh R, Maganti RJ, Fleming S, Wall SM, Everett LA, Green ED, Wangemann P. Macrophage invasion contributes to degeneration of stria vascularis in Pendred syndrome mouse model. BMC Med. 2006;4:37.

Jamesdaniel S, Hu B, Kermany MH, Jiang H, Ding D, Coling D, Salvi R. Noise induced changes in the expression of p38/MAPK signaling proteins in the sensory epithelium of the inner ear. J Proteomics. 2011;75:410–24.

Jiang Z-G, Shi X-R, Guan B-C, Zhao H, Yang Y-Q. Dihydropyridines inhibit acetylcholine-induced hyperpolarization in cochlear artery via blockade of intermediate-conductance calcium-activated potassium channels. J Pharmacol Exp Ther. 2007;320:544–51.

Juhn SK, Rybak LP. Labyrinthine barriers and cochlear homeostasis. Acta Otolaryngol. 1981;91:529–34.

Juhn SK, Hunter BA, Odland RM. Blood-labyrinth barrier and fluid dynamics of the inner ear. Int Tinnitus J. 2001;7:72–83.

Kamogashira T, Fujimoto C, Yamasoba T. Reactive oxygen species, apoptosis, and mitochondrial dysfunction in hearing loss. Biomed Res Int. 2015;2015:617207.

Karasawa T, Steyger PS. Intracellular mechanisms of aminoglycoside-induced cytotoxicity. Integr Biol. 2011;3:879–86.

Karasawa T, Wang Q, Fu Y, Cohen DM, Steyger PS. TRPV4 enhances the cellular uptake of aminoglycoside antibiotics. J Cell Sci. 2008;121:2871–9.

Kaur T, Mukherjea D, Sheehan K, Jajoo S, Rybak LP, Ramkumar V. Short interfering RNA against STAT1 attenuates cisplatin-induced ototoxicity in the rat by suppressing inflammation. Cell Death Dis. 2011;2:e180.

Kellerhals B. Acoustic trauma and cochlear microcirculation. An experimental and clinical study on pathogenesis and treatment of inner ear lesions after acute noise exposure. Adv Otorhinolaryngol. 1972;18:91.

Kikuchi T, Adams JC, Miyabe Y, So E, Kobayashi T. Potassium ion recycling pathway via gap junction systems in the mammalian cochlea and its interruption in hereditary nonsyndromic deafness. Med Electron Microsc. 2000;33:51–6.

Kim SH, Kim JY, Lee HJ, Gi M, Kim BG, Choi JY. Autoimmunity as a candidate for the etiopathogenesis of Meniere's disease: detection of autoimmune reactions and diagnostic biomarker candidate. PLoS One. 2014;9(10):e111039.

Kim JM, Hong K-S, Song WK, Bae D, Hwang I-K, Kim JS, Chung H-M. Perivascular progenitor cells derived from human embryonic stem cells exhibit functional characteristics of pericytes and improve the retinal vasculature in a rodent model of diabetic retinopathy. Stem Cells Transl Med. 2016;5:1268–76.

Kohn S, Nir I, Fradis M, Podoshin L, David YB, Zidan J, Robinson E. Toxic effects of cisplatin alone and in combination with gentamicin in stria vascularis of guinea pigs. Laryngoscope. 1991;101:709–16.

Komune S, Nakagawa T, Hisashi K, Kimituki T, Uemura T. Movement of monovalent ions across the membranes of marginal cells of the stria vascularis in the guinea pig cochlea. ORL J Otorhinolaryngol Relat Spec. 1993;55:61–7.

Koo J-W, Wang Q, Steyger PS. Infection-mediated vasoactive peptides modulate cochlear uptake of fluorescent gentamicin. Audiol Neurootol. 2011;16:347–58.

Kujawa SG, Liberman MC. Synaptopathy in the noise-exposed and aging cochlea: primary neural degeneration in acquired sensorineural hearing loss. Hear Res. 2015;330:191–9.

Kurata N, Schachern PA, Paparella MM, Cureoglu S. Histopathologic evaluation of vascular findings in the cochlea in patients with presbycusis. JAMA Otolaryngol Head Neck Surg. 2016;142(2):173–8.

Lamm K, Arnold W. Successful treatment of noise-induced cochlear ischemia, hypoxia, and hearing loss. Ann N Y Acad Sci. 1999;884:233–48.

Lamm K, Arnold W. The effect of blood flow promoting drugs on cochlear blood flow, perilymphatic pO(2) and auditory function in the normal and noise-damaged hypoxic and ischemic guinea pig inner ear. Hear Res. 2000;141:199–219.

Laurell G, Viberg A, Teixeira M, Sterkers O, Ferrary E. Blood-perilymph barrier and ototoxicity: an in vivo study in the rat. Acta Otolaryngol. 2000;120:796–803.

Le Prell CG, Dolan DF, Schacht J, Miller JM, Lomax MI, Altschuler RA. Pathways for protection from noise induced hearing loss. Noise Health. 2003;5:1–17.

Le Prell CG, Hughes LF, Miller JM. Free radical scavengers vitamins A, C, and E plus magnesium reduce noise trauma. Free Radic Biol Med. 2007a;42:1454–63.

Le Prell CG, Yamashita D, Minami SB, Yamasoba T, Miller JM. Mechanisms of noise-induced hearing loss indicate multiple methods of prevention. Hear Res. 2007b;226:22–43.

Li H, Kachelmeier A, Furness DN, Steyger PS. Local mechanisms for loud sound-enhanced aminoglycoside entry into outer hair cells. Front Cell Neurosci. 2015;9:130.

Liberman LD, Suzuki J, Liberman MC. Erratum to: dynamics of cochlear synaptopathy after acoustic overexposure. J Assoc Res Otolaryngol. 2015;16:221.

Lin DW, Trune DR. Breakdown of stria vascularis blood-labyrinth barrier in C3H/lpr autoimmune disease mice. Otolaryngol Head Neck Surg. 1997;117:530–4.

Liu H, Ren J-G, Cooper WL, Hawkins CE, Cowan MR, Tong PY. Identification of the antivasopermeability effect of pigment epithelium-derived factor and its active site. Proc Natl Acad Sci U S A. 2004;101:6605–10.

Liu JT, Chen YL, Chen WC, Chen HY, Lin YW, Wang SH, Man KM, Wan HM, Yin WH, Liu PL. Role of pigment epithelium-derived factor in stem/progenitor cell-associated neovascularization. J Biomed Biotechnol. 2012a;2012:871272.

Liu S, Agalliu D, Yu C, Fisher M. The role of pericytes in blood-brain barrier function and stroke. Curr Pharm Des. 2012b;18:3653–62.

Maddox DE, Shibata S, Goldstein IJ. Stimulated macrophages express a new glycoprotein receptor reactive with Griffonia simplicifolia I-B4 isolectin. Proc Natl Acad Sci U S A. 1982;79:166–70.

Marcus DC, Marcus NY, Thalmann R. Changes in cation contents of stria vascularis with ouabain and potassium-free perfusion. Hear Res. 1981;4:149–60.

Meech RP, Campbell KC, Hughes LP, Rybak LP. A semiquantitative analysis of the effects of cisplatin on the rat stria vascularis. Hear Res. 1998;124:44–59.

Miettinen S, Laurell G, Andersson A, Johansson R, Laurikainen E. Blood flow-independent accumulation of cisplatin in the guinea pig cochlea. Acta Otolaryngol. 1997;117:55–60.

Mijovic T, Zeitouni A, Colmegna I. Autoimmune sensorineural hearing loss: the otology–rheumatology interface. Rheumatology. 2013;52(5):780–9. https://doi.org/10.1093/rheumatology/ket009.

Miller JM, Brown JN, Schacht J. 8-iso-prostaglandin F2α, a product of noise exposure, reduces inner ear blood flow. Audiol Neurotol. 2003;8:207–21.

Miller JM, Dengerink H. Control of inner ear blood flow. Am J Otolaryngol. 1988;9:302–16.

Misrahy G, Shinabarger E, Arnold J. Changes in cochlear endolymphatic oxygen availability, action potential, and microphonics during and following asphyxia, hypoxia, and exposure to loud sounds. J Acoust Soc Am. 1958;30:701–4.

Miyao M, Firestein GS, Keithley EM. Acoustic trauma augments the cochlear immune response to antigen. Laryngoscope. 2008;118:1801–8.

Moon S-K, Moon S-K, Park R, Moon S-K, Park R, Lee H-Y, Nam G-J, Cha K, Andalibi A, Lim DJ. Spiral ligament fibrocytes release chemokines in response to otitis media pathogens. Acta Otolaryngol. 2006;126:564–9.

Mouadeb DA, Ruckenstein MJ. Antiphospholipid inner ear syndrome. Laryngoscope. 2005;115:879–83.

Mudar RA, Husain FT. Neural alterations in acquired age-related hearing loss. Front Psychol. 2016;7:828.

Murillo-Cuesta S, Contreras J, Zurita E, Cediel R, Cantero M, Varela-Nieto I, Montoliu L. Melanin precursors prevent premature age-related and noise-induced hearing loss in albino mice. Pigment Cell Melanoma Res. 2010;23:72–83.

Nair TS, Kozma KE, Hoefling NL, Kommareddi PK, Ueda Y, Gong T-W, Lomax MI, Lansford CD, Telian SA, Satar B. Identification and characterization of choline transporter-like protein 2, an inner ear glycoprotein of 68 and 72 kDa that is the target of antibody-induced hearing loss. J Neurosci. 2004;24:1772–9.

Nakai Y, Masutani H, Moriguchi M, Matsunaga K, Kato A, Maeda H. Microvasculature of normal and hydropic labyrinth. Scanning Microsc. 1992;6:1097–103; discussion 1103–4.

Nakamoto T, Mikuriya T, Sugahara K, Hirose Y, Hashimoto T, Shimogori H, Takii R, Nakai A, Yamashita H. Geranylgeranylacetone suppresses noise-induced expression of proinflammatory cytokines in the cochlea. Auris Nasus Larynx. 2012;39:270–4.

Nakashima T. Autoregulation of cochlear blood flow. Nagoya J Med Sci. 1999;62:1–9.

Nakashima T, Suzuki T, Iwagaki T, Hibi T. Effects of anterior inferior cerebellar artery occlusion on cochlear blood flow–a comparison between laser-Doppler and microsphere methods. Hear Res. 2001;162:85–90.

Nakashima T, Naganawa S, Sone M, Tominaga M, Hayashi H, Yamamoto H, Liu X, Nuttall AL. Disorders of cochlear blood flow. Brain Res Rev. 2003;43:17–28.

Neng L, Zhang F, Kachelmeier A, Shi X. Endothelial cell, pericyte, and perivascular resident macrophage-type melanocyte interactions regulate cochlear intrastrial fluid–blood barrier permeability. J Assoc Res Otolaryngol. 2013a;14:175–85.

Neng L, Zhang W, Hassan A, Zemla M, Kachelmeier A, Fridberger A, Auer M, Shi X. Isolation and culture of endothelial cells, pericytes and perivascular resident macrophage-like melanocytes from the young mouse ear. Nat Protoc. 2013b;8:709–20.

Neng L, Zhang J, Yang J, Zhang F, Lopez IA, Dong M, Shi X. Structural changes in thestrial blood–labyrinth barrier of aged C57BL/6 mice. Cell Tissue Res. 2015;361(3):685–96.

Nuttall AL. Techniques for the observation and measurement of red blood cell velocity in vessels of the guinea pig cochlea. Hear Res. 1987;27:111–9.

Nuttall AL. Sound-induced cochlear ischemia/hypoxia as a mechanism of hearing loss. Noise Health. 1999;2:17.

Nuttall AL, Lawrence M. Endocochlear potential and scala media oxygen tension during partial anoxia. Am J Otolaryngol. 1980;1:147–53.

O'Farrell FM, Attwell D. A role for pericytes in coronary no-reflow. Nat Rev Cardiol. 2014;11(7):427–32.

O'Malley JT, Nadol JB Jr, McKenna MJ. Anti CD163+, Iba1+, and CD68+ cells in the adult human inner ear: normal distribution of an unappreciated class of macrophages/microglia and implications for inflammatory otopathology in humans. Otol Neurotol. 2016;37:99–108.

Oberman B, Patel V, Cureoglu S, Isildak H. The aetiopathologies of Ménière's disease: a contemporary review. Acta Otorhinolaryngol Ital. 2017;37(4):250–63.

Oh G-S, Kim H-J, Choi J-H, Shen A, Kim C-H, Kim S-J, Shin S-R, Hong S-H, Kim Y, Park C. Activation of lipopolysaccharide–TLR4 signaling accelerates the ototoxic potential of cisplatin in mice. J Immunol. 2011;186:1140–50.

Ohlemiller KK, Gagnon PM. Genetic dependence of cochlear cells and structures injured by noise. Hear Res. 2007;224:34–50.

Ohlemiller KK, Rice MER, Gagnon PM. Strial microvascular pathology and age-associated endocochlear potential decline in NOD congenic mice. Hear Res. 2008;244:85–97.

Ohlemiller KK, Rybak Rice ME, Lett JM, Gagnon PM. Absence of strial melanin coincides with age-associated marginal cell loss and endocochlear potential decline. Hear Res. 2009;249:1–14.

Oishi N, Talaska AE, Schacht J. Ototoxicity in dogs and cats. Vet Clin North Am Small Anim Pract. 2012;42:1259–71.

Ottaviani F, Cadoni G, Marinelli L, Fetoni AR, De Santis A, Romito A, Vulpiani P, Manna R. Anti-endothelial autoantibodies in patients with sudden hearing loss. Laryngoscope. 1999;109:1084–7.

Pender D. Endolymphatic hydrops and Ménière's disease: a lesion meta-analysis. J Laryngol Otol. 2014;128:859–65.

Penha R, O'Neill M, Goyri ONJ, Esperanca PJ. Ultrastructural aspects of the microvasculature of the cochlea: the internal spiral network. Otolaryngol Head Neck Surg. 1999;120:725.

Peppiatt CM, Howarth C, Mobbs P, Attwell D. Bidirectional control of CNS capillary diameter by pericytes. Nature. 2006;443:700–4.

Pfister F, Feng Y, Vom Hagen F, Hoffmann S, Molema G, Hillebrands J-L, Shani M, Deutch U, Hammes H-P. Pericyte migration: a novel mechanism of pericyte loss in experimental diabetic retinopathy. Diabetes. 2008;57:2495–502.

Plonka P, Passeron T, Brenner M, Tobin D, Shibahara S, Thomas A, Slominski A, Kadekaro A, Hershkovitz D, Peters E. What are melanocytes really doing all day long…? Exp Dermatol. 2009;18:799–819.

Prat A, Biernacki K, Wosik K, Antel JP. Glial cell influence on the human blood-brain barrier. Glia. 2001;36:145–55.

Prazma J, Carrasco VN, Butler B, Waters G, Anderson T, Pillsbury HC. Cochlear microcirculation in young and old gerbils. Arch Otolaryngol Head Neck Surg. 1990;116:932.

Qu C, Liang F, Smythe NM, Schulte BA. Identification of ClC-2 and CIC-K2 chloride channels in cultured rat type IV spiral ligament fibrocytes. J Assoc Res Otolaryngol. 2007;8:205–19.

Quaegebeur A, Segura I, Carmeliet P. Pericytes: blood-brain barrier safeguards against neurodegeneration? Neuron. 2010;68:321–3.

Quintanilla-Dieck L, Larrain B, Trune D, Steyger PS. Effect of systemic lipopolysaccharide-induced inflammation on cytokine levels in the murine cochlea: a pilot study. Otolaryngol Head Neck Surg. 2013. https://doi.org/10.1177/0194599813491712.

Quirk W, Laurikainen E, Avinash G, Nuttall A, Miller J. The role of endothelin on the regulation of cochlear blood flow. Assoc Res Otolaryngol. 1992;15:37.

Rehm HL, Zhang D-S, Brown MC, Burgess B, Halpin C, Berger W, Morton CC, Corey DP, Chen Z-Y. Vascular defects and sensorineural deafness in a mouse model of Norrie disease. J Neurosci. 2002;22:4286–92.

Reif R, Zhi Z, Dziennis S, Nuttall AL, Wang RK. Changes in cochlear blood flow in mice due to loud sound exposure measured with Doppler optical microangiography and laser Doppler flowmetry. Quant Imaging Med Surg. 2013;3:235.

Ruckenstein MJ, Hu L. Antibody deposition in the stria vascularis of the MRL-Fas lpr mouse. Hear Res. 1999;127:137–42.

Rybak LP, Ramkumar V. Ototoxicity. Kidney Int. 2007;72:931–5.

Rybak LP, Whitworth CA, Mukherjea D, Ramkumar V. Mechanisms of cisplatin-induced ototoxicity and prevention. Hear Res. 2007;226:157–67.

Sakaguchi N, Spicer SS, Thomopoulos GN, Schulte BA. Immunoglobulin deposition in thickened basement membranes of aging strial capillaries. Hear Res. 1997a;109:83–91.

Sakaguchi N, Spicer SS, Thomopoulos GN, Schulte BA. Increased laminin deposition in capillaries of the stria vascularis of quiet-aged gerbils. Hear Res. 1997b;105:44–56.

Salt AN, Mleichar I, Thalmann R. Mechanisms of endocochlear potential generation by stria vascularis. Laryngoscope. 1987;97:984–91.

Sara S, Teh B, Friedland P. Bilateral sudden sensorineural hearing loss: review. J Laryngol Otol. 2014;128:S8–S15.

Sato E, Shick HE, Ransohoff RM, Hirose K. Repopulation of cochlear macrophages in murine hematopoietic progenitor cell chimeras: the role of CX3CR1. J Comp Neurol. 2008;506:930–42.

Schacht J, Talaska AE, Rybak LP. Cisplatin and aminoglycoside antibiotics: hearing loss and its prevention. Anat Rec. 2012;295:1837–50.

Scheibe F, Haupt H, Ludwig C. Intensity-related changes in cochlear blood flow in the guinea pig during and following acoustic exposure. Eur Arch Otorhinolaryngol. 1993;250:281–5.

Schulte BA, Schmiedt RA. Lateral wall Na, K-ATPase and endocochlear potentials decline with age in quiet-reared gerbils. Hear Res. 1992;61:35–46.

Seidman MD, Quirk WS, Shirwany NA. Mechanisms of alterations in the microcirculation of the cochlea, ototoxicity: basic science and clinical applications. Ann N Y Acad Sci. 1999;884:226–32.

Shaddock LC, Hamernik RP, Axelsson A. Cochlear vascular and sensorycell changes induced by elevated temperature and noise. Am J Otolaryngol. 1984;5:99–107.

Shepro D, Morel N. Pericyte physiology. FASEB J. 1993;7:1031–8.

Shi X. Cochlear pericyte responses to acoustic trauma and the involvement of hypoxia-inducible factor-1alpha and vascular endothelial growth factor. Am J Pathol. 2009;174:1692–704.

Shi X. Resident macrophages in the cochlear blood-labyrinth barrier and their renewal via migration of bone-marrow-derived cells. Cell Tissue Res. 2010;342:21–30.

Shi X. Physiopathology of the cochlear microcirculation. Hear Res. 2011;282:10–24.

Shi X. Pathophysiology of the cochlear intrastrial fluid-blood barrier (review). Hear Res. 2016;338:52–63.

Shi X, Nuttall AL. The demonstration of nitric oxide in cochlear blood vessels in vivo and in vitro: the role of endothelial nitric oxide in venular permeability. Hear Res. 2002;172:73–80.

Shi X, Nuttall AL. Upregulated iNOS and oxidative damage to the cochlear stria vascularis due to noise stress. Brain Res. 2003;967:1–10.

Shi X, Nuttall AL. Expression of adhesion molecular proteins in the cochlear lateral wall of normal and PARP-1 mutant mice. Hear Res. 2007;224:1–14.

Shi X, Han W, Yamamoto H, Tang W, Lin X, Xiu R, Trune DR, Nuttall AL. The cochlear pericytes. Microcirculation. 2008;15:515–29.

Sims DE. Diversity within pericytes. Clin Exp Pharmacol Physiol. 2000;27:842–6.

Slominski A. Neuroendocrine activity of the melanocyte. Exp Dermatol. 2009;18:760–3.

Slominski A, Zmijewski MA, Pawelek J. L-tyrosine and L-dihydroxyphenylalanine as hormone-like regulators of melanocyte functions. Pigment Cell Melanoma Res. 2012;25:14–27.

Spicer SS, Schulte BA. Differentiation of inner ear fibrocytes according to their ion transport related activity. Hear Res. 1991;56:53–64.

Spicer SS, Schulte BA. The fine structure of spiral ligament cells relates to ion return to the stria and varies with place-frequency. Hear Res. 1996;100:80–100.

Spicer SS, Schulte BA. Spiral ligament pathology in quiet-aged gerbils. Hear Res. 2002;172:172–85.

Spicer SS, Schulte BA. Novel structures in marginal and intermediate cells presumably relate to functions of apical versus basal strial strata. Hear Res. 2005;200:87–101.

Spiess AC, Lang H, Schulte BA, Spicer S, Schmiedt RA. Effects of gap junction uncoupling in the gerbil cochlea. Laryngoscope. 2002;112:1635–41.

Sprinzl G, Riechelmann H. Current trends in treating hearing loss in elderly people: a review of the technology and treatment options–a mini-review. Gerontology. 2010;56:351–8.

Steel K, Barkway C. Another role for melanocytes: their importance for normal stria vascularis development in the mammalian inner ear. Development. 1989;107:453–63.

Steel K, Davidson DR, Jackson I. TRP-2/DT, a new early melanoblast marker, shows that steel growth factor (c-kit ligand) is a survival factor. Development. 1992;115:1111–9.

Sulaimon SS, Kitchell BE. The biology of melanocytes. Vet Dermatol. 2003;14:57–65.

Suzuki M, Yamasoba T, Kaga K. Development of the blood-labyrinth barrier in the rat. Hear Res. 1998;116:107–12.

Suzuki M, Yamasoba T, Ishibashi T, Miller JM, Kaga K. Effect of noise exposure on blood–labyrinth barrier in guinea pigs. Hear Res. 2002;164:12–8.

Tagaya M, Yamazaki M, Teranishi M, Naganawa S, Yoshida T, Otake H, Nakata S, Sone M, Nakashima T. Endolymphatic hydrops and blood–labyrinth barrier in Meniere's disease. Acta Otolaryngol. 2011;131:474–9.

Takahashi M, Harris JP. Anatomic distribution and localization of immunocompetent cells in normal mouse endolymphatic sac. Acta Otolaryngol. 1988;106:409–16.

Tavanai E, Mohammadkhani G. Role of antioxidants in prevention of age-related hearing loss: a review of literature. Eur Arch Otorhinolaryngol. 2017;274(4):1821–34.

Thomopoulos GN, Spicer SS, Gratton MA, Schulte BA. Age-related thickening of basement membrane in stria vascularis capillaries. Hear Res. 1997;111:31–41.

Tornabene SV, Sato K, Pham L, Billings P, Keithley EM. Immune cell recruitment following acoustic trauma. Hear Res. 2006;222:115–24.

Toubi E, Halas K, Ben-David J, Sabo E, Kessel A, Luntz M. Immune-mediated disorders associated with idiopathic sudden sensorineural hearing loss. Ann Otol Rhinol Laryngol. 2004;113:445–9.

Trowe M-O, Maier H, Schweizer M, Kispert A. Deafness in mice lacking the T-box transcription factor Tbx18 in otic fibrocytes. Development. 2008;135:1725–34.

Trune DR, Nguyen-Huynh A. Vascular pathophysiology in hearing disorders, Semin Hear. Thieme Medical Publishers; 2012. p. 242–50.

Trune DR, Kempton JB, Mitchell CR, Hefeneider SH. Failure of elevated heat shock protein 70 antibodies to alter cochlear function in mice. Hear Res. 1998;116:65–70.

Ueda S, Yamagishi S-I, Okuda S. Anti-vasopermeability effects of PEDF in retinal-renal disorders. Curr Mol Med. 2010;10:279–83.

Varol C, Mildner A, Jung S. Macrophages: development and tissue specialization. Annu Rev Immunol. 2015;33:643–75.

Wakaoka T, Motohashi T, Hayashi H, Kuze B, Aoki M, Mizuta K, Kunisada T, Ito Y. Tracing Sox10-expressing cells elucidates the dynamic development of the mouse inner ear. Hear Res. 2013;302:17–25.

Wang Y, Hirose K, Liberman MC. Dynamics of noise-induced cellular injury and repair in the mouse cochlea. J Assoc Res Otolaryng. 2002;3:248–68.

Wang Q, Steyger PS. Trafficking of systemic fluorescent gentamicin into the cochlea and hair cells. J Assoc Res Otolaryngol. 2009;10:205–19.

Wangemann P. Cochlear blood flow regulation. Adv Otorhinolaryngol. 2002;59:51–7.

Wangemann P, Liu J. Osmotic water permeability of capillaries from the isolated spiral liga-ment: new in-vitro techniques for the study of vascular permeability and diameter. Hear Res. 1996;95:49–56.

Wong ML, Young JS, Nilaver G, Morton JI, Trune DR. Cochlear IgG in the C3H/lpr autoimmune strain mouse. Hear Res. 1992;59:93–100.

Wood MB, Zuo J. The contribution of immune infiltrates to ototoxicity and cochlear hair cell loss. Front Cell Neurosci. 2017;11:106.

Wu T, Marcus DC. Age-related changes in cochlear endolymphatic potassium and potential in CD-1 and CBA/CaJ mice. J Assoc Res Otolaryngol. 2003;4:353–62.

Yamamoto H, Omelchenko I, Shi X, Nuttall AL. The influence of NF-kappaB signal-transduction pathways on the murine inner ear by acoustic overstimulation. J Neurosci Res. 2009;87:1832–40.

Yamane H, Nakai Y, Konishi K, Sakamoto H, Matsuda Y, Iguchi H. Strial circulation impairment due to acoustic trauma. Acta Otolaryngol. 1991;111:85–93.

Yang Y, Dai M, Wilson TM, Omelchenko I, Klimek JE, Wilmarth PA, David LL, Nuttall AL, Gillespie PG, Shi X. Na+/K+-ATPase alpha1 identified as an abundant protein in the blood-labyrinth barrier that plays an essential role in the barrier integrity. PLoS One. 2011;6:e16547.

Yehudai D, Shoenfeld Y, Toubi E. The autoimmune characteristics of progressive or sudden senso-rineural hearing loss. Autoimmunity. 2006;39:153–8.

Yorgason JG, Luxford W, Kalinec F. In vitro and in vivo models of drug ototoxicity: studying the mechanisms of a clinical problem. Expert Opin Drug Metab Toxicol. 2011;7:1521–34.

Yoshida N, Kristiansen A, Liberman MC. Heat stress and protection from permanent acoustic injury in mice. J Neurosci. 1999;19:10116–24.

Young J, Morton J, Nilaver G, Trune D. Distribution of IgG in the inner ear of C3H/lpr autoim-mune disease mice. Abst Assoc Res Otolaryngol. 1988;225.

Zhang W, Dai M, Fridberger A, Hassan A, Degagne J, Neng L, Zhang F, He W, Ren T, Trune D, Auer M, Shi X. Perivascular-resident macrophage-like melanocytes in the inner ear are essential for the integrity of the intrastrial fluid-blood barrier. Proc Natl Acad Sci U S A. 2012;109:10388–93.

Zhang F, Dai M, Neng L, Zhang JH, Zhi Z, Fridberger A, Shi X. Perivascular macrophage-like melanocyte responsiveness to acoustic trauma—a salient feature of strial barrier associated hearing loss. FASEB J. 2013;27:3730–40.

Zhang J, Chen S, Hou Z, Cai J, Dong M, Shi X. Lipopolysaccharide-induced middle ear inflam-mation disrupts the cochlear intra-strial fluid-blood barrier through down-regulation of tight junction proteins. PLoS One. 2015;10(3):e0122572.

Chapter 5
Cochlear Inflammation Associated with Noise-Exposure

Elizabeth M. Keithley

Abstract While we know a great deal about the anatomical and physiological changes that occur within the cochlea as a result of noise exposure of various spectra, intensities and durations, we know relatively little about the inflammatory response to these noises. Some cochlear cells up-regulate their expression of inflammatory mediators in response to noise and presumably thereby, recruit circulating macrophages into the cochlea or activate resident cells. The mechanisms that mediate these process are not yet known. The value of the inflammatory response in terms of cochlear repair is not known. Investigators have described immune responses within the stria vascularis, the spiral ligament, the mesothelial cells below the basilar membrane and the epithelial cells of the organ of Corti. The cooperation and/or interactions among these various cells are not known. This chapter is an attempt to identify what is known of the inflammatory response and stimulate new research to clarify the response and its function.

Keywords Inflammation · Cochlea · Basilar membrane · Spiral ligament · Organ of Corti · Stria vascularis · Macrophages

1 Cochlear Response to Noise Exposure

It has been known for centuries that exposure to loud sounds results in hearing loss. Early studies of human temporal bones from hunters revealed that the auditory sensory cells within the basal turn of the organ of Corti were completely degenerated (Johnsson and Hawkins 1976). Systematic evaluation of the effects of noise exposure of various intensities on cochlear anatomy and physiology was begun in the 1970s (Eldredge et al. 1973; Henderson et al. 1974; Liberman and Kiang 1978; Liberman 1978) and from these experiments it became clear that auditory thresholds were permanently elevated after very loud noise exposures, but less loud sound

E. M. Keithley (✉)
Division of Otolaryngology/Head and Neck Surgery, University of California, San Diego, La Jolla, CA, USA
e-mail: ekeithley@ucsd.edu

© Springer International Publishing AG, part of Springer Nature 2018
V. Ramkumar, L. P. Rybak (eds.), *Inflammatory Mechanisms in Mediating Hearing Loss*, https://doi.org/10.1007/978-3-319-92507-3_5

exposures resulted in only temporary threshold shifts. The morphological changes underlying the hearing loss were seen in cells comprising the cochlear sensory epithelium, spiral ganglion neurons, stria vascularis, spiral ligament, and spiral limbus depending on the sound spectrum, amplitude and exposure duration (Liberman and Dodds 1987; Pujol et al. 1993; Wang et al. 2002; Miller et al. 1996). Even relatively low level noise exposure can lead to delayed onset hearing loss (Kujawa and Liberman 2006).

Noise exposure also affects blood flow within the cochlea and stria vascularis including decreased blood flow, increased vascular permeability and ischemia (Hillerdal et al. 1987; Quirk et al. 1992; Thorne and Nuttall 1989; Scheibe et al. 1993; Seidman et al. 1999; Miller et al. 2003; Shi 2011, 2016). Consistent with these changes, there is morphological damage to the marginal, intermediate and basal cells forming the stria vascularis, and the endocochlear potential is reduced as a result of loud noise exposure (Syka et al. 1981; Ohlemiller and Gagnon 2007; Zhang et al. 2013).

Since many of these structures are involved in maintaining the endocochlear potential and controlling potassium ion concentrations through recycling (Spicer and Schulte 1991; Wangemann 2006) the difficulty of this task during prolonged, high intensity noise may underlie the noise induced damage (Wang et al. 2002; Hirose and Liberman 2003). The immune response then is likely to be directed at these cellular structures.

2 Immune Aspects of the Normal Cochlea

The epithelial cells of the skin, gastrointestinal tract and respiratory system are constantly exposed to pathogens and, therefore, well-adapted to protecting themselves. The cochlea, on the other hand, is less often exposed to pathogens, from the systemic circulation or from the middle ear through the round window membrane. Certainly, it has the capability of defending itself, but the identification of the cells and molecular mechanisms for providing this protection has only recently begun to be explored (Harris 1983). In addition to pathogens, stress and cellular damage products (damage-associated molecular patterns—DAMPs) which interact with toll-like receptors (TLRs), RAGE (receptor for advanced glycation end-products) and other receptors, can also activate an innate immune response (Bianchi 2007; Knoops et al. 2016). Loud noise exposure that results in cellular degeneration produces stress and DAMPs within the cochlea and initiates an innate immune response (Hirose et al. 2005; Tornabene et al. 2006). It is not surprising, though, that the inflammatory and immune processes in the cochlea are tailored to this location because of the inability of the mammalian cochlear cells to regenerate after any injury and "bystander cellular-injury" from inflammation is well documented.

Resident, tissue macrophages provide the front-line surveillance for immune function throughout mammalian organisms. They have various shapes in different organs and have been given different names by anatomists, such as Kupffer cells in the liver and microglia in the central nervous system. They constantly sample the molecular environment, where they serve in the innate immune response by phagocytosing foreign molecules and cells and then, can initiate an adaptive immune response through antigen-presentation to T-lymphocytes. They coordinate and initiate cellular responses by secreting cytokines.

The resident macrophages in the cochlea do not have sufficiently distinct morphological features to make them identifiable in hematoxylin and eosin stains or other basophilic stains generally used to assess cochlear micro-anatomy and pathology. It was not, therefore, until the availability of antibodies that could be used in immunohistochemical assays and the ability to inject green-fluorescent-protein (GFP) labeled, bone-marrow derived monocytes into irradiated mice, that these cells were described in the cochlea. One of the earliest identifications of cochlear-resident macrophages was in the avian cochlea where they were identified by immunohistochemical assays. In an avian, organ-culture system, macrophages were seen to assist in hair cell regeneration following laser damage to the sensory cells (Warchol 1997).

Cochlear Vasculature and Leukocyte Extravasation

As in all organs, the vascular pattern of the cochlea is related to its specialized function (Axelsson 1968). The spiral modiolar artery enters the cochlea near the basal turn and spirals apically within the modiolus with radiating arterioles periodically branching off and arching over the scala vestibuli toward the lateral wall. Within the spiral ligament these vessels branch to form the capillaries of the stria vascularis whose endothelial cell tight-junctions create the blood-labyrinthine barrier, and the capillaries of the spiral ligament whose endothelial cells are not connected by tight junctions. As in all other tissues, the capillaries rejoin to form post-capillary venules. The venules are within the spiral ligament at the level of the spiral prominence and below the level of the junction of the basilar membrane and the ligament. This is the type IV fibrocyte region of the ligament (Spicer and Schulte 1991) and where intercellular adhesion molecule-1 (ICAM-1) is expressed (Tornabene et al. 2006). Post-capillary venules are generally the location of immune cell extravasation. Macrophage-induced expression of adhesion molecules by the venular endothelial cells enables binding of leukocytes prior to extravasation into tissues. Once extravasated, the cells migrate within tissues through the interaction of integrins with extracellular matrix proteins and adhesion molecules like ICAM-1. The cochlea, with its fluid-filled spaces and sensory epithelium that is devoid of extracellular matrix (Santi et al. 2016), presumably imposes limitations on this system, likely preventing leukocyte migration among the sensory epithelial cells and requiring the formation of extracellular matrix in the scalae to enable leukocyte migration (Keithley and Harris 1996).

Resident Macrophages of the Spiral Ligament

Although the first studies using immunohistochemical assays of the leukocyte markers, Mac-1, (Takahashi and Harris 1988) and CD-45 (Hashimoto et al. 2005) in mice cochleas did not report seeing any labeling of these cells in the spiral ligament of normal animals, it is now clear that immune cells are present in the normal cochlea. Antibodies to the pan-leukocyte, cell surface marker, CD-45, a receptor tyrosine phosphatase, reveal macrophages within the spiral ligament in the region of the type II and IV fibrocytes (Hirose et al. 2005; Sato et al. 2008; Tornabene et al. 2006) (Fig. 5.1a), in the stria vascularis (Du et al. 2011) and on the scala tympani side of the basilar membrane (Yang et al. 2015). They vary from stellate shaped to spherical. The reported number of cells in the spiral ligament per section is related to the section thickness. Hirose et al. (2005) reported cutting 30 μm frozen sections with 5 CD45+ cells per section, while Tornabene et al. (2006) cut 6 μm thick sections and found 0.3 CD45+ cells per section.

Several studies have used radiation to deplete mice of their existing leukocyte populations, and then injected GFP+ bone-marrow-derived leukocytes to determine whether these cells extravasate to provide cochlear tissues with macrophages or whether the resident cells divide when activated. These experiments have demonstrated that in the spiral ligament (Lang et al. 2006a; Okano et al. 2008; Sato et al. 2008) and spiral ganglion (Okano et al. 2008), macrophages originate from the bone marrow and turn over every 3 or 4 months in normal mammals. The blood vessels of the normal spiral ligament express the ICAM-1 (Tornabene et al. 2006; Shi and Nuttall 2007) that is involved in recruiting and enabling extravasation of leukocytes into tissues from the vasculature and likely, is involved in this slow turnover of

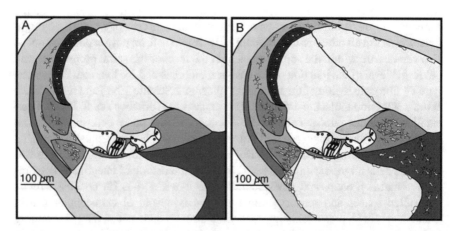

Fig. 5.1 Schematic drawing of a cross-section of the cochlear duct illustrating the location of resident and infiltrated macrophages. (**a**) Control ear with few resident macrophages concentrated in the lower spiral ligament and scattered in the upper spiral ligament. (**b**) Cochlear duct 7 days after noise damage, with large numbers of inflammatory cells in the lateral wall, spiral ganglion, and spiral limbus and marching into the perilymph-filled space of the scala tympani and scala vestibuli. (Reproduced with permission from Hirose et al. 2005)

CD45+ cells in the normal spiral ligament. The chemokine CCL2 and its receptor CCR2 do not seem to be involved, however (Sautter et al. 2006; Sato et al. 2008). There are still many unanswered questions concerning the immune regulatory activity in the spiral ligament because Lang et al. (2006a) concluded that the majority of GFP+, bone-marrow-derived cells that infiltrate the spiral ligament differentiate into fibrocytes, not macrophages, based on the infiltrated GFP+ cells non-reactivity with the leukocyte marker, CD45, and their positive immunoreactivity with Na-K-ATPase, an enzyme used by fibrocytes for potassium recycling. These authors did, however, also report that a few GFP+ hematopoietic cells, identified in the type II and type IV fibrocyte region of the spiral ligament, did immunolabel for the CD45 leukocyte marker. So perhaps, depending on signals of the spiral ligament cells, the infiltrating cells may differentiate into fibrocytes or macrophages.

Resident Macrophages of the Stria Vascularis

Resident macrophages are also present in association with the complex capillary system of the stria vascularis (Zhang et al. 2012, 2013; Shi 2016). The cell bodies are physically very close to the capillary endothelial cells and pericytes with multiple processes partially enveloping a vessel (Fig. 5.2). These cells immuno-label

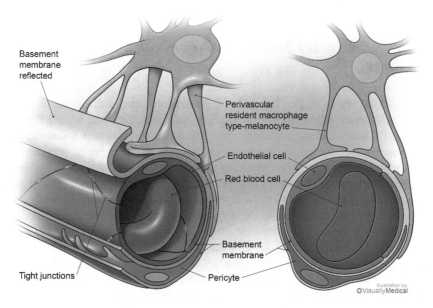

Fig. 5.2 Illustration of a stria vascularis capillary in cross-section showing the major components of the intrastrial fluid-blood barrier or blood-labyrinthine barrier. The vessel lumen is formed by endothelial cells on a dense basement membrane shared with pericytes. The endothelial cells are connected by tight junctions. Perivascular resident macrophages-like melanocytes have end-feet that cover a large portion of the capillary surface external to the pericytes. (Reproduced with permission from Shi 2016)

with antibodies against macrophages including F4/80, CD68, and CD11b. They express scavenger-receptor molecules on their cell membranes and they are slowly renewed from bone-marrow derived cells (Shi 2010). They also express markers for melanocytes (Zhang et al. 2012). As such, they have characteristics of both macrophages and melanocytes and have been named "perivascular resident macrophage-like melanocytes" by Zhang et al. (2012). Unlike the macrophages in the spiral ligament and below the basilar membrane, the physiology of these cells is under active investigation and they are considered to participate in a "cochlear-vascular unit" (Shi 2016). In the normal cochlea, it appears that they serve to modulate cochlear blood-flow and maintain the endocochlear potential through control of endothelial-cell tight-junctions and adhesive junction protein expression (Shi 2010; Zhang et al. 2012, 2013). One factor involved in this control is pigment epithelium-derived factor (PEDF), a 50 kDa glycoprotein. When its expression is inhibited, the tight junction proteins of the stria-vascularis, capillary-endothelial cells are down-regulated and the endocochlear potential is reduced. The shape and dendritic arborization of the "perivascular resident macrophage-like melanocytes" are also affected as the cells lose their arborizations and become more spherical. Loss of these cells is associated with blood-cochlear-barrier breakdown and edema of the stria vascularis (Zhang et al. 2013). The inclusion of physiological studies to the examination of immune function in the cochlea shines light on the importance of these newly described cells to normal cochlear function.

Resident Cells of the Sensory Epithelium with Macrophage Attributes

Based on the argument that circulating inflammatory cells enter the cochlea from the venules in the spiral ligament or scala tympani and therefore, are not likely to directly influence the remodeling of the organ of Corti that occurs following loud noise exposure, Cai et al. (2014) examined the cells of the organ of Corti itself to determine if these cells had some capacity to act as immunocompetent cells. This is a reasonable hypothesis because it appears that inflammatory cells do not migrate into the sensory epithelium. Consistent with their hypothesis, Cai et al. (2014) found at least 45 genes related to inflammatory pathways in normal cells of the sensory-epithelium on the basilar membrane. Toll-like receptor signaling pathway genes were among those identified with the strongest expression levels. Four proteins that are related to the Toll-like receptor pathway, who's mRNA was detected in the gene analysis, were chosen for analysis in immunohistochemical assays. Labeling of the normal organ of Corti demonstrated strong TLR4 expression in the inner hair cell membranes, but no label of the outer hair cells or Deiter's cells. TLR3, however, was not identified in any of the cells, although the mRNA was seen in the PCR assay. IRF7 (interferon regulatory factor 7) and STAT1 (signal transducer and activator of transcription 1) are expressed in supporting cells (Hensen cells and Deiter's cells), but not in the sensory cells. STAT1 immunoreactivity was localized in the phalangeal processes of the Deiter's cells that form the reticular lamina and IRF7 was in the tubulin containing phalangeal processes that extend from the cell body up to the reticular lamina (Cai et al. 2014). In addition to molecules related to Toll-like receptors, Mif (macrophage migration inhibitory factor) and TNFα (tumor necrosis factor-alpha), both of which are pro-inflammatory cytokines, are also constitutively expressed in the organ of Corti. Mif is involved in regulating macrophage function and TNFα has been shown to induce recruitment of inflammatory cells to the cochlea (Keithley et al. 2008). The work of Cai and his colleagues is quite novel and opens a new path for exploring the effects of loud noise on cochlear degeneration, hearing loss and the auditory sensory apparatus to establish whether pharmaceutical assistance can prevent permanent hearing loss.

Resident Macrophages of the Basilar Membrane

The physically closest macrophages to the cochlear sensory cells are those that reside on the scala tympani side of the basilar membrane among the tympanic lamellar cells. Evaluation of these cells was performed following dissection of the basilar membrane to create surface preparations of the membrane and sensory epithelium (Yang et al. 2015). The morphology of these CD45+ cells underneath the mouse basilar membrane varies from apex to base of the cochlea, with the cells in the apical third having a stellate shape with multiple long dendritic projections. In approximately the middle cochlear turn, the cells have fewer and shorter projections, while in the basal turn, the cells are amoeboid-shaped with few projections. In addition to expressing CD45, all these cells express F4/80, the antigen expressed by macrophages, but not other leukocytes. In addition to these cells, there are smaller, spherical shaped cells along the whole length of the basilar membrane. F4/80 labeling ranged from strong to not-detectable in the spherical cells suggesting that they may not all be macrophages. In total, 95 ± 17 macrophages per cochlea were counted along the entire length of the basilar membrane. They are evenly distributed with no clustering of cells (Yang et al. 2015). The relative paucity of these cells explains why they were not described by previous researchers who have looked at radially sectioned material using antibodies against CD45 and F4/80.

Further characterization of the dendritic-like cells under the basilar membrane revealed that the apical cells do not immunolabel with antibodies to the dendritic cell markers CD11c, CD14 or MHC class II antigens. While in the middle and basal regions, CD45-positive cells also label with CD11c, CD14 consistent with the basal cells being dendritic cells, while the apical cells are not. MHC II immunoreactivity was restricted to only about ten or fewer cells per cochlea scattered throughout the whole length of the basilar membrane (Yang et al. 2015).

Lymphatic Drainage of the Cochlea

In all mammalian tissues the resident immune cells work in collaboration with the lymphatic system which is largely composed of the lymph nodes and the lymphatic vessels that drain organs and return extracellular fluid to the heart. The inner ear, like the central nervous system, has always been considered independent of this system. As the inflammatory capability of these organs, however, has become apparent, it seems reasonable to reconsider their interactive capability with draining cervical lymph nodes. Using newly available lymphatic markers, connections between cerebral spinal fluid and cervical lymph nodes have been identified (Louveau et al. 2015; Aspelund et al. 2015).

Yimtae et al. (2001) addressed this issue for the inner ear by slowly injecting a large molecular weight antigen that could be immuno-labeled in tissue sections into the scala tympani of normal guinea pigs. The cervical lymph nodes (mandibular, parotid, superficial ventral and deep cranial) and the spleen as well as the temporal bones were collected shortly after the injection. Immuno-labeled-antigen in the cochlea showed that it had been absorbed by cells lining the scala tympani and the inferior portion of the spiral ligament in the region of the type IV fibrocytes where the resident macrophages are located. In some cases, antigen was also seen in the modiolus and the suprastrial portion of the spiral ligament. The important finding is that antigen was identified in the superficial ventral cervical lymph nodes and the spleen 15 min after injection into the scala tympani. It is very clear that antigens present in the cochlea are monitored by systematic immune cells at all times, whether through lymphatics within the spiral ligament and modiolus or through the cochlear aqueduct to the cerebral meninges. What goes on in the inner ear does not stay in the inner ear. With the experimental tools available today, it seems possible to determine whether there are lymphatic vessels within the cochlea or whether the cochlea is connected to the cervical lymph nodes via the cochlear aqueduct and the cerebral spinal fluid.

Conclusions

It seems clear from this summary that the "resting", normal cochlea has all the cellular components to provide rapid immunological responses to invasions by pathogens or signals from cellular damage (damage associated molecular patterns, DAMPS) and stress. It is interesting that there is great heterogeneity among the resident cochlear macrophages including their locations: the spiral ligament and spiral limbus, stria vascularis and basilar membrane; and therefore, perhaps there is a broad range of physiological responses to various stress situations. Physical limitations to immune activity in the cochlea are imposed by the lack of blood vessels and extracellular matrix within the organ of Corti and obviously within the fluid filed scalae. This lack of matrix in the organ of Corti prevents leukocytes from moving among the epithelial cells comprising this tissue. An obvious question, then, is whether either signaling molecules and/or cell debris from damaged cells can diffuse from the epithelium across the basilar membrane to the resident macrophages below it, or are the basilar membrane macrophages solely responsible for surveillance of the perilymph? Are the supporting cells of the organ of Corti endowed with the immune capability to protect the epithelium? Likewise, are the macrophages of the spiral ligament, which are clearly related to the post-capillary-venule system of the cochlea, restricted to respond to local events in the spiral ligament? Since the fibrocytes are involved in antigen absorption from the scalae (Yimtae et al. 2001) and potassium recycling (Spicer and Schulte 1996) they are likely stressed by loud noise exposure (Wang et al. 2002). And finally, are the macrophages associated with the strial capillaries solely charged with maintaining oxygen and other vascular responsibilities?

3   Inflammatory Response to Acoustic Trauma

While the anatomical and physiological changes to the cochlea in response to noise exposures of various intensities, frequencies and durations have been very well defined, the cochlear inflammatory responses to noise are just beginning to be investigated. As suggested above, it was not until the advent of commercially available antibodies against cell surface markers of specific leukocyte antigens that studies of inflammatory responses could be undertaken at all. At this point, there is no sense of the relationship between the quality and magnitude of the inflammatory response and the noise stimulus parameters. The various studies that have been published have used different mammalian species (mice and rats of different strains and chinchillas), noise exposures and post-noise exposure evaluation times. The following discussion then must be considered general and may not apply to all situations. Certainly, in the future systematic studies of cochlear inflammation relative to noise exposure will define the responses and how they relate to cochlear pathology and repair.

   Given that the damaging effects of loud noise are present in the sensory epithelium, the vasculature of the stria vascularis and the spiral ligament, it seems reasonable to look at the inflammatory responses in all the cellular structures separately.

Inflammatory Response of the Spiral Ligament

Hirose et al. (2005) were the first to observe that in response to noise exposure in the mammalian cochlea, there were leukocytes, immunolabeled with antibodies to the CD45, common leukocyte cell surface antigen, recruited to the spiral ligament, spiral limbus, and scala tympani where they are adherent to the bony wall of the cochlear duct (Fig. 5.1b). The cells within the spiral ligament are first seen in the region of the type IV fibrocytes, but over time they expand into the region of the type I fibrocytes. This is also the location of the post-capillary venules for the

Fig. 5.3 Macrophages, immuno-labeled with antibody to F4/80 antigen, in the spiral ligament, stria vascularis and below the basilar membrane from a mouse exposed to an 8–16 kHz noise at 118 dB for 2 h and sacrificed 7 days post-exposure (Tornabene et al. 2006). *SM* scala media, *ST* scala tympani. Scale bar 50 μm

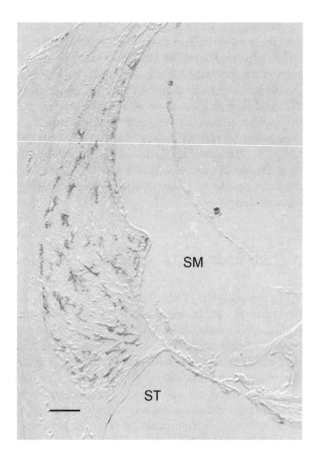

vasculature of the lateral wall (Axelsson 1968), the general site for leukocyte extravasation into tissues. These CD45+ cells were observed by 1 day post-noise exposure at noise levels that resulted in permanent threshold shifts. At 7 and 14 days after noise exposure, 20–30 cells per 30-μm section were still present in the region of the cochlea where the greatest damage to hair cells occurred. Similar results were subsequently reported using antibodies to F/480, a macrophage-specific marker by Tornabene et al. (2006) and Tan et al. (2008) who point out that the maximum number of infiltrated cells was at 3 days after the noise exposure (Fig. 5.3).

By 7 days post-exposure, the cochleas exposed to noise that resulted in only a temporary threshold shift also experienced an influx of CD45+ cells (Hirose et al. 2005). This result might be interpreted to indicate that at lower sound levels, signals that recruit circulating cells are slower to be activated, but once activated, produce a similar response. After 14 days, the number of labeled cells was reduced in the mice exposed to 106 or 112 dB noise, but continued to increase in the mice exposed to the 120 dB noise. Although the quantitative analysis was done with an immunohistochemical assay using the CD45 antibody, making it impossible to differentiate between influx of cells or division of resident cells, a transgenic mouse line with GFP+ circulating monocytes/macrophages was used to demonstrate that the leukocytes did extravasate from the systemic circulation and were derived from the bone marrow monocyte lineage (Hirose et al. 2005; Tan et al. 2008).

An understanding of the signals that induce the recruitment of circulating macrophages to the spiral ligament and limbus is not yet defined; however, the DNA transcription factor, NFκ-B, (nuclear factor kappa-light-chain-enhancer of activated B cells) is a protein complex that controls DNA transcription related to inflammatory mediators and cell survival, and is activated 2–6 h following noise exposure (Masuda et al. 2006), so is a potential candidate. Sound levels that cause only a temporary threshold shift in mice, but no hair cell degeneration (92–112 dB, 8–16 kHz noise for 2 h) activate NFκB, and in transgenic, NFκB-reporter mice the activated cells were seen predominantly among the type I fibrocytes in the spiral ligament and in the spiral limbus. There were also a few activated cells in the type II fibrocyte region of the spiral ligament as early as 24 h after the termination of noise exposure (Adams et al. 2009). As the sound level was increased, the number of cells showing NFκB activation increased. These cells are not, however, located adjacent to the venules of the inferior portion of the spiral ligament which presumably support the extravasation of macrophages, thus complicating our understanding of the relevant mechanisms.

Another set of likely candidates for initiating the immune response are chemokines that are involved in chemotaxis of leukocytes to tissues. It is reasonable to expect that expression of at least one of the known chemokines is involved in the infiltration of macrophages to the spiral ligament in response to noise exposure. In accordance with this expectation, Sato et al. (2008) developed chimeric mice with GFP+ bone-marrow cells and no gene for producing the chemokine receptor, CX3CR1, known to function in the central nervous system, and exposed the mice to noise (8–16 kHz, 112 dB, for 2 h). They subsequently evaluated the mice for macrophage migration into the spiral ligament. There was no difference in the number of infiltrated cells between the CX3CR1-null mice and normal mice, suggesting that CX3CR1, generally expressed on tissue macrophages, is not necessary for chemotaxis into the spiral ligament (Sato et al. 2008). Another chemokine, CXCL12, or SDF-1 (stromal-cell derived factor-1) is up-regulated immediately after noise exposure in mice (Dai et al. 2010) and was immunohistochemically identified to be expressed on cells in the type II fibrocyte region of the spiral ligament 3 days after noise exposure with reduced expression after that time (Tan et al. 2008). This expression pattern is consistent with the timing for the recruitment of macrophages. CXCL12 is expressed by cells that are involved in potassium recycling that are likely stressed by loud noise exposure, so this chemokine clearly deserves further experimental examination to define its involvement in chemotaxis.

Another approach, used by most investigators addressing this issue, is to evaluate cochlear tissues for mRNA expression by PCR or protein detection assays like ELISA or Western blot following noise exposure. Numerous studies describe the results of investigation of several possible chemokines, cytokines or other pro-inflammatory molecules. Several different animals have been examined using a variety of noise exposures and survival times. The results of these assays are shown in Table 5.1. Given the existing data, it seems that CCL2 is the most highly expressed

Table 5.1 Chemokines and Inflammatory Molecules expressed in the cochlea following noise exposure

Investigation	Animal	Noise exposure	Time after noise	mRNA, ELISA or Western Blot Assays											
				CCL2/ MCP-1	CCL4/ MIP1β	CCL7	CCL12/ MCP-5	CXCL10	CXCL12/ SDF-1α	IL-1β	IL-6	TNF-α	TGF-β	ICAM-1	inos
Acute															
Tournabene et al. (2006)	NIH Swiss mice	8–16 kHz, 118 dB, 2 h	2 h	XX	0		XX							X	
Fujioka et al. (2006)	rat	4 kHz octave band, 124 dB, 2 h	3 h							XX	XX	X			
Nakamoto et al. (2012)	CBA mice	4 kHz octave band, 130 dB, 3 h	3 h							XX	XX				
Wakabayashi et al. (2010)	B6 mice	4 kHz octave band, 124 dB, 2 h	6 h								XX				
Gratton et al. (2011)	129S1 mice	7–14 kHz, 105 dB, 1 h	6 h					X							
Tan et al. (2016)	B6 mice	8–16 kHz, 100 dB, 24 h	6 h	XXXX						X		XXX		XX	
Tournabene et al. (2006)	NIH Swiss mice	8–16 kHz, 118 dB, 2 h	8 h	XX	X		XX							X	
Tournabene et al. (2006)	NIH Swiss mice	8–16 kHz, 118 dB, 2 h	18 h	XXX	0		XX							X	
Murillo-Cuesta et al. (2015)	B6, CBA, MF1 mice	2–20 kHz, 100–120 dB, 0.5 or 12 h	24 h										X		
Tan et al. (2016)	B6 mice	8–16 kHz, 100 dB, 24 h	24 h	XXXX						0		0		X	

Tournabene et al. (2006)	NIH Swiss mice	8–16 kHz, 118 dB, 2 h	48 h	X	X	X	X				X
										X	X
Tan et al. (2016)	B6 mice	8–16 kHz, 100 dB, 24 h	3 days	0				X	X		X
Cai et al. (2014)	B6, CBA mice, sensory epithelium	1–7 kHz, 120 dB, 1 h	1 and 4 days			0		0	0		
Tan et al. (2016)	B6 mice	8–16 kHz, 100 dB, 24 h	7 days	X	X	X		X	X	XX	XXX
Sautter et al. (2006)	B6 mice	8–16 kHz, 112 dB, 2 h	7 days	XXX	XX	XX					
Chronic											
Dai et al. (2010)	B6 mice, lateral wall	Broadband, 120 dB, 3 h/day to 2 days	0–48 h			X	X				X
Arslan et al. (2012)	Wistar rats	White noise, 115 dB, 3 h/day for 10 days	7 days			0		0	0	0	
Tan et al. (2016)	B6 mice	8–16 kHz, 90 dB, 2 h/day, 1–4 weeks	14 days	XX	X	X	X		X	X	X

0—The molecule was not found by the investigator. The number of X's represents an estimate of the quantity of the molecules relative to other molecules within the experimenters' sample chemokine ligand among those tested (Tornabene et al. 2006; Sautter et al. 2006; Tan et al. 2016). It is not necessary for chemotaxis, however, because knockout mice that express no CCL2 are still able to recruit macrophages to the cochlea in response to noise exposure (Sautter et al. 2006). These experiments illustrate the complexity of the signaling pathways involved in the recruitment of leukocytes to the spiral ligament.

Pro-inflammatory cytokines are expressed by infiltrated inflammatory cells, as well as cochlear fibrocytes, during induced immune responses (Satoh et al. 2002). Noise exposure also results in rapid and transitory expression of the pro-inflammatory cytokines IL-1β and IL-6 (Fujioka et al. 2006; Wakabayashi et al. 2010; Nakamoto et al. 2012). TNF-α is also up-regulated (Fujioka et al. 2006; Tan et al. 2016) although perhaps slightly later than IL-1β and IL-6. As shown in Table 5.1, it is not clear what cytokine expression patterns occur with chronic noise exposure and longer times after the cessation of the noise (Arslan et al. 2012; Tan et al. 2016). The data of Cai et al. (2014), that reflect expression in the sensory epithelium, with the lateral wall excluded, suggest that the organ of Corti epithelial cells are not the source of the cytokine expression. Unfortunately, these authors looked long after the noise exposure, so this may also explain the lack of cytokine expression in their samples. TGFβ was also identified by PCR as a key player in inflammatory reactions to noise exposure in CBA mice exposed to broad-band noise (2–20 kHz) at 100 or 120 dB for 3 or 24 h (Murillo-Cuesta et al. 2015). In mice exposed to 105 dB noise (7–14 kHz) for 1 h the membranous labyrinth and lateral wall tissues showed up-regulation of Socs-3 (Suppressor of cytokine signaling 3) and CXCL10 mRNA at 6 h after noise, even though this exposure was not enough to cause auditory threshold shifts or sensory cell damage (Gratton et al. 2011). While Socs-3 is thought to suppress cytokine signaling, CXCL10 is thought to contribute to recruitment of macrophages and deserves further evaluation for its relevance to immune activity following noise exposure.

Finally, some studies have used immunohistochemical techniques that reveal the location of the chemokines or other molecules involved in macrophage recruitment or initiation of inflammation. Fujioka et al. (2006) used immunohistochemistry to demonstrate the location of the pro-inflammatory cytokine, IL-6, in and around the type IV fibrocytes shortly after noise exposure. The expression pattern became more diffuse over time after the exposure. ICAM-1 (also known as CD54) expression on vascular endothelial cells contributes to the capture of leukocytes prior to extravasation. Following noise exposure of various levels and frequencies, the expression of ICAM-1 in the inferior portion of the spiral ligament has been consistently observed by several different investigators (Tornabene et al. 2006; Miyao et al. 2008; Tan et al. 2016; Shi and Nuttall 2007). PECAM-1 and P-selectin were also evaluated by immunohistochemistry and found to be expressed in the blood vessels of the inferior portion of the spiral ligament (Shi and Nuttall 2007). Shi and Nuttall (2007) have pursued the significance and importance of the ICAM-1 expression for the recruitment of macrophages to the spiral ligament by using poly (ADP-ribose) polymerase-1 (PARP-1) mutant mice. The PARP-1 null mice are unable to up-regulate PECAM-1, P-selectin or ICAM-1 because of the absence of PARP-1 which modulates the expression of intercellular adhesion molecules. When these mice were exposed to noise and examined for macrophages in the spiral ligament using antibodies to CD45, there were none present. PARP-1 null mice seem like a very useful model for investigating the role of the inflammatory response to acoustic trauma and whether it contributes to threshold shifts or helps repair the spiral ligament assisting in recovery from temporary threshold shifts. Unfortunately, there were no measurements of the hearing losses in these mice strains, so it is not known how the absence of macrophage infiltration affects threshold recovery after noise exposure. In a separate study, antibodies to ICAM-1 were systemically injected into rats before and after noise exposure that produces only a temporary shift. Relative to control rats that were not injected with antibody, the animals with the anti-ICAM-1 antibodies had less auditory brainstem threshold shift after the exposure. No histology was reported, so it is not known whether or not macrophages entered the spiral ligament at the sound levels used to induce trauma (Seidman et al. 2009). It can be concluded, though, that at least one or all of the examined adhesion molecules is required for macrophage infiltration of the spiral ligament, and that the enzyme, poly (ADP-ribose) polymerase-1, is necessary for their expression.

Inflammatory Response of the Stria Vascularis

As discussed above, the stria vascularis includes a resident population of immuno-competent cells given the name "perivascular resident, macrophage-like melano-cytes" because of their characteristics, reminiscent of both macrophages and melanocytes. These cells contribute to the normal capillary structure by having a cell body that is outside the endothelial cell basement membrane with processes and "end-feet" in contact with that membrane. The remaining cell type participating in the "cochlear-vascular unit" is the pericyte that ensheaths the capillary wall outside the endothelial cell basement membrane (Shi 2010; Zhang et al. 2012; Shi 2016).

They are very sensitive to noise and retract their processes and appear less tightly bound to the capillary after broadband noise exposure (120 dB, 3 h/day for 2 days). Their production of the signaling molecule, PEDF, which contributes to their dendritic-arbor morphology, and to the integrity of endothelial-cell, tight-junction proteins that create the blood-labyrinthine barrier, is reduced at both the mRNA and protein levels. One result of the reduction of PEDF is a breakdown of the blood-labyrinthine barrier thus allowing the influx of serum proteins among other mole-cules, and a decline in the endocochlear potential by 30–50% (Zhang et al. 2013). As if to compensate for the morphological changes of the resident macrophage-like melanocytes, circulating CD45+ and F4/80+ cells are recruited to the stria vascu-laris (Du et al. 2011). This process was examined in detail using mice with GFP+ circulating bone marrow cells (Dai et al. 2010). These investigators demonstrated that immediately following noise exposure as above, macrophages begin to enter the stria vascularis where they assist in repair of the blood-labyrinth barrier. The number of macrophages per unit length of the stria vascularis increased from about

5 in mice sacrificed immediately following the noise to 20 or 25 at 1 and 2 weeks and then declined to slightly fewer at 4 weeks which was the final time of analysis because by that time the infiltrated cells had coalesced with the strial vessels and cells were no longer undergoing extravasation.

The strial perivascular pericytes respond to noise exposure that results in permanent threshold shifts by dramatically up-regulating inducible nitric oxide synthase (iNOS) (Shi et al. 2003) and the chemokine, stromal-cell-derived factor-1α (SDF-1α) (Dai et al. 2010). Because these molecules are likely to be involved in the macrophage recruitment process, Dai et al. (2010) tested the hypothesis that iNOS, its downstream product, SDF-1α, and the SDF-1α receptor, CXCR4, are, in fact, involved. In a series of experiments, the investigators used iNOS wild-type mice and iNOS-null-mice with GFP+ bone marrow cells and a blocker of CXCR4. The iNOS-null mice showed significantly less SDF-1α expression and macrophage infiltration following the noise exposure than did the wild-type mice, illustrating the importance of iNOS in the recruitment process. The mice injected with the CXCR4 blocker also experienced few infiltrated bone-marrow cells following noise exposure, implying that the SDF-1α is also required for extravasation. It seems then, that these molecules are involved in macrophage recruitment to the stria vascularis following loud noise exposure. As in the spiral ligament, the adhesion molecule, ICAM-1 is also expressed in the vessels of the stria vascularis after noise exposure and is likely involved in the recruitment process (Shi and Nuttall 2007).

Unlike other cochlear structures, it seems that the vasculature within the stria vascularis is repaired following noise exposure and this repair is accomplished with the assistance of the infiltrated circulating bone-marrow derived cells (Shi 2016). By 4 weeks after noise exposure, Dai et al. (2010) describe that the infiltrated GFP+ cells have differentiated into macrophages, pericytes and vascular endothelial cells, and reformed the normal vascular pattern of the stria vascularis. Zhang et al. (2013) demonstrated that injection of PEDF into mice after the noise-induced damage of the macrophages-like melanocytes had occurred that the endocochlear potential could be regained. It seems then, that following naturally occurring repair of the damaged stria vascularis that, if the growth factor is secreted by the newly infiltrated cells, at least the endocochlear potential component of noise induced hearing loss should be repaired.

Inflammatory Response of the Sensory Epithelium

The sensory cells of the organ of Corti were the first to be identified as those that are damaged by loud noise exposure, and until recently were not considered to possess any immune function or even be protected by immune surveillance. As described above, no macrophage-like cells have been identified through the use of cell surface markers within the organ of Corti or sensory epithelium following noise exposure even though hair cell debris is rapidly eliminated and the reticular lamina is very quickly repaired following noise-induced damage (Raphael and Altschuler 1991).

This remarkable repair is achieved by Deiter's cells (Abrashkin et al. 2006; Anttonen et al. 2014), but the signals to activate the process may relate to innate immunity. Based on this assumption, Cai et al. (2014) have examined the cochlear, sensory epithelium for its immune activation capability in response to noise exposure and discovered that, in fact, the hair cells and supporting cells of the organ of Corti react to the exposure by up-regulating immune-related genes. A subsequent study comparing mice and rat cochleas exposed to noise and using the same dissection technique, also showed up-regulation of numerous genes related to inflammatory pathways in both species (Yang et al. 2016). To further investigate gene expression as a result of noise exposure, Cai et al. (2014) used B6 and CBA mice exposed for 1 h to broadband noise (1–7 kHz, 120 dB) that were sacrificed 1 or 4 days after the termination of the noise. This exposure causes permanent threshold shifts of about 30 dB and hair cell degeneration mostly in the basal half of the cochlea. mRNA analysis of gene expression showed that several immune-related genes were up-regulated including *Irf7*, interferon regulatory factor 7, *Ddx58* and *Stat1*, signal transducer and activator of transcription 1, 1 day after noise exposure with a return to normal expression levels by 4 days. The gene for TLR3 was also slightly up-regulated at 1 day post exposure. *Irf7* showed the largest increase in mRNA expression with about a 70-fold increase. All these genes are involved in innate immunity, the detection of viruses in tissues and the initiation of a response to eliminate the pathogens. IRF7 has been shown to play a role in the activation of interferon related genes. Ddx58 functions as a pattern recognition receptor for viruses. STAT1 is a transcription factor involved in up-regulating genes in response to signals from interferons. And finally, the ligand for TLR3 is double-stranded RNA of viruses. The bound receptor subsequently activates interferon pathways that participate in destroying viral pathogens. Assuming that there was no undetected, viral infection of the mouse cochleas examined by these investigators, one interpretation of these findings is that cell-damage from noise exposure activates the same pathways generally used for viral elimination. In fact, there is evidence that DAMPS do bind to TLR4 and activate an innate immune response (Peri and Calabrese 2014). The heat shock protein, HSP60, generally expressed during cellular stress and death, may also be a natural ligand for the bacterial endotoxin receptor, TLR4 (Ohashi et al. 2000). This hypothesis provides a novel approach to understanding the reorganization and repair of the noise-damaged organ of Corti.

Using immunohistochemistry to localize TLR3 proteins in the sensory epithelium 1 day following noise-exposure, Cai et al. (2014) described scattered inner and outer hair cells labeled with antibodies to TLR3. In the region of the organ of Corti with the greatest amount of outer hair cell damage, the greatest number of outer hair cells was labeled. There was also a correlation with cells undergoing apoptosis and TLR3 immunoreactivity. It was concluded that TLR3 expression is, in fact, related to a damaged hair cell. No TLR3 label was seen in normal animals. Unlike TLR3, TLR4 immunoreactivity was seen in some scattered, non-noise-exposed cochlear inner hair cells and Hensen cells. As with TLR3 immunoreactivity, TLR4 expression was clearly seen in Deiter's cells below damaged outer hair cells (Cai et al. 2014). In a subsequent study, Vethanayagam et al. (2016) exposed TLR4 null mice to the same broadband noise (1–7 kHz, 120 dB, 1 h) and these mice had less hair cell damage and auditory threshold shifts than wild-type mice without the TLR4 deletion, supporting the interpretation that TLR4 activates immune activity in the sensory epithelial cells and that this activation results in greater sensory cell degeneration.

Another piece of evidence in support of sensory cell participation in immune activation comes from a study in which transgenic-mouse, cochlear-sensory cells were deleted in the absence of any other cochlear injury. Following the death of hair cells, the supporting cells expressed the chemokine, CX3CL1, and circulating macrophages were recruited to the basilar membrane (Kaur et al. 2015).

Inflammatory Response of the Basilar Membrane

Among the cells described as mesothelial cells below the basilar membrane are phagocytic cells that label with the leukocyte marker, CD45 (Hirose et al. 2005; Tornabene et al. 2006; Kaur et al. 2015). Yang et al. (2015) evaluated these cells after exposure to broadband noise (1–7 kHz, 120 dB, 1 h). At 1 day following the exposure, the number of monocyte/macrophages in the basal half of the cochlea did not change. By 4 days after the noise, however, there were significantly more CD45+ cells and they were both below the tunnel of Corti and at the junction of the basilar membrane and the spiral ligament where it appeared cells had migrated from the venules and were adherent to the basilar membrane. The location along the length of the cochlea of the influx of monocytes was related to the location of greatest hair cell loss. The surface preparations and fluorescent labels used by these investigators made the shapes of the cells easy to assess (Yang et al. 2015). The shapes were different on different days and the authors interpreted these shape changes to reflect maturation of infiltrated monocytes to mature macrophages. After 10 days, the labeling pattern had returned to normal. TLR4-null mice, which have less sensory cell loss and threshold shifts than wild-type mice, still had the same pattern of macrophages recruitment to the basilar membrane, suggesting that TLR4 activation is not required for the recruitment or maturation of these macrophages (Vethanayagam et al. 2016).

The mature macrophages showed evidence of antigen-presentation ability and T-cells, the lymphocytes that respond to antigen presentation and initiate an adaptive immune response, were identified among the infiltrated circulating cells. This observation is the most provocative. It implies that cochlear proteins released from cells as a result of damage caused by acoustic trauma have the potential to initiate an auto-immune response resulting in auto-immune hearing loss (Ryan et al. 2001). For the majority of cases, however, this would not occur since the presented proteins would be recognized as "self". The proteins would also have to travel through the basilar membrane to reach the inflammatory cells. The observation does, however, help explain the result that acoustic trauma exacerbates the cochlear response to foreign antigens (Miyao et al. 2008). It also implies that if there are bacterial or viral antigens in the cochlea at the time of acoustic trauma, the response may be more destructive than just the noise trauma alone.

Conclusions

Over the last 30 years the ability of the cochlea to mount an immune response has been demonstrated and more recently the immune response induced by noise exposure has been examined. Based on the data presented above, noise of sufficient amplitude and duration to cause permanent threshold shifts and cellular damage has effects related to immunity on the sensory epithelium, resident macrophages on the scala tympani side of the basilar membrane, the fibrocytes of the spiral ligament, and the epithelial cells and capillaries of the stria vascularis. Each of these locations has been studied separately, by different groups of investigators using different noise trauma parameters and experimental assays. Hopefully in the future, a systematic and holistic approach will provide a better understanding of whether these cells work together or whether they are independent of each other, whether they are activated simultaneously and work in synchrony in response to the same stimulus level or whether they respond independently of each other and are activated at different sound levels.

It seems that the recruitment of macrophages is very specific, related to the type of injury, and not generalized to the whole cochlea. The next phase of this research should address the relationship between stimulus parameters, hearing loss and tissue damage relative to the activation of the immune response. As we begin to understand how noise damages cochlear cells, this should provide understanding of what immune responses may be generated. An insight into this was provided by Kaur et al. (2015), who used a transgenic mouse model in which hair cells can be killed by injection of diphtheria toxin with no other cochlear cell damage. The treatment induces recruitment of macrophages below the basilar membrane, but not the spiral ligament. This result supports the idea that potassium recirculation, which is presumably not challenged in this model, may be the signal for spiral ligament macrophages activation. This model would also be useful for examining the immune response of the Deiter's cells following the death of the hair cells.

At this point it is not clear whether the cochlear immune response is beneficial or harmful to the cochlea and hearing. The data of Shi and her colleagues, discussed above, strongly suggests that the infiltrated macrophages provide a benefit to the repair of the cochlear vasculature within the stria vascularis (Shi 2016). The data of Vethanayagam et al. (2016), on the other hand, using TLR4 null mice suggests that the organ of Corti is less damaged when the immune activity of the supporting cells is inhibited. And in general, cochlear inflammation is damaging to cochlear structures (Satoh et al. 2002), but in the case of acoustic trauma, it seems reasonable to assume the response is intended to repair the damage although there may be collateral damage. Transgenic mice that do not express the *Mif* gene coding for the pro-inflammatory cytokine, macrophage migration inhibitory factor, that is constitutively expressed by organ of Corti cells (Cai et al. 2014) showed less hearing recovery after noise exposure than control mice, supporting the idea that the inflammatory response is, in fact, beneficial (Kariya et al. 2015). While this study provided little information about the actual immune response in the Mif−/− mice, it serves as a starting point for investigating this question. Another initiator of inflammation, NFκB, which is activated by low level noise, appeared to have a beneficial effect on the spiral ganglion neurons. Transgenic mice lacking the p50 subunit of NFκB (p50−/−) are more susceptible to trauma from low level noise exposure (Lang et al. 2006b), again supporting the idea that the immune response is beneficial.

Most of the cochlear duct structures including the spiral ligament and spiral limbus fibrocytes participate in potassium ion recycling related to inner hair cell depolarization (Spicer and Schulte 1991, 1996, 1998; Wangemann 2002, 2006). Fibrocytes may also be involved in glutamate homeostasis which is also likely challenged during noise exposure (Furness et al. 2009). As previously suggested, the energy requirement and cellular stress related to performing the potassium and glutamate recycling functions during prolonged, loud noise exposure are likely very great (Wang et al. 2002). It is not surprising then, that NFκB is activated (Adams et al. 2009) and inflammatory signals are generated by type I spiral ligament fibrocytes and macrophages are recruited. What is hard to explain is why the type IV fibrocytes are the cells in the spiral ligament that degenerate in response to even relatively low level noise (Wang et al. 2002; Hirose et al. 2005). Apparently these cells are not involved in the potassium recycling, but they do produce connective tissue growth factor, CTGF, (Adams 2009) that interacts with transforming growth factor β, TGFβ, and facilitates leukocyte migration within tissues (Lau 2016). They seem likely then, to interact with the macrophages that enter the spiral ligament in response to noise. The activity of the macrophages, once in the spiral ligament, is unknown. Whether they are involved in the repair of the ligament following the degeneration of the type IV fibrocytes has not been examined.

The function of resident macrophages below the basilar membrane is yet to be defined. They do not pass through the basilar membrane to facilitate the repair of the organ of Corti after acoustic trauma; however, it is possible that they act through paracrine signaling to orchestrate the Deiter's cell phagocytosis of cellular debris and repair of the reticular lamina (Kaur et al. 2015). On the other hand, these macrophages may serve to provide surveillance and clean-up of the perilymph. In a very early study of cochlear inflammation, virus injected into the scala tympani was identified in the mesothelial cells below the basilar membrane and adhering to the walls of the scala tympani very shortly after injection (Keithley et al. 1988). Given what we know now, it is likely that the cells containing viral antigen were phagocytic macrophages involved in destroying the virus and recruiting immunocompetent cells to the cochlea. A modification of this response is likely a component of the response to cellular stress induced by acoustic trauma.

We are just beginning to investigate the molecular stress responses within the cochlea that occur as a result of acoustic trauma and what role infiltrating inflammatory cells play in that response. Hopefully, this chapter has clearly presented what is known about the immune responses and stimulated thoughts about how we can improve our understanding of the inflammatory response to acoustic trauma.

Acknowledgements I thank Allen F. Ryan, Ph.D. for reviewing and providing meaningful discussions concerning this manuscript, Jeffery P. Harris, M.D., Ph.D. for introducing me to immunity in the inner ear, and Gary S. Firestein, M.D. and Joe C. Adams, Ph.D. for their many years of collaboration and contributions towards gaining an understanding of inflammatory mechanisms in the inner ear. This work was supported in part by VA Merit grant BX001205 and NIH/NIDCD grant DC012595.

References

Abrashkin KA, Izumikawa M, Miyazawa T, Wang CH, Crumling MA, Swiderski DL, Beyer LA, Gong TW, Raphael Y. The fate of outer hair cells after acoustic or ototoxic insults. Hear Res. 2006;218:20–9.

Adams JC. Immunocytochemical traits of type IV fibrocytes and their possible relations to cochlear function and pathology. J Assoc Res Otolaryngol. 2009;10:369–82.

Adams JC, Seed B, Lu N, Landry A, Xavier RJ. Selective activation of nuclear factor kappa B in the cochlea by sensory and inflammatory stress. Neuroscience. 2009;160:530–9.

Anttonen T, Belevich I, Kirjavainen A, Laos M, Brakebusch C, Jokitalo E, Pirvola U. How to bury the dead: elimination of apoptotic hair cells from the hearing organ of the mouse. J Assoc Res Otolaryngol. 2014;15:975–92.

Arslan HH, Satar B, Serdar MA, Ozler M, Yilmaz E. Effects of hyperbaric oxygen and dexamethasone on proinflammatory cytokines of rat cochlea in noise-induced hearing loss. Otol Neurotol. 2012;33:1672–8.

Aspelund A, Antila S, Proulx ST, Karlsen TV, Karaman S, Detmar M, Wiig H, Alitalo K. A dural lymphatic vascular system that drains brain interstitial fluid and macromolecules. J Exp Med. 2015;212:991–9.

Axelsson A. The vascular anatomy of the cochlea in the guinea pig and in man. Acta Otolaryngol. 1968;Suppl 243:1–134.

Bianchi ME. DAMPs, PAMPs and alarmins: all we need to know about danger. J Leukoc Biol. 2007;81:1–5.

Cai Q, Vethanayagam RR, Yang S, Bard J, Jamison J, Cartwright D, Dong Y, Hu BH. Molecular profile of cochlear immunity in the resident cells of the organ of Corti. J Neuroinflammation. 2014;11:173.

Dai M, Yang Y, Omelchenko I, Nuttall AL, Kachelmeier A, Xiu R, Shi X. Bone marrow cell recruitment mediated by inducible nitric oxide synthase/stromal cell-derived factor-1alpha signaling repairs the acoustically damaged cochlear blood-labyrinth barrier. Am J Pathol. 2010;177:3089–99.

Du X, Choi CH, Chen K, Cheng W, Floyd RA, Kopke RD. Reduced formation of oxidative stress biomarkers and migration of mononuclear phagocytes in the cochleae of chinchilla after antioxidant treatment in acute acoustic trauma. Int J Otolaryngol. 2011;2011:612690.

Eldredge DH, Mills JH, Bohne BA. Anatomical, behavioral, and electrophysiological observations on chinchillas after long exposures to noise. Adv Otorhinolaryngol. 1973;20:64–81.

Fujioka M, Kanzaki S, Okano HJ, Masuda M, Ogawa K, Okano H. Proinflammatory cytokines expression in noise-induced damaged cochlea. J Neurosci Res. 2006;83:575–83.

Furness DN, Lawton DM, Mahendrasingam S, Hodierne L, Jagger DJ. Quantitative analysis of the expression of the glutamate-aspartate transporter and identification of functional glutamate uptake reveal a role for cochlear fibrocytes in glutamate homeostasis. Neuroscience. 2009;162:1307–21.

Gratton MA, Eleftheriadou A, Garcia J, Verduzco E, Martin GK, BL L–M, Vázquez AE. Noise-induced changes in gene expression in the cochleae of mice differing in their susceptibility to noise damage. Hear Res. 2011;277:211–26.

Harris JP. Immunology of the inner ear: response of the inner ear to antigen challenge. Otolaryngol Head Neck Surg. 1983;91:18–32.

Hashimoto S, Billings P, Harris JP, Firestein GS, Keithley EM. Innate immunity contributes to cochlear adaptive immune responses. Audiol Neurootol. 2005;10:35–43.

Henderson D, Hamernik RP, Sitler RW. Audiometric and histological correlates of exposure to 1-msec noise impulses in the chinchilla. J Acoust Soc Am. 1974;56:1210–21.

Hillerdal M, Jansson B, Engstrom B, Hultcrantz E, Borg E. Cochlear blood flow in noise-damaged ears. Acta Otolaryngol. 1987;104:270–8.

Hirose K, Liberman MC. Lateral wall histopathology and endocochlear potential in the noise-damaged mouse cochlea. J Assoc Res Otolaryngol. 2003;4:339–52.

Hirose K, Discolo C, Keasler J, Ransohoff R. Mononuclear phagocytes migrate into the murine cochlea after acoustic trauma. J Comp Neurol. 2005;489:180–94.

Johnsson LG, Hawkins JE Jr. Degeneration patterns in human ears exposed to noise. Ann Otol Rhinol Laryngol. 1976;85:725–39.

Kariya S, Okano M, Maeda Y, Hirai H, Higaki T, Noyama Y, Haruna T, Nishihira J, Nishizaki K. Macrophage migration inhibitory factor deficiency causes prolonged hearing loss after acoustic overstimulation. Otol Neurotol. 2015;36:1103–8.

Kaur T, Zamani D, Tong L, Rubel EW, Ohlemiller KK, Hirose K, Warchol ME. Fractalkine signaling regulates macrophage recruitment into the cochlea and promotes the survival of spiral ganglion neurons after selective hair cell lesion. J Neurosci. 2015;35:15050–61.

Keithley EM, Harris JP. Late sequelae of cochlear infection. Laryngoscope. 1996;106:341–5.

Keithley EM, Sharp P, Woolf NK, Harris JP. Temporal sequence of viral antigen expression in the cochlea induced by cytomegalovirus. Acta Otolaryngol. 1988;106:46–54.

Keithley EM, Wang X, Barkdull GC. Tumor necrosis factor alpha can induce recruitment of inflammatory cells to the cochlea. Otol Neurotol. 2008;29:854–9.

Knoops B, Argyropoulou V, Becker S, Ferté L, Kuznetsova O. Multiple roles of peroxiredoxins in inflammation. Mol Cells. 2016;39:60–4.

Kujawa SG, Liberman MC. Acceleration of age-related hearing loss by early noise exposure: evidence of a misspent youth. J Neurosci. 2006;26:2115–23.

Lang H, Ebihara Y, Schmiedt RA, Minamiguchi H, Zhou D, Smythe N, Liu L, Ogawa M, Schulte BA. Contribution of bone marrow hematopoietic stem cells to adult mouse inner ear: mesenchymal cells and fibrocytes. J Comp Neurol. 2006a;496:187–201.

Lang H, Schulte BA, Zhou D, Smythe N, Spicer S, Schmiedt RA. Nuclear factor kappaβ deficiency is associated with auditory nerve degeneration and increased noise-induced hearing loss. J Neurosci. 2006b;26:3541–50.

Lau LF. Cell surface receptors for CCN proteins. J Cell Commun Signal. 2016;10:121–7.

Liberman MC. Auditory-nerve response from cats raised in a low-noise chamber. J Acoust Soc Am. 1978;63:442–55.

Liberman MC, Dodds LW. Acute ultrastructural changes in acoustic trauma: serial-section reconstruction of stereocilia and cuticular plates. Hear Res. 1987;26:45–64.

Liberman MC, Kiang NY. Acoustic trauma in cats. Cochlear pathology and auditory-nerve activity. Acta Otolaryngol Suppl. 1978;358:1–63.

Louveau A, Smirnov I, Keyes TJ, Eccles JD, Rouhani SJ, Peske JD, Derecki NC, Castle D, Mandell JW, Lee KS, Harris TH, Kipnis J. Structural and functional features of central nervous system lymphatic vessels. Nature. 2015;523:337–41.

Masuda M, Nagashima R, Kanzaki S, Fujioka M, Ogita K, Ogawa K. Nuclear factor-kappa B nuclear translocation in the cochlea of mice following acoustic overstimulation. Brain Res. 2006;1068:237–47.

Miller JM, Ren T-Y, Dengerink HA, Nuttall AL. Cochlear blood flow changes with short sound stimulation. In: Axelsson A, Borchgrevink H, Hamernik RP, Hellstrom P-A, Henderson D,

Salvi RJ, editors. Scientific basis of noise-induced hearing loss. New York: Thieme; 1996. p. 95–109.

Miller JM, Brown JN, Schacht J. 8-Iso-Prostaglandin F_2, a product of noise exposure, reduces inner ear blood flow. Audiol Neurootol. 2003;8:207–21.

Miyao M, Firestein GS, Keithley EM. Acoustic trauma augments the cochlear immune response to antigen. Laryngoscope. 2008;118(10):1801–8.

Murillo-Cuesta S, Rodríguez-de la Rosa L, Contreras J, Celaya AM, Camarero G, Rivera T, Varela-Nieto I. Transforming growth factor β1 inhibition protects from noise-induced hearing loss. Front Aging Neurosci. 2015;20:7–32.

Nakamoto T, Mikuriya T, Sugahara K, Hirose Y, Hashimoto T, Shimogori H, Takii R, Nakai A, Yamashita H. Geranylgeranylacetone suppresses noise-induced expression of proinflammatory cytokines in the cochlea. Auris Nasus Larynx. 2012;39:270–4.

Ohashi K, Burkart V, Flohé S, Kolb H. Cutting edge: heat shock protein 60 is a putative endogenous ligand of the toll-like receptor-4 complex. J Immunol. 2000;164:558–61.

Ohlemiller KK, Gagnon PM. Genetic dependence of cochlear cells and structures injured by noise. Hear Res. 2007;224:34–50.

Okano T, Nakagawa T, Kita T, Kada S, Yoshimoto M, Nakahata T, Ito J. Bone marrow-derived cells expressing Iba1 are constitutively present as resident tissue macrophages in the mouse cochlea. J Neurosci Res. 2008;86:1758–67.

Peri F, Calabrese V. Toll-like Receptor 4 (TLR4) modulation by synthetic and natural compounds: an update. J Med Chem. 2014;57:3612–22.

Pujol R, Puel JL, Gervais d'Aldin C, Eybalin M. Pathophysiology of the glutamatergic synapses in the cochlea. Acta Otolaryngol. 1993;113:330–4.

Quirk WS, Avinash G, Nuttall AL, Miller JM. The influence of loud sound on red blood cell velocity and blood vessel diameter in the cochlea. Hear Res. 1992;63:102–7.

Raphael Y, Altschuler RA. Reorganization of cytoskeletal and junctional proteins during cochlear hair cell degeneration. Cell Motil Cytoskeleton. 1991;18:215–27.

Ryan AF, Keithley EM, Harris JP. Autoimmune inner ear disorders. Curr Opin Neurol. 2001;14:35–40.

Santi PA, Aldaya R, Brown A, Johnson S, Stromback T, Cureoglu S, Rask-Andersen H. Scanning electron microscopic examination of the extracellular matrix in the decellularized mouse and human cochlea. J Assoc Res Otolaryngol. 2016;17:159–71.

Sato E, Shick HE, Ransohoff RM, Hirose K. Repopulation of cochlear macrophages in murine hematopoietic progenitor cell chimeras: the role of CX3CR1. J Comp Neurol. 2008;506:930–42.

Satoh H, Firestein GS, Billings PB, Harris JP, Keithley EM. Tumor necrosis factor-alpha, an initiator, and etanercept, an inhibitor of cochlear inflammation. Laryngoscope. 2002;112:1627–34.

Sautter NB, Shick EH, Ransohoff RM, Charo IF, Hirose K. CC chemokine receptor 2 is protective against noise-induced hair cell death: studies in CX3CR1(+/GFP) mice. J Assoc Res Otolaryngol. 2006;7:361–72.

Scheibe F, Haupt H, Ludwig C. Intensity-related changes in cochlear blood flow in the guinea pig during and following acoustic exposure. Eur Arch Otorhinolaryngol. 1993;250:281–5.

Seidman MD, Quirk WS, Shirwany NA. Mechanisms of alterations in the microcirculation of the cochlea. Ann N Y Acad Sci. 1999;884:226–32.

Seidman MD, Tang W, Shirwany N, Bai U, Rubin CJ, Henig JP, Quirk WS. Anti-intercellular adhesion molecule-1 antibody's effect on noise damage. Laryngoscope. 2009;119:707–12.

Shi X. Resident macrophages in the cochlear blood-labyrinth barrier and their renewal via migration of bone-marrow-derived cells. Cell Tissue Res. 2010;342:21–30.

Shi X. Physiopathology of the cochlear microcirculation. Hear Res. 2011;282:10–24.

Shi X. Pathophysiology of the cochlear intrastrial fluid-blood barrier (review). Hear Res. 2016;338:52–63.

Shi X, Nuttall AL. Expression of adhesion molecular proteins in the cochlear lateral wall of normal and PARP-1 mutant mice. Hear Res. 2007;224:1–14.

Shi X, Dai C, Nuttall AL. Altered expression of inducible nitric oxide synthase (iNOS) in the cochlea. Hear Res. 2003;177:43–52.

Spicer SS, Schulte BA. Differentiation of inner ear fibrocytes according to their ion transport related activity. Hear Res. 1991;56:53–64.

Spicer SS, Schulte BA. The fine structure of spiral ligament cells relates to ion return to the stria and varies with place-frequency. Hear Res. 1996;100:80–100.

Spicer SS, Schulte BA. Evidence for a medial K+ recycling pathway from inner hair cells. Hear Res. 1998;118:1–12.

Syka J, Melichar I, Ulehlová L. Longitudinal distribution of cochlear potentials and the K+ concentration in the endolymph after acoustic trauma. Hear Res. 1981;4:287–98.

Takahashi M, Harris JP. Analysis of immunocompetent cells following inner ear immunostimulation. Laryngoscope. 1988;98:1133–8.

Tan BTG, Lee MMG, Ruan R. Bone marrow-derived cells that home to acoustic deafened cochlea preserved their hematopoietic identity. J Comp Neurol. 2008;509:167–79.

Tan WJT, Thorne PR, Vlajkovic SM. Characterization of cochlear inflammation in mice following acute and chronic noise exposure. Histochem Cell Biol. 2016;146:219–30.

Thorne PR, Nuttall AL. Alterations in oxygenation of cochlear endolymph during loud sound exposure. Acta Otolaryngol (Stockh). 1989;107:71–9.

Tornabene SV, Sato K, Pham L, Billings P, Keithley EM. Immune cell recruitment following acoustic trauma. Hear Res. 2006;222:115–24.

Vethanayagam RR, Yang W, Dong Y, Hu BH. Toll-like receptor 4 modulates the cochlear immune response to acoustic injury. Cell Death Dis. 2016;7:e2245.

Wakabayashi K, Fujioka M, Kanzaki S, Okano HJ, Shibata S, Yamashita D, Masuda M, Mihara M, Ohsugi Y, Ogawa K, Okano H. Blockade of interleukin-6 signaling suppressed cochlear inflammatory response and improved hearing impairment in noise-damaged mice cochlea. Neurosci Res. 2010;66:345–52.

Wang Y, Hirose K, Liberman MC. Dynamics of noise-induced cellular injury and repair in the mouse cochlea. J Assoc Res Otolaryngol. 2002;3:248–68.

Wangemann P. K+ cycling and the endocochlear potential. Hear Res. 2002;165:1–9.

Wangemann P. Supporting sensory transduction: cochlear fluid homeostasis and the endocochlear potential. J Physiol. 2006;576:11–21.

Warchol ME. Macrophage activity in organ cultures of the avian cochlea: demonstration of a resident population and recruitment to sites of hair cell lesions. J Neurobiol. 1997;33:724–34.

Yang W, Vethanayagam RR, Dong AY, Cai Q, Hu BH. Activation of the antigen presentation function of mononuclear phagocyte populations associated with the basilar membrane of the cochlea after acoustic overstimulation. Neuroscience. 2015;303:1–15.

Yang S, Cai Q, Vethanayagam RR, Wang J, Yang W, Hu BH. Immune defense is the primary function associated with the differentially expressed genes in the cochlea following acoustic trauma. Hear Res. 2016;333:283–94.

Yimtae K, Song H, Billings P, Harris JP, Keithley EM. Connection between the inner ear and the lymphatic system. Laryngoscope. 2001;111:1631–5.

Zhang W, Dai M, Fridberger A, Hassan A, Degagne J, Neng L, Zhang F, He W, Ren T, Trune D, Auer M, Shi X. Perivascular-resident macrophage-like melanocytes in the inner ear are essential for the integrity of the intrastrial fluid-blood barrier. Proc Natl Acad Sci U S A. 2012;109:10388–93.

Zhang F, Dai M, Neng L, Zhang JH, Zhi Z, Fridberger A, Shi X. Perivascular macrophage-like melanocyte responsiveness to acoustic trauma--a salient feature of strial barrier associated hearing loss. FASEB J. 2013;27:3730–40.

Chapter 6
Middle Ear Infection and Hearing Loss

Arwa Kurabi, Daniel Schaerer, and Allen F. Ryan

Abstract The primary inflammatory disease of the middle ear (ME) is otitis media (OM), a common pediatric infection that accounts for more office visits and surgeries than any other pediatric condition. It also affects adults to a lesser degree. The presence of inflammatory mediators and cells is one of the hallmarks of OM. It is mediated primarily by innate immune receptors, which interact with molecules from the bacteria that cause ME infections without the need for prior sensitization. Chronic and recurrent ME infections in children lead to hearing loss during critical periods of language acquisition and learning, causing delays in reaching developmental milestones and if left untreated, have the potential risks of permanent damage to the middle and inner ear. In this review, we document the presence of inflammation in the ME during OM, discuss current evidence implicating innate immunity in the generation and regulation of ME inflammation, and review the effects of ME inflammation and infection on hearing, auditory processing, the acquisition of language and learning.

Keywords Otitis media · Innate immunity · Inflammatory mediators

A. Kurabi (✉)
Division of Otolaryngology, Department of Surgery, University of California San Diego, La Jolla, CA, USA

Veterans Administration San Diego Healthcare System, San Diego, CA, USA
e-mail: akurabi@ucsd.edu

D. Schaerer
Division of Otolaryngology, Department of Surgery, University of California San Diego, La Jolla, CA, USA

A. F. Ryan
Division of Otolaryngology, Department of Surgery, University of California San Diego, La Jolla, CA, USA

Department of Neurosciences, University of California San Diego, La Jolla, CA, USA

Veterans Administration San Diego Healthcare System, San Diego, CA, USA

© Springer International Publishing AG, part of Springer Nature 2018
V. Ramkumar, L. P. Rybak (eds.), *Inflammatory Mechanisms in Mediating Hearing Loss*, https://doi.org/10.1007/978-3-319-92507-3_6

1   Introduction

Otitis media (OM) is a common infectious disease in children worldwide, resulting in substantial health care expenditures and burden (Acuin 2004; Ahmed and Shapiro 2014). In the USA, more than 90% of children experience OM before age 5, making it the most common condition warranting medical therapy for children in this age group (Teele et al. 1989; Thomas and Brook 2014). While acute, uncomplicated OM tends to be self-limiting, 10–20% of children experience persistent, recurrent or chronic OM (Daly et al. 1999). The long-lasting forms of this condition can result in significant conductive hearing loss. The peak of incidence of OM is from 6 months to 3 years (MacIntyre et al. 2010). Because this is a critical period for early language acquisition (Gervain 2015), persistent OM can result in delayed speech and communication development. Chronic forms of OM also carry a risk of permanent damage to the middle and inner ears, resulting in hearing loss that does not recover. OM has much more serious complications in developing countries (Bluestone 1998; Kubba et al. 2000; Ahmed et al. 2014). It is estimated that one half of the world's burden of significant hearing loss is related to undertreated OM (WHO 2004). Currently, treatment for uncomplicated OM consists of watchful waiting or antibiotics (Forgie et al. 2009; Bascelli and Losh 2001). Tympanostomy tubes for middle ear (ME) ventilation are often recommended for recalcitrant cases (Qureishi et al. 2014; Ahmmed et al. 2001; Ambrosio and Brigger 2014). Insertion of tympanostomy tubes is the most common ambulatory surgery performed on children in the US; with 670,000 insertions, costing over four billion dollars in 2009 (Rosenfeld et al. 2013).

The etiology of OM is generally thought to be multifactorial. OM incidence has been found to be influenced by variation in the infecting pathogens, host anatomy, immunological status and prior viral infection. However, at the molecular level, OM can largely be defined by ME inflammatory responses. These responses occur via the activation of pro-inflammatory transcription factors, followed by production and release of inflammatory mediators (Hernandez et al. 2015). Mucosal hyperplasia, leukocytic infiltration into the ME cavity, and secretion of mucus-rich effusions are also characteristic sequelae of OM (Allen et al. 2014). These host responses can largely be attributed to the ME response to bacteria, which ascend into the ME from the nasopharynx through the Eustachian tube. The nasopharyngeal commensals *Streptococcus pneumonia*, nontypeable *Haemophilus influenza* (NTHi), and *Moraxella catarrhalis* are the three most common pathogens causing ME infection (Daly et al. 1999; Vergison 2008). However, OM is often preceded by respiratory viral infections, and viruses are often detected in ME effusions (Heikkinen and Chonmaitree 2003; Barenkamp 2014). Viruses can enhance nasopharyngeal bacterial load, alter the function of the Eustachian tube, and modulate host immunity, any of which can increase the probability of ME infection.

Generally, children are thought to be more prone to OM than adults for several reasons. The Eustachian tube is shorter, oriented more horizontally and may function less efficiently compared to adults, potentially allowing easier access of nasopharyngeal bacteria to the ME (Fuchs et al. 2014; Alles et al. 2001). In addition, since their cognate immune systems are immature, they are less able to react efficiently to pathogens, to which they are immunologically naïve (Faden 2001; Sharma and Pichichero 2013). Still, most children rapidly resolve OM without the need for treatment. Indeed, only 10–20% exhibit recurrent/chronic disease. Children who experience more than three episodes of acute OM (AOM) within 6 months are considered otitis-prone, and are likely to require tympanostomy tube insertion (Qureishi et al. 2014; Ryan et al. 2001).

Why some children progress to persistent/recurrent OM, while the majority experience no or only a few OM episodes, is not well understood. However, epidemiologic studies indicate that OM-proneness in humans receives contributions from Eustachian tube dysfunction, economic and health care status, exposure to cigarette smoke, prior exposure to upper respiratory viral infections, and frequent day care attendance (Ambrosio and Brigger 2014; Vergison 2008; Rye et al. 2012). It is also clear that genetics play a significant role, as indicated by twin studies (Rye et al. 2012; Goodwin and Post 2002; Post 2011; Kvaerner et al. 1996). Current data indicate that OM proneness is polygenic, and there is evidence implicating genes involved in craniofacial structure as well as in various aspects of immune defense (Fuchs et al. 2014; Daly et al. 2004). Craniofacial anomalies that disrupt the normal structure and function of the Eustachian tube can alter ME pressure and enhance access of bacteria to the tympanic cavity (Bluestone 1998). However, most children with chronic/recurrent OM do not have overt craniofacial defects, implicating more subtle Eustachian tube dysfunction or deficiencies in immunity. Regarding the latter, two distinct defense strategies can protect a host from infectious damage: alleviating the pathogenic burden by increasing host resistance, or reducing pathological impact by enhancing host tolerance (Medzhitov et al. 2012). Changes in the balance of these two fundamental defense mechanisms (resistance versus tolerance) can be linked to OM proneness and chronicity. Immune factors include mutations or polymorphisms in innate immune and inflammatory genes, defects in cellular processes that reduce infection such as phagocytosis, and changes in factors that initiate and regulate tissue repair and recovery after inflammation and injury. A recent survey of the transcriptome of otitis-prone children found that many genes in these categories are altered during OM (Liu et al. 2013).

It should also be noted that multivalent vaccines against *Streptococcus pneumoniae* have decreased the prevalence of *S. pneumonia* OM (Pletz et al. 2008; Eskola et al. 2001; Sabirov and Metzger 2006). Overall incidence of OM has also been decreased to some extent (Taylor et al. 2012), although OM due to other pathogens and to uncovered *S. pneumonia* strains has increased (Cripps and Otczyk 2006). Fortunately, the covered *S. pneumonia* strains are among the most invasive and dangerous, which has decreased the severity of OM due to this pathogen (Taylor et al. 2012).

2 Inflammatory Mediators in the ME

The ME is a closed cavity, with the only means of egress being the Eustachian tube. Since the tubal orifice is above the floor of the ME, any fluid which accumulates in the tympanic cavity has a tendency to be cleared very slowly. The closed nature of the ME means that any inflammation that occurs is not only slow to resolve, but can also recruit additional pro-inflammatory mediators and inflammatory cells, resulting in a large number of mediators and prolonged pathology. This may explain the wealth of inflammatory mediators that have been isolated from the MEs of patient with diseases such as OM and cholesteatoma.

Virtually all pro-inflammatory mediators have been detected in ME effusion from children with OM. This includes a large number of cytokines (e.g. Yellon et al. 1991; Smirnova et al. 2002), with IL1b, IL-6 and TNFa being highly expressed. Similarly, many chemokines have been detected, including those that are chemotactic for neutrophils and macrophages (e.g. Kaur et al. 2015). The inflammatory complement activation products C3a and C5a are present (Bernstein 1976; He et al. 2013), as have many arachidonic acid metabolites (e.g. Brodsky et al. 1991).

While human studies document the mediators of mature disease, animal studies have documented not only the presence and kinetics of these mediators, (e.g. Melhus and Ryan 2000; Hernandez et al. 2015) but also allow interventional studies. Trune et al. (2015) and Hernandez et al. (2015) found a great majority of all cytokine genes are expressed during experimental OM, with most peaking within 3–6 h of bacterial infection. While many chemoattractant chemokines are similarly regulated (Leichtle et al. 2010), CXCLs 5, 9 and 17 and CCL9 are expressed for several days after ME infection, including the resolution phase of OM. These chemokines are associated with vascular development, T-cell chemoattraction and dendritic cell chemoattraction, events that would be expected to occur over a longer timespan than acute inflammation. Complement activation peaks at 24 h after infection (Li et al. 2012), consistent with leukocyte chemoattraction. Immune-induced ME inflammation is also reduced by prior complement depletion (Ryan et al. 1986). Arachidonic acid metabolites peak at 24 h after ME infection. Leukotriene inhibition strongly inhibited for formation of effusion in experimental OM (Jung et al. 2004), indicating a role in vascular permeability.

3 Initiation and Regulation of Inflammation During ME Infection

The detection of inflammatory mediators in the ME of patients with disease, and in animal models, says little about their initiation. To understand the origins of inflammation, we rely primarily upon animal models that allow us to evaluate the earliest stages of OM and the functional role of genes in producing and regulating ME responses to infection.

Innate Immunity and ME Inflammation

In the normal child, uncomplicated OM resolves in only a few days, even in the absence of antibiotic therapy (Thomas and Brook 2014). In fact, antibiotic therapy only reduces the recovery time by approximately 24 h, leading to the recommendation that antibiotics not be prescribed for simple, uncomplicated OM (e.g. Venekamp et al. 2015). The normal recovery period for OM is thus too short for the development of cognate immunity to play a significant role in resolution. This strongly suggests that the innate immune system, which is activated without prior sensitization, is the major effector of uncomplicated OM resolution.

The innate immune system is comprised of pattern recognition receptors (PRRs) that respond to pathogen associated molecular patterns (PAMPs). The receptors of innate immunity include the Toll-like receptors (TLRs), Nod-like receptors (NLRs), Rig-I-like receptors (RLRs), C-type lectin receptors (CLRs), and nucleic acid receptors (Beutler 2004; Kawai and Akira 2009). Activated PRRs recruit adaptor molecules, which in turn initiate signaling cascades. The majority of these cascades end in the activation of transcription factors, including NFκB, AP-1, and interferon-regulatory factors (IRFs), which localize to the nucleus and up-regulate the expression of many pro-inflammatory genes. They also stimulate the production of chemokines that attract inflammatory leukocytes, primarily neutrophils and macrophages, into the ME cavity (Hernandez et al. 2015; Kurabi et al. 2016).

Because inflammation has a high potential for bystander damage, anti-inflammatory genes are also up-regulated. Innate immune responses are thus balanced between pro-inflammatory responses that fight infection and anti-inflammatory responses protecting against host tissue damage and initiating repair and healing. This reflects the balance between resistance and tolerance as discussed earlier. For example, NFκB induces both pro-inflammatory genes and genes that limit the duration and magnitude of the inflammatory response. Innate immunity also plays a critical role in the development of cognate immunity by recruiting and activating lymphocytes, as well as macrophages and other cells involved in antigen presentation. Thus, innate immunity affects not only initial defense against infection, but also the development of immunologic memory (Akira et al. 2006).

Innate immune responses to the bacterial pathogens of OM can be predicted based upon their known ligands. Herein, we will discuss how the activation of innate immune signaling pathways, in particular by the TLRS and NLRs, may be a common pathway for OM pathogenesis and recovery. The expression of PRRs in the human ME has been examined by Granath et al. (2011) and others (Ryan et al. 2001; Kim·et al. 2010; Kim et al. 2015; Hernandez et al. 2015) to name a few. Several are expressed at significant levels in the normal ME cavity and, following infection, these and many additional PRRs are upregulated. Triggering these PRRs activates inflammatory and immune responses leading to efficient resolution of invading pathogens but also causing disease pathogenesis.

The TLR Family

The TLR family, the first PRRs to be described in depth, consists of ten human receptors. Other species have different numbers of TLR genes. For example, there are 13 TLR genes in mice. However, those in common behave similarly between the two species (Takeda et al. 2003). TLRs evolved to recognize non-host PAMPs from fungi, bacteria, viruses and other parasites (Medzhitov and Janeway 2000). TLR1, TLR2, TLR4, TLR5 and TLR6 are cell surface receptors that primarily recognize lipoproteins, lipopolysaccharides (LPS) and flagellins in the extracellular environment. In contrast, TLR3, TLR7, TLR8 and TLR9 are cytosolic receptors located on endolysosomes and they primarily detect bacterial and viral nucleic acids (Mogensen 2009). Activated TLRs recruit and activate adaptor molecules, which initiate and amplify down-stream signaling cascades. Most TLRs utilize the adaptor myeloid differentiation factor-88 (MyD88) (Takeda et al. 2003). TLR3 uses the TIR domain containing adaptor inducing IFN-β (TRIF), while TLR4 can utilize either the MyD88-dependent or TRIF-dependent pathways (Kawai and Akira 2009). MyD88 activation of the down-stream IRAK1/4 (interleukin-1 receptor-associated kinase 1/4) scaffold results in phosphorylation and binding of TRAF6 (TNF receptor-associated factor 6). A resulting cascade activates a variety of effectors including NFκB and c-Jun terminal kinase (JNK), leading to the expression of interleukin, TNFα and leukocyte recruitment genes. TRIF can activate not only TRAF6, but also TBK1 (tank-binding kinase-1) leading to IRF activation and interferon expression. Subsequent inflammatory responses mediated primarily by the cytokines, chemokines and interferons follow. Among the TLR subtypes, TLR2 and TLR4 play a crucial role in determining the outcome of infection during OM (Hernandez et al. 2008; Shuto et al. 2001; Leichtle et al. 2009a; Hirano et al. 2009; Lim et al. 2008). TLR2 is involved in the recognition of a wide variety of microbial ligands like peptidoglycans, lipoteichoic acid and lipoproteins. On the other hand, TLR4 primarily binds to LPS specifically associated gram-negative bacteria.

Several animal studies have demonstrated a crucial role for TLRs and TLR-dependent signaling cascades in the generation and persistence of inflammation as well as the resolution of OM (Lim et al. 2008; Hirano et al. 2007; Kawano et al. 2013). Both TLR2 and TLR4 deficient mice exhibit persistent inflammation with impaired bacterial clearance and delays in OM recovery. In addition, genetic deletions of downstream signaling molecules, MyD88, TRIF, TNF, JNK1 and JNK2 also result in prolonged inflammation and abnormal OM recovery (Leichtle et al. 2008; Han et al. 2009; Leichtle et al. 2009b; Yao et al. 2014). In particular, mouse model studies have shown that in the absence of TLR2 or MyD88, initial neutrophil and macrophage recruitments are significantly delayed (decreasing host resistance and initial leukocyte-mediated inflammation), but also leading to host inability to clear bacteria with resultant prolonged inflammatory responses (up to 42 days, as compared to 5 days in wild-type mice). In particular, TLR2-deficient mice were susceptible to *S. pneumoniae* infection of the ME and show decreased expression of NOD2, IL1, NFκB, TNFα, MIP1α, Muc5ac and Muc5b, encumbering timely bacterial clearance (Han et al. 2009). In turn, TLR4 induces early activation of TLR2 in OM (Leichtle et al. 2009a) and plays a role in acquired adaptive mucosal immunity in the ME (Hirano et al. 2009). Moreover, both TLR2 and TLR4 are critical to the up-regulation of the key pro-inflammatory mediator TNFα early in OM (Leichtle et al. 2009a). This up-regulation induces the downstream target chemokine CCL3, promoting macrophage recruitment and bacterial phagocytosis and clearance. Conversely, IRAK-M, a negative regulator of the TLR family, has recently been shown to suppress the production of pro-inflammatory mediators in the lung during NTHi infection (Miyata et al. 2015).

Along with the activation of pro-inflammatory mediators, many anti-inflammatory mediators are up-regulated with similar kinetics. These include IL-4 and IL-10, interleukin-1 receptor antagonist, the MyD88 signaling inhibitors SOCs3 and IRAK3, the TRAF6 inhibitor A20, and the IRF inhibitor ATF3 (Hernandez et al. 2015). These tend to blunt the ME inflammatory response, limiting host tissue damage and balancing the inflammatory responses. For instance, IL10$^{-/-}$ mice show a reduced number of goblet and mucin-producing cells during OM. While IL10 plays an anti-inflammatory role by inhibiting the NFκB-dependent pathway down-stream of the TLRs, it is also known to regulate the ME mucosal hyperplastic response during OM via TLR2 (Tsuchiya et al. 2008).

In humans, polymorphisms in genes encoding TLR2, TLR4, and the TLR4 binding partner CD14, have been identified in genome-wide association studies of OM susceptibility, as they have effector genes TNFα and IL1R (reviewed in Rye et al. 2011). In addition, changes in the mRNA levels of TLR2, TLR4, TLR5, TLR7, and TLR9 in addition to cytokines like IL1, IL6, IL8, and IL10 and chemokines like CCL2, CCL3, CXCR3 in the ME infiltrate and inflamed mucosa in patients have also been documented clinically (Si et al. 2014; Lee et al. 2013a, b; Emonts et al. 2007; Granath et al. 2011; Kaur et al. 2015). Taken together with the animal data, these results provide strong evidence that the TLRs via innate immunity play a significant role in generating ME inflammation, but also in OM recovery.

In addition to sensing pathogens, several TLRs also serve as receptors for damage-associated molecular patterns (DAMPs), which are released from or produced by damaged or necrotic host cells. The cellular responses to DAMPs are similar to those evoked by PAMPs. Since infection and inflammation frequently produce host cell damage, DAMPs provide an additional source of ME inflammation during OM, which can augment the response to bacteria (El Mezayen et al. 2007).

The NLR Family

The NOD-like receptors (NLRs) comprise an additional family of PRRs. All of the NLRs are located intracellularly, and detect bacterial and viral molecules in the cytoplasm. NLR activation leads to the secretion of pro-inflammatory cytokines, including TNFα, interleukins and interferons. Members of this family include NOD1 and NOD2, as well as the pyrins (NLRP1–14), and the NLR family CARD domain containing protein 4 (NLRC4). The NLRs organize large signaling complexes such as NOD signalosomes and NLRP inflammasomes (Rathinam et al. 2012). Activated NODs signal via the adaptor RIP2 (receptor interacting protein 2) that, similar to the TLR/MyD88 cascade, leads to the activation of NFκB and MAPK pathways (Kim et al. 2008). Alternatively, they can interact with MAVS (mitochondrial antiviral-signaling), leading to type I interferon production (Lupfer and Kanneganti 2013). Recent studies in patients with chronic OM revealed that NOD1 and NOD2 mRNA was reduced in the ME effusions of otitis-prone children when compared to non-otitis-prone individuals (Kim et al. 2010). However, mucosal tissue samples from patients with chronic OM revealed that NOD2 expression is elevated when compared to normal mucosa (Granath et al. 2011). Increased expression of mucosal NOD2 has also been shown to participate in NTHi-induced β-defensin2 production and in turn to modulate the recruitment of inflammatory cells and bacterial clearance from the ME cavity in animal studies (Woo et al. 2014), highlighting the interplay between inflammatory leukocyte infiltration and mucosal defense of the ME.

Activation of the NLRPs results in the formation of inflammasomes, multi-molecular signaling platforms consisting of an active NLRP, the bipartite adaptor protein ASC (inflammasome adaptor protein apoptosis-associated speck-like protein containing CARD), and Caspase1 (Brodsky and Monack 2009). Inflammasomes, involving NLRPs, NLRC4, the DNA-sensing AIM2 (absent in melanoma-2) and RIG1 (retinoic acid-inducible gene-1) receptors sense a wide range of PAMPs and hence are involved in the development of acute and chronic inflammatory responses. Inflammasomes cleave pro-Caspase-1 to active Caspase1, a key mediator of inflammatory processes. Active caspase1 in turn cleaves pro-IL1β and pro-IL18 into their active forms. Caspase1 can also execute a rapid, programmed cell death response termed pyroptosis. Pyroptosis is a Caspase1 dependent inflammatory process of cell self-destruction to eliminate cellular niches that favor microbial growth (Denes et al. 2012). However, cell damage can release DAMPs, providing another pathway leading to inflammation.

We, and others, have previously demonstrated that IL1 is important for the recruitment of inflammatory cells to the ME (Watanabe et al. 2001; Catanzaro et al. 1991). Moreover, mice deficient in the key inflammasome component ASC, which modulates Caspase1 recruitment and activation, are more susceptible to ME infection than wild-type mice (Kurabi et al. 2015). This heightened susceptibility is associated with decreased activation of IL1β due to a reduction in active Caspase1 (Denes et al. 2012). These data indicate that the inflammasome and the NLRPs play a significant role in recovery from OM.

Nucleic Acid Receptors

PRRs that can detect pathogen DNA and RNA include the RLRs, AIM2 and TLR9. In addition, Pol-III can transcribe bacterial DNA into RNA, which can be sensed by RIG1 or MDA5 receptors which function as cytosolic sensors alerting innate immunity to viral and bacterial RNA. These receptors act via proteins associated with the mitochondria (i.e., IPS1; interferon-β promoter stimulator 1) and the endoplasmic reticulum (ER) to induce IRF3 and IRF7 activation, leading to IFN production inflammation. RLRs also can cross-talk with TLR signaling through MyD88-dependent and IPS-1-dependent pathways (Kawai and Akira 2009). As noted above, AIM2 receptors sensing bacterial DNA complex with the adaptor molecule ASC and pro-Caspase1 to form an inflammasome, leading to IL1β and IL18 processing (Jin et al. 2012). In animal models, genes encoding many of the nucleic acid sensing molecules are significantly regulated during OM (Leichtle et al. 2010). Experimentally, RIG1 gene expression was found to be enhanced shortly after NTHi induced infection, as was AIM2 mRNA (Kurabi et al. 2015), meanwhile mice deficient in TLR9 exhibit a prolonged OM phenotype (Leichtle et al. 2012). In OM-prone patients, the expression levels of TLR9 and RIG1 mRNA were found to be lower than in normal patients (Kim et al. 2010), possibly linking these receptors to OM reoccurrence.

Additional PRRs

Besides TLRs, NLRs and RLRs, inflammatory responses can be triggered by additional innate immune receptors for residual DNA or carbohydrate molecules from invading pathogens or damaged/dying cells. Those receptors include members of the PGRP (peptidoglycan recognition proteins) and C-type lectin receptors (CLRs). PGRPs act both as sensors and scavengers of peptidoglycan and modulate the level of the host immune response to the presence of infectious agents (Dziarski 2003). The CLRs comprise a large family of receptors that recognize carbohydrate residues. They are divided into 17 groups based on their structural features and homology (Geijtenbeek and Gringhuis 2009). The roles of PGRPs and CLRs in the pathogenesis of OM have only begun to be elucidated. Recent clinical data from patients with OME and chronic OM showed increased mRNA expression of CLRs and CLR related molecules such as DECTIN-1, MR1, MR2, MINCLE, SYK, CARD9, BCL10, MALT1, SRC, DEC205, GALECTIN1, TIM3, TREM1, and DAP12 in ME effusions (Lee et al. 2013a, b). CLRs also function as DAMP receptors, responding to cytosolic carbohydrates.

There are additional receptors which respond to DAMPs, such as advanced glycation end-products, high-mobility group box 1 (HMGB1), S100A8 and S100A9, and serum amyloid A, but have no or a lesser role in innate immunity. These include receptor for advanced glycation end-products (RAGE) (Bierhaus et al. 2005), triggering receptor expressed on myeloid cells 1 (TREM-1), TREM-2 (Weber et al. 2014) and Mincle (Yamasaki et al. 2008).

4   Hearing Consequences of ME Infection and Inflammation

In the most common form of ME inflammation, AOM, any hearing loss is typically conductive, due to the presence of ME effusion. ME may result in no hearing loss, if EM fluid does not reduce the ME volume significantly. Higher levels of serous fluid in the ME can result in mild (25 dB) hearing loss (Guan and Gan 2011). As the inflammation of uncomplicated OM typically resolves in a few days, this hearing loss is temporary.

However, for 10–20% of children, OM can be chronic, persistent or recurrent. This may result in persistent, mild hearing loss. However, if the effusion occupies more of the ME, the conductive hearing loss may be more significant, reaching 40 dB. If the fluid is viscous, as in mucoid OM, purulent OM or glue ear, somewhat more loss may be present, especially at low frequencies, due to decreased compliance of the tympanic membrane and ME conductive apparatus. However, the effects of even very high viscosity generally do not exceed 5–10 dB (Thornton et al. 2013).

The impact of hearing loss due to persistent or recurrent OM is controversial. There is little question that children with multiple episodes of OM, which typically peaks at 6–24 months during the period of early language acquisition, display delayed language (e.g. Friel-Patti et al. 1982; Friel-Patti and Finitzo 1990; Rvachew et al. 1999). Moreover, surveys of language and learning in school-age children have reported significant deficits for those with a history of OM (Holm and Kunze 1969; Luotonen et al. 1996; Winskel 2006). However, other studies found that any effects on language were temporary, disappearing by the time children reached grade school age (Grievink et al. 1993; Johnson et al. 2008; Zumach et al. 2010). A rigorous meta-analysis of the best data (Roberts et al. 2004) concluded that the evidence did not support an effect of early OM on school-age language, but left open the possibility of a minor effect. More recently, studies of higher-order auditory processing have detected deficits in school-age children (Haapala et al. 2014; Villa and Zanchetta 2014) raising the possibility that not all effects of a history of OM recover with age.

A more significant issue with chronic/persistent OM, especially purulent OM, is the possibility of permanent damage to the conductive apparatus of the ME. Fibrosis, a common consequence of inflammation, can also affect ME structures subsequent to OM. Scarring of the tympanic membrane consistent with prior OM was observed in 25% of Danish teenagers (Stangerup et al. 1994). Adhesions limiting the mobility of the ossicular chain is less common, but affects a small percentage of children after OM (Forsen 2000). These changes in hearing do not recover without surgery, and can have lasting impacts.

Inflammation in the ME from OM also has the potential to influence the delicate tissues of the inner ear. TNFα has been shown to directly damage cochlear hair cells (Haake et al. 2009), while interferon g induces apoptosis receptor expression in the organ of Corti (Bodmer et al. 2002). Inflammatory mediators can enter the cochlear via the round window membrane or the annular ligament (Juhn et al. 2008). High-frequency hearing loss reflecting the loss of basal turn hair cells has been reported after persistent or recurrent OM, especially in the case of purulent OM (Mittal et al. 2015).

Mastoiditis and cholesteatoma are serious conditions that occur as sequelae of chronic, suppurative OM (Minovi and Dazert 2014). Mastoiditis is an inflammation and infection of the mastoid air cells. While involvement of the mastoid occurs in virtually all OM, infection and inflammation can persist in the mastoid even when the ME has cleared. It can lead to erosion of bone in the mastoid, ME and inner ear, leading to conductive and sensorineural hearing loss. Cholesteatoma is an ingrowth of the keratinized epithelium on the external surface of the tympanic membrane, through to be caused due to negative ME pressure inducing a retraction pocket in the membrane. The resulting structure of keratinized epithelium within a capsule of mucosal tissue induces inflammation and grows destructively within the ME. If untreated, this can result in destruction of the ME conductive apparatus and even the inner ear. The treatment for chronic mastoiditis and cholesteatoma is typically surgical.

The possibility of permanent hearing loss due to OM are greatly magnified in certain areas of the developing world, especially sub-Saharan Africa and Southeast Asia. Due to the combination of low socioeconomic status, reduced opportunities for personal hygiene and limited access to health care, the consequences of OM are dramatically more severe. The WHO estimates that one half of the world's burden of serious hearing loss (approximately 175 million individuals) is the result of OM, and that 28,000 children die from complications of undertreated OM each year (WHO 2004).

5   Conclusions

The results of the studies reviewed above indicate that inflammation plays a central role in the pathobiology and resolution of OM. A wide variety of inflammatory mediators have been identified in the effusions of OM patients. This reflects not only initial inflammatory events evoked by pathogens, but also the ability of inflammation to recruit additional inflammatory pathways via cytokine receptors and tissue damage. Inflammatory mediators are critical for the resolution of OM, as evidenced by the association of polymorphisms in TNFα gene with OM proneness in humans, and the dramatic persistence of experimental OM in animals deficient in this cytokine (Leichtle et al. 2010). However, it is axiomatic that inflammation is a double-edged sword, contributing also to OM pathogenesis.

The source of these mediators in AOM appears to be the innate immune system, although cognate immunity can also generate inflammation if a pathogen is subject to immunologic memory (Ryan et al. 1986). Deficiencies in any of the innate immune receptors and signaling molecules that have been examined to date in animals have resulted in the persistence of bacterial OM beyond the period normally observed in wild-type animals. Virtually all of the innate immune pathways examined to date appear to be required for appropriate resolution of OM. Some pathways do, however, appear to be more central than others. A prominent example is the deletion of the TLR adaptor molecule MyD88 in mice which results in persistence of NTHi-induced OM for several weeks, as compared to a few days for wild-type mice (Hernandez et al. 2008). In contrast, deletion of the alternative TLR4 adaptor TRIF results in a much milder OM phenotype (Leichtle et al. 2009b) likely due to the fact that MyD88 operates downstream from many more TLRs than does TRIF. Nonetheless, all deficiencies appear to decrease the host tolerance mechanisms and increase susceptibility to inflammation.

While many individual PRRs appear to be required for normal OM resolution, the results from animal studies also make it clear that there is significant collaboration, complementation and some redundancy in innate sensing through the multiple PRRs present. For example, deletion of the inflammasome adaptor ASC hindered the normal timely OM resolution in mouse models due to decreased IL1β levels. However, all mice eventually cleared the ME infection. Such an outcome is to be expected since there are other PRR pathways which can redundantly, albeit less efficiently, activate many of the same effector genes as those whose expression is most effectively induced by the inflammasomes. However, it is apparent that these alternative pathways are not enough to mediate normal resolution on their own.

Innate immunity also involves contributions from phagocytic cells like neutrophils and macrophages capable of direct migration, microbial uptake and clearance. Pro-inflammatory molecules like TNFα, IL1β, and CCL3 appear to play a key role in promoting cellular migration into the ME and/or activation of the cells for microbial clearance. Addition of recombinant TNFα to the guinea pig ME was sufficient to induce an inflammatory response in the absence of microbial infection (Catanzaro et al. 1991), while a knockout mouse OM model showed that lack of TNF reduced both neutrophilic migration into the ME and macrophage phagocytosis. Interestingly, addition of recombinant CCL3, a chemokine expressed at high levels during the course of OM, abolished infection by restoring normal macrophage phagocytic function in the knockout mouse. Similarly, exogenous CCL3 aided the return of phagocytic function in TLR2-, MyD88- or TNF-deficient macrophages. Future therapeutic approaches to OM should focus on small molecules that can boost key innate immune mechanisms in addition to factors that block microbial virulence at the host level. For instance, the use of infliximab (monoclonal TNFα antibody) reduced the inflammatory activity of OM in animal models (Kariya et al. 2013; Lee et al. 2008). It would be interesting to see if the use of antibodies like canakinumab (monoclonal anti-IL1β antibody) would have similar effects. While anti-inflammatory therapies have the potential to reduce pathology in OM and hearing loss due to ME inflammation, any such therapies must be applied carefully, since they also have the potential to interfere with OM recovery.

OM inflammation can induce high fever, headaches, irritable pain and diminished hearing. While ultimately resulting in the resolution of ME infection, the pathology induced by inflammation produces temporary and permanent hearing loss. This has been shown to lead to delays in speech and language acquisition and learning. For most children in developed countries, this hearing loss is temporary. As fluid within the ME resolves, their hearing returns to normal. Even for otitis-prone children, deficits in language typically recover over time, so that by age 7 they perform normally. However, it should be noted that some recent studies showing an association of deficits in higher-order auditory processing with a history of OM raise the possibility of more lasting changes. Of course, more severe OM can produce permanent damage to ME conduction or inner ear sensorineural structures. Recent results from a large population study identified significant hearing loss in Scandinavian adults who had been diagnosed with chronic suppurative OM as children (Aarhus et al. 2015). As noted above, the problem of permanent hearing loss is particularly acute in developing countries ME infection and inflammation are the greatest single cause of hearing loss at the world level (WHO 2004), providing great need for more effective treatment.

References

Aarhus L, Tambs K, Kvestad E, Engdahl B. Childhood otitis media: a cohort study with 30-year follow-up of hearing (The HUNT study). Ear Hear. 2015;36(3):302–8.

Acuin J. Chronic suppurative otitis media: burden of illness and management options. Geneva, Switzerland: World Health Organization; 2004. Available: http://www.who.int/pbd/publications/Chronicsuppurativeotitis_media.pdf

Ahmed S, Shapiro NL. Bhattacharyya N. Incremental health care utilization and costs for acute otitis media in children. Laryngoscope. 2014;124(1):301–5.

Ahmed S, Arjmand E, Sidell D. Role of obesity in otitis media in children. Curr Allergy Asthma Rep. 2014;14(11):469.

Ahmmed AU, Curley JW, Newton VE, Mukherjee D. Hearing aids versus ventilation tubes in persistent otitis media with effusion: a survey of clinical practice. J Laryngol Otol. 2001;115(4):274–9.

Akira S, Uematsu S, Takeuchi O. Pathogen recognition and innate immunity. Cell. 2006;124(4):783–801.

Allen EK, Manichaikul A, Sale MM. Genetic contributors to otitis media: agnostic discovery approaches. Curr Allergy Asthma Rep. 2014;14(2):411.

Alles R, Parikh A, Hawk L, Darby Y, Romero JN, Scadding G. The prevalence of atopic disorders in children with chronic otitis media with effusion. Pediatr Allergy Immunol. 2001;12(2):102–6.

Ambrosio A, Brigger MT. Surgery for otitis media in a universal health care model: socioeconomic status and race/ethnicity effects. Otolaryngol Head Neck Surg. 2014;151(1):137–41.

Barenkamp SJ. Editorial commentary: respiratory viruses and otitis media in young children. Clin Infect Dis. 2014;60(1):10–1.

Bascelli LM, Losh DP. How does a "wait and see" approach to prescribing antibiotics for acute otitis media (AOM) compare with immediate antibiotic treatment? J Fam Pract. 2001;50(5):469.

Bernstein JM. Biological mediators of inflammation in middle ear effusions. Ann Otol Rhinol Laryngol. 1976;85(2 Suppl 25 Pt 2):90–6.

Beutler B. Innate immunity: an overview. Mol Immunol. 2004;40(12):845–59.

Bierhaus A, et al. Understanding RAGE, the receptor for advanced glycation end products. J Mol Med. 2005;83(11):876–86.

Bluestone CD. Epidemiology and pathogenesis of chronic suppurative otitis media: implications for prevention and treatment. Int J Pediatr Otorhinolaryngol. 1998;42(3):207–23.

Bodmer D, Brors D, Pak K, Keithley EM, Mullen L, Ryan AF, Gloddek B. Inflammatory signals increase Fas ligand expression by inner ear cells. J Neuroimmunol. 2002;129(1–2):10–7.

Brodsky IE, Monack D. NLR-mediated control of inflammasome assembly in the host response against bacterial pathogens. Semin Immunol. 2009;21(4):199–207.

Brodsky L, Faden H, Bernstein J, Stanievich J, DeCastro G, Volovitz B, Ogra PL. Arachidonic acid metabolites in middle ear effusions of children. Ann Otol Rhinol Laryngol. 1991;100(7):589–92.

Catanzaro A, Ryan A, Batcher S, Wasserman SI. The response to human rIL-1, rIL-2, and rTNF in the middle ear of guinea pigs. Laryngoscope. 1991;101(3):271–5.

Cripps AW, Otczyk DC. Prospects for a vaccine against otitis media. Expert Rev Vaccines. 2006;5(4):517–34.

Daly KA, Hunter LL, Giebink GS. Chronic otitis media with effusion. Pediatr Rev. 1999;20(3):85–93. quiz 4

Daly KA, Brown WM, Segade F, Bowden DW, Keats BJ, Lindgren BR, Levine SC, Rich SS. Chronic and recurrent otitis media: a genome scan for susceptibility loci. Am J Hum Genet. 2004;75(6):988–97.

Denes A, Lopez-Castejon G, Brough D. Caspase-1: is IL-1 just the tip of the ICEberg? Cell Death Dis. 2012;3:e338.

Dziarski R. Recognition of bacterial peptidoglycan by the innate immune system. Cell Mol Life Sci. 2003;60(9):1793–804.

El Mezayen R, et al. Endogenous signals released from necrotic cells augment inflammatory responses to bacterial endotoxin. Immunol Lett. 2007;111(1):36–44.

Emonts M, Veenhoven RH, Wiertsema SP, Houwing-Duistermaat JJ, Walraven V, de Groot R, Hermans PW, Sanders EA. Genetic polymorphisms in immunoresponse genes TNFA, IL6, IL10, and TLR4 are associated with recurrent acute otitis media. Pediatrics. 2007;120(4):814–23.

Eskola J, Kilpi T, Palmu A, Jokinen J, Eerola M, Haapakoski J, Herva E, Takala A, et al. Efficacy of a pneumococcal conjugate vaccine against acute otitis media. N Engl J Med. 2001;344(6):403–9.

Faden H. The microbiologic and immunologic basis for recurrent otitis media in children. Eur J Pediatr. 2001;160(7):407–13.

Forgie S, Zhanel G, Robinson J. Management of acute otitis media. Paediatr Child Health. 2009;14(7):457–64.

Forsen TW. Chronic disorders of the middle ear and mastoid. In: Wetmore RF, Muntz HR, TJ MG, et al., editors. Pediatric otolaryngology: principles and practice pathways. Stuttgart, Germany: Thieme; 2000. p. 281–304.

Friel-Patti S, Finitzo T. Language learning in a prospective study of otitis media with effusion in the first two years of life. J Speech Hear Res. 1990;33(1):188–94.

Friel-Patti S, Finitzo-Hieber T, Conti G, Brown KC. Language delay in infants associated with middle ear disease and mild, fluctuating hearing impairment. Pediatr Infect Dis. 1982;1(2):104–910.

Fuchs JC, Linden JF, Baldini A, Tucker AS. A defect in early myogenesis causes otitis media in two mouse models of 22q11.2 deletion syndrome. Hum Mol Genet. 2014;24(7):1869–82.

Geijtenbeek TB, Gringhuis SI. Signalling through C-type lectin receptors: shaping immune responses. Nat Rev Immunol. 2009;9(7):465–79.

Gervain J. Plasticity in early language acquisition: the effects of prenatal and early childhood experience. Curr Opin Neurobiol. 2015;35:13–20.

Goodwin JH, Post JC. The genetics of otitis media. Curr Allergy Asthma Rep. 2002;2(4):304–8.

Granath A, Cardell LO, Uddman R, Harder H. Altered Toll- and Nod-like receptor expression in human middle ear mucosa from patients with chronic middle ear disease. J Infect. 2011;63(2):174–6.

Grievink EH, Peters SA, van Bon WH, Schilder AG. The effects of early bilateral otitis media with effusion on language ability: a prospective cohort study. J Speech Hear Res. 1993;36(5):1004–12.

Guan X, Gan RZ. Effect of middle ear fluid on sound transmission and auditory brainstem response in guinea pigs. Hear Res. 2011;277(1–2):96–106.

Haake SM, Dinh CT, Chen S, Eshraghi AA, Van De Water TR. Dexamethasone protects auditory hair cells against TNFalpha-initiated apoptosis via activation of PI3K/Akt and NFkappaB signaling. Hear Res. 2009;255(1–2):22–32.

Haapala S, Niemitalo-Haapola E, Raappana A, Kujala T, Suominen K, Kujala T, Jansson-Verkasalo E. Effects of recurrent acute otitis media on cortical speech-sound processing in 2-year old children. Ear Hear. 2014;35(3):e75–83.

Han F, Yu H, Tian C, Li S, Jacobs MR, Benedict-Alderfer C, Zheng QY. Role for toll-like receptor 2 in the immune response to streptococcus pneumoniae infection in mouse otitis media. Infect Immun. 2009;77(7):3100–8.

He Y, Scholes MA, Wiet GJ, Li Q, Clancy C, Tong HH. Complement activation in pediatric patients with recurrent acute otitis media. Int J Pediatr Otorhinolaryngol. 2013;77(6):911–7.

Heikkinen T, Chonmaitree T. Importance of respiratory viruses in acute otitis media. Clin Microbiol Rev. 2003;16(2):230–41.

Hernandez M, Leichtle A, Pak K, Ebmeyer J, Euteneuer S, Obonyo M, Guiney DG, Webster NJ, Broide D, Ryan AF, Wasserman SI. Myeloid differentiation primary response gene 88 is required for the resolution of otitis media. J Infect Dis. 2008;198(12):1862–9.

Hernandez M, Leichtle A, Pak K, Webster NJ, Wasserman SI, Ryan AF. The transcriptome of a complete episode of acute otitis media. BMC Genomics. 2015;16:259.

Hirano T, Kodama S, Fujita K, Maeda K, Suzuki M. Role of Toll-like receptor 4 in innate immune responses in a mouse model of acute otitis media. FEMS Immunol Med Microbiol. 2007;49(1):75–83.

Hirano T, Kodama S, Moriyama M, Kawano T, Suzuki M. The role of Toll-like receptor 4 in eliciting acquired immune responses against nontypeable Haemophilus influenzae following intranasal immunization with outer membrane protein. Int J Pediatr Otorhinolaryngol. 2009;73(12):1657–65.

Holm VA, Kunze LH. Effect of chronic otitis media on language and speech development. Pediatrics. 1969;43(5):833–9.

Jin T, Perry A, Jiang J, Smith P, Curry JA, Unterholzner L, et al. Structures of the HIN domain:DNA complexes reveal ligand binding and activation mechanisms of the AIM2 inflammasome and IFI16 receptor. Immunity. 2012;36(4):561–71.

Johnson DL, McCormick DP, Baldwin CD. Early middle ear effusion and language at age seven. J Commun Disord. 2008;41(1):20–32.

Juhn SK, Jung MK, Hoffman MD, Drew BR, Preciado DA, Sausen NJ, Jung TT, Kim BH, Park SY, Lin J, Ondrey FG, Mains DR, Huang T. The role of inflammatory mediators in the pathogenesis of otitis media and sequelae. Clin Exp Otorhinolaryngol. 2008;1(3):117–38.

Jung TT, Park SK, Rhee CK. Effect of inhibitors of leukotriene and/or platelet activating factor on killed H. influenzae induced experimental otitis media with effusion. Int J Pediatr Otorhinolaryngol. 2004;68(1):57–63.

Kariya S, Okano M, Higaki T, Makihara S, Haruna T, Eguchi M, Nishizaki K. Neutralizing antibody against granulocyte/macrophage colony-stimulating factor inhibits inflammatory response in experimental otitis media. Laryngoscope. 2013;123(6):1514–8.

Kaur R, Casey J, Pichichero M. Cytokine, chemokine, and Toll-like receptor expression in middle ear fluids of children with acute otitis media. Laryngoscope. 2015;125(1):E39–44.

Kawai T, Akira S. The roles of TLRs, RLRs and NLRs in pathogen recognition. Int Immunol. 2009;21(4):317–37.

Kawano T, Hirano T, Kodama S, Mitsui MT, Ahmed K, Nishizono A, Suzuki M. Pili play an important role in enhancing the bacterial clearance from the middle ear in a mouse model of acute otitis media with Moraxella catarrhalis. Pathog Dis. 2013;67(2):119–31.

Kim Y-G, Park J-H, Shaw MH, Franchi L, Inohara N, Nunez G. The cytosolic sensors Nod1 and Nod2 are critical for bacterial recognition and host defense after exposure to Toll-like receptor ligands. Immunity. 2008;28(2):246–57.

Kim MG, Park DC, Shim JS, Jung H, Park MS, Kim YI, Lee JW, Yeo SG. TLR-9, NOD-1, NOD-2, RIG-I and immunoglobulins in recurrent otitis media with effusion. Int J Pediatr Otorhinolaryngol. 2010;74(12):1425–9.

Kim SH, Cha SH, Kim YI, Byun JY, Park MS, Yeo SG. Age-dependent changes in pattern recognition receptor and cytokine mRNA expression in children with otitis media with effusion. Int J Pediatr Otorhinolaryngol. 2015;79(2):229–34.

Kubba H, Pearson JP, Birchall JP. The aetiology of otitis media with effusion: a review. Clin Otolaryngol Allied Sci. 2000;25(3):181–94.

Kurabi A, Lee J, Wong C, Pak K, Hoffman HM, Ryan AF, Wasserman SI. The inflammasome adaptor ASC contributes to multiple innate immune processes in the resolution of otitis media. Innate Immun. 2015;21(2):203–14.

Kurabi A, Pak K, Ryan AF, Wasserman SI. Innate immunity: orchestrating inflammation and resolution of otitis media. Curr Allergy Asthma Rep. 2016;16:6.

Kvaerner KJ, Tambs K, Harris JR, Magnus P. The relationship between otitis media and intrauterine growth: a co-twin control study. Int J Pediatr Otorhinolaryngol. 1996;37(3):217–25.

Lee DH, Yeo SW, Chang KH, Park SY, Oh JH, Seo JH. Effect of infliximab on experimentally induced otitis media in rats. Ann Otol Rhinol Laryngol. 2008;117(6):470–6.

Lee HY, Chung JH, Lee SK, Byun JY, Kim YI, Yeo SG. Toll-like receptors, cytokines & nitric oxide synthase in patients with otitis media with effusion. Indian J Med Res. 2013a;138(4):523–30.

Lee JH, Park DC, Oh IW, Kim YI, Kim JB, Yeo SG. C-type lectin receptors mRNA expression in patients with otitis media with effusion. Int J Pediatr Otorhinolaryngol. 2013b;77(11):1846–51.

Leichtle A, Wasserman SI, Hernandez M, Pak K, Ryan A. TNF and MyD88 are critical to the clearance of nontypeable nontypable haemophilus influenzae (NTHi)-induced otitis media via inflammatory cell recruitment and phagocytosis. J Allergy Clin Immunol. 2008;121(2):S268–9.

Leichtle A, Hernandez M, Pak K, Yamasaki K, Cheng C-F, Webster NJ, Ryan AF, Wasserman SI. TLR4-mediated induction of TLR2 signaling is critical in the pathogenesis and resolution of otitis media. Innate Immun. 2009a;15(4):205–15.

Leichtle A, Hernandez M, Pak K, Webster NJ, Wasserman SI, Ryan AF. The Toll-like receptor adaptor TRIF contributes to otitis media pathogenesis and recovery. BMC Immunol. 2009b;10:45.

Leichtle A, Hernandez M, Ebmeyer J, Yamasaki K, Lai Y, Radek K, Choung YH, Euteneuer S, Pak K, Gallo R, Wasserman SI, Ryan AF. CC chemokine ligand 3 overcomes the bacteriocidal and phagocytic defect of macrophages and hastens recovery from experimental otitis media in TNF−/− mice. J Immunol. 2010;184(6):3087–97.

Leichtle A, Hernandez M, Lee J, Pak K, Webster NJ, Wollenberg B, Wasserman SI, Ryan AF. The role of DNA sensing and innate immune receptor TLR9 in otitis media. Innate Immun. 2012;18(1):3–13.

Li Q, Li YX, Douthitt K, Stahl GL, Thurman JM, Tong HH. Role of the alternative and classical complement activation pathway in complement mediated killing against Streptococcus pneumoniae colony opacity variants during acute pneumococcal otitis media in mice. Microbes Infect. 2012;14(14):1308–18.

Lim JH, Ha U, Sakai A, Woo CH, Kweon SM, Xu H, Li JD. Streptococcus pneumoniae synergizes with nontypeable Haemophilus influenzae to induce inflammation via upregulating TLR2. BMC Immunol. 2008;9:40.

Liu K, Chen L, Kaur R, Pichichero ME. Transcriptome signature in young children with acute otitis media due to non-typeable Haemophilus influenzae. Int Immunol. 2013;25(6):353–61.

Luotonen M, Uhari M, Aitola L, Lukkaroinen AM, Luotonen J, Uhari M, Korkeamäki RL. Recurrent otitis media during infancy and linguistic skills at the age of nine years. Pediatr Infect Dis J. 1996;15(10):854–8.

Lupfer C, Kanneganti TD. The expanding role of NLRs in antiviral immunity. Immunol Rev. 2013;255(1):13–24.

MacIntyre EA, Karr CJ, Koehoorn M, et al. Otitis media incidence and risk factors in a population-based birth cohort. Paediatr Child Health. 2010;15(7):437–42.

Medzhitov R, Janeway C Jr. Innate immunity. N Engl J Med. 2000;343(5):338–44.

Medzhitov R, Schneider DS, Soares MP. Disease tolerance as a defense strategy. Science. 2012; 335:936–41.

Melhus A, Ryan AF. Expression of cytokine genes during pneumococcal and nontypeable Haemophilus influenzae acute otitis media in the rat. Infect Immun. 2000;68(7):4024–31.

Minovi A, Dazert S. Diseases of the middle ear in childhood. GMS Curr Top Otorhinolaryngol Head Neck Surg. 2014;13:Doc11.

Mittal R, Lisi CV, Gerring R, Mittal J, Mathee K, Narasimhan G, Azad RK, Yao Q, Grati M, Yan D, Eshraghi AA, Angeli SI, Telischi FF, Liu XZ. Current concepts in the pathogenesis and treatment of chronic suppurative otitis media. J Med Microbiol. 2015;64(10):1103–16.

Miyata M, Lee J-Y, Susuki-Miyata S, Wang WY, Xu H, Kai H, Kobayashi KS, Flavell RA, Li JD. Glucocorticoids suppress inflammation via the upregulation of negative regulator IRAK-M. Nat Commun. 2015;6:6062.

Mogensen TH. Pathogen recognition and inflammatory signaling in innate immune defenses. Clin Microbiol Rev. 2009;22(2):240–73.

Pletz MW, Maus U, Krug N, Welte T, Lode H. Pneumococcal vaccines: mechanism of action, impact on epidemiology and adaption of the species. Int J Antimicrob Agents. 2008;32(3):199–206.

Post JC. Genetics of otitis media. Adv Otorhinolaryngol. 2011;70:135–40.

Qureishi A, Lee Y, Belfield K, Birchall JP, Daniel M. Update on otitis media - prevention and treatment. Infect Drug Resist. 2014;7:15–24.

Rathinam VA, Vanaja SK, Fitzgerald KA. Regulation of inflammasome signaling. Nat Immunol. 2012;13(4):333–42.

Roberts JE, Rosenfeld RM, Zeisel SA. Otitis media and speech and language: a meta-analysis of prospective studies. Pediatrics. 2004;113(3 Pt 1):e238–48.

Rosenfeld RM, Schwartz SR, Pynnonen MA, Tunkel DE, Hussey HM, Fichera JS, Grimes AM, et al. Clinical practice guideline: Tympanostomy tubes in children. Otolaryngol Head Neck Surg. 2013;149(1 Suppl):S1–35.

Rvachew S, Slawinski EB, Williams M, Green CL. The impact of early onset otitis media on babbling and early language development. J Acoust Soc Am. 1999;105(1):467–75.

Ryan AF, Catanzaro A, Wasserman SI, Harris JP, Vogel CW. The effect of complement depletion on immunologically mediated middle ear effusion and inflammation. Clin Immunol Immunopathol. 1986;40(3):410–21.

Ryan R, Harkness P, Fowler S, Topham J. Management of paediatric otitis media with effusion in the UK: a survey conducted with the guidance of the Clinical Effectiveness Unit at the Royal College of Surgeons of England. J Laryngol Otol. 2001;115(6):475–8.

Rye MS, Bhutta MF, Cheeseman MT, Burgner D, Blackwell JM, Brown SD, Jamieson SE. Unraveling the genetics of otitis media: from mouse to human and back again. Mamm Genome. 2011;22(1–2):66–82.

Rye MS, Blackwell JM, Jamieson SE. Genetic susceptibility to otitis media in childhood. Laryngoscope. 2012;122(3):665–75.

Sabirov A, Metzger DW. Mouse models for the study of mucosal vaccination against otitis media. Vaccine. 2006;26(12):1501–24.

Sharma SK, Pichichero ME. Cellular immune response in young children accounts for recurrent acute otitis media. Curr Allergy Asthma Rep. 2013;13(5):495–500.

Shuto T, Xu H, Wang B, Han J, Kai H, Gu XX, Murphy TF, Lim DJ, Li JD. Activation of NF-kappa B by nontypeable Hemophilus influenzae is mediated by toll-like receptor 2-TAK1-dependent NIK-IKK alpha/beta-I kappa B alpha and MKK3/6-p38 MAP kinase signaling pathways in epithelial cells. Proc Natl Acad Sci U S A. 2001;98(15):8774–9.

Si Y, Zhang ZG, Chen SJ, Zheng YQ, Chen YB, Liu Y, Jiang H, Feng LQ, Huang X. Attenuated TLRs in middle ear mucosa contributes to susceptibility of chronic suppurative otitis media. Hum Immunol. 2014;75(8):771–6.

Smirnova MG, Kiselev SL, Gnuchev NV, Birchall JP, Pearson JP. Role of the pro-inflammatory cytokines tumor necrosis factor-alpha, interleukin-1 beta, interleukin-6 and interleukin-8 in the pathogenesis of the otitis media with effusion. Eur Cytokine Netw. 2002;13(2):161–72.

Stangerup SE, Tos M, Arnesen R, Larsen P. A cohort study of point prevalence of eardrum pathology in children and teenagers from age 5 to age 16. Eur Arch Otorhinolaryngol. 1994;251(7):399–403.

Takeda K, Kaisho T, Akira S. Toll-like receptors. Annu Rev Immunol. 2003;21:335–76.

Taylor S, Marchisio P, Vergison A, Harriague J, Hausdorff WP, Haggard M. Impact of pneumococcal conjugate vaccination on otitis media: a systematic review. Clin Infect Dis. 2012;54(12):1765–73.

Teele DW, Klein JO, Rosner B. Epidemiology of otitis media during the first seven years of life in children in greater Boston: a prospective, cohort study. J Infect Dis. 1989;160(1):83–94.

Thomas NM, Brook I. Otitis media: an update on current pharmacotherapy and future perspectives. Expert Opin Pharmacother. 2014;15(8):1069–83.

Thornton JL, Chevallier KM, Koka K, Gabbard SA, Tollin DJ. Conductive hearing loss induced by experimental middle-ear effusion in a chinchilla model reveals impaired tympanic membrane-coupled ossicular chain movement. J Assoc Res Otolaryngol. 2013;14(4):451–64.

Trune DR, Kempton B, Hausman FA, Larrain BE, MacArthur CJ. Correlative mRNA and protein expression of middle and inner ear inflammatory cytokines during mouse acute otitis media. Hear Res. 2015;326:49–58.

Tsuchiya K, Komori M, Zheng QY, Ferrieri P, Lin J. Interleukin 10 is an essential modulator of mucoid metaplasia in a mouse otitis media model. Ann Otol Rhinol Laryngol. 2008;117(8):630–6.

Venekamp RP, Sanders SL, Glasziou PP, Del Mar CB, Rovers MM (2015) Antibiotics for acute otitis media in children. Cochrane Database Syst Rev 23(6):CD000219.

Vergison A. Microbiology of otitis media: a moving target. Vaccine. 2008;26(Suppl 7):G5–10.

Villa PC, Zanchetta S. Auditory temporal abilities in children with history of recurrent otitis media in the first years of life and persistent in preschool and school ages. Codas. 2014;26(6):494–502.

Watanabe T, Hirano T, Suzuki M, Kurono Y, Mogi G. Role of interleukin-1beta in a murine model of otitis media with effusion. Ann Otol Rhinol Laryngol. 2001;110(6):574–80.

Weber B, Schuster S, Zysset D, Rihs S, Dickgreber N, Schürch C, Riether C, Siegrist M, Schneider C, Pawelski H, Gurzeler U, Ziltener P, Genitsch V, Tacchini-Cottier F, Ochsenbein A, Hofstetter W, Kopf M, Kaufmann T, Oxenius A, Reith W, Saurer L, Mueller C. TREM-1 deficiency can attenuate disease severity without affecting pathogen clearance. PLoS Pathog. 2014;10(1):e1003900.

WHO. 2004. http://www.who.int/pbd/publications/Chronicsuppurativeotitismedia.pdf

Winskel H. The effects of an early history of otitis media on children's language and literacy skill development. Br J Educ Psychol. 2006;76(Pt 4):727–44.

Woo JI, Oh S, Webster P, Lee YJ, Lim DJ, Moon SK. NOD2/RICK-dependent beta-defensin 2 regulation is protective for nontypeable Haemophilus influenzae-induced middle ear infection. PLoS One. 2014;9(3):e90933.

Yamasaki S, et al. Mincle is an ITAM-coupled activating receptor that senses damaged cells. Nat Immunol. 2008;9(10):1179–88.

Yao W, Frie M, Pan J, Pak K, Webster N, Wasserman SI, Ryan AF. C-Jun N-terminal kinase (JNK) isoforms play differing roles in otitis media. BMC Immunol. 2014;15:46.

Yellon RF, Leonard G, Marucha PT, Craven R, Carpenter RJ, Lehmann WB, Burleson JA, Kreutzer DL. Characterization of cytokines present in middle ear effusions. Laryngoscope. 1991;101(2):165–9.

Zumach A, Gerrits E, Chenault M, Anteunis L. Long-term effects of early-life otitis media on language development. J Speech Lang Hear Res. 2010;53(1):34–43.

Chapter 7
Inflammation Potentiates Cochlear Uptake of Ototoxins and Drug-Induced Hearing Loss

Peter S. Steyger

Abstract Serious bacterial infections are often treated with aminoglycosides, especially when the cause of systemic infection is unknown. Severe infections trigger specific systemic inflammatory response pathways. Aminoglycosides are primarily trafficked across the cochlear blood-labyrinth barrier into the stria vascularis, prior to clearance into endolymph and entry into hair cells with subsequent cytotoxicity and loss of auditory function: cochleotoxicity. Systemic inflammation potentiates cochlear uptake of aminoglycosides and increases the risk of hearing loss in both preclinical models and human studies. Here, we review the data that establishes the above narrative, and articulate the need for translational studies to promote ototoxicity monitoring in neonatal intensive care units and cystic fibrosis clinics.

Keywords Aminoglycosides · Ototoxicity · Stria vascularis · Drug trafficking · Infection

1 Infection and Inflammation

Any person can potentially become infected by bacteria, viruses, fungi or parasites. These infections can trigger specific systemic inflammatory responses that facilitate recovery to good health. Severe infections cause obvious systemic symptoms including fever, chills, lethargy, and decrease cognitive function—as the initial inflammatory response is vastly amplified to combat the infection. Contemporaneously, however, our sense of hearing and vision is rarely compromised, except when infections are localized to the middle or inner ear, or within the eye. Until recently, the inner ear has been considered, systemically, an immunologically-privileged site, as few components of the inflammatory response (e.g., immune cells, antibodies) are detected within the inner ear, excluded by the blood-labyrinth barrier that is akin to the blood-brain barrier (Oh et al. 2012). This immunologically-isolated view of

P. S. Steyger (✉)
Oregon Hearing Research Center, Oregon Health & Science University, Oregon, USA
e-mail: steygerp@ohsu.edu

© Springer International Publishing AG, part of Springer Nature 2018
V. Ramkumar, L. P. Rybak (eds.), *Inflammatory Mechanisms in Mediating Hearing Loss*, https://doi.org/10.1007/978-3-319-92507-3_7

the inner ear has been displaced in recent years by several pioneering studies that show the inner ear actively participating in classical local and systemic inflammatory mechanisms, with unexpected and unintended consequences.

Middle ear infections degrade the ability to hear low level sound, primarily through impaired conductive transmission of acoustic stimuli. Recent studies show that middle ear infections can trigger intra-cochlear inflammatory responses, disrupting cochlear homeostasis, and initiate cochlear tissue remodeling, all of which transiently or permanently impact auditory function (Trune et al. 2015). Middle ear inflammation increases the permeability of the round window to macromolecules, enabling pro-inflammatory signals and bacterial endotoxins in the middle ear to penetrate through the round window into the perilymphatic scala tympani of the cochlea (Kawauchi et al. 1989, Ikeda et al. 1990). Spiral ligament fibrocytes lining the scala tympani respond to these immunogenic signals, releasing inflammatory chemokines that attract immune cells to migrate across the blood-labyrinth barrier into the inner ear (Oh et al. 2012, Kaur et al. 2015b). Inner ear recruitment of systemic immune cells is also evident after selective hair cell death (Kaur et al. 2015a, b). In addition, macrophage-like cells are localized adjacent to blood vessels (perivascular macrophages) within the inner ear (Zhang et al. 2012); and supporting cells in the organ of Corti exhibit glial-like (anti-inflammatory) properties by phagocytosing cellular debris following sensory hair cell death (Monzack et al. 2015, Francis and Cunningham 2017). These data imply that inner ear tissues can mount a local response to inflammatory signals similar to that accomplished in other tissues following sterile induction, e.g., after a crushing injury resulting in necrotic cell death (Rock et al. 2010), and more specifically, by noise-induced cochlear cell death (Fujioka et al. 2014, Hirose et al. 2005). Yet, it is not fully understood how cochlear inflammation induced by middle ear infections could cause auditory dysfunction. Middle ear infections and inflammation can down-regulate the expression of cochlear gene products for ion channels and transporters in the stria vascularis that are crucial for cochlear fluid homeostasis, and the integrity of the blood-labyrinth barrier (BLB). Remarkably, glucocorticoids can partially reverse this dysregulation and restore ion homeostasis and improve auditory function (MacArthur et al. 2011, 2015), indicating that these drugs can modulate cochlear physiology, as well as having anti-inflammatory effects.

Three meningeal membranes envelop cerebrospinal fluid (CSF) to protect the brain and spinal cord, and are nourished by the highly-vascularized blood-brain barrier. Infection of the meningeal membranes—meningitis—has long been known to induce labyrinthitis, cochlear fibrosis and ossification that can impair optimal cochlear implantation procedures (Caye-Thomasen et al. 2012). Strikingly, meningogenic bacteria migrate from CSF through the cochlear aqueduct into the perilymphatic scala tympani at the base of the cochlea (Takumida and Anniko 2004), and can induce temporary (when treated rapidly with non-ototoxic antibiotics) or permanent hearing loss (Richardson et al. 1997, Perny et al. 2016, Bhatt et al. 1993). Over time, bacteria spread through perilymph via the helicotrema to the basal region of the scala vestibuli before entering cochlear endolymph and the vestibular apparatus, inducing widespread inflammation (Takumida and Anniko 2004). Preclinical models with untreated meningitis frequently develop hearing loss, and this was closely correlated with rapid elevation of markers for inflammation in CSF (Bhatt et al. 1993, Perny et al. 2016). Local infusion of bacterial endotoxin into the cochlea also induced a dose-dependent increase in inflammatory infiltrates and hearing loss (Darrow et al. 1992, Tarlow et al. 1991).

In contrast to these direct inflammatory challenges to the cochlea, systemic infections or inflammation do not generally modulate auditory function, as shown experimentally (Hirose et al. 2014b, Koo et al. 2015). Nonetheless, systemic inflammation changes cochlear physiology. Systemic administration of immunogenic stimuli (e.g., bacterial lipopolysaccharides, LPS) triggers cochlear recruitment of mononuclear phagocytes into the spiral ligament over several days (Hirose et al. 2014b). While LPS and inflammation typically vasodilate blood vessels, facilitating greater extravasation of plasma and immune cells into the interstitial fluids, the tight junctions between endothelial cells of cochlear capillaries are thought to remain intact (unpublished data). Yet, systemic LPS-induced inflammation altered the permeability of the blood-perilymph barrier, with a two- to threefold increase in fluorescein in perilymph (Hirose et al. 2014a). In addition, systemic LPS also increased cochlear levels of inflammatory markers (Koo et al. 2015, Quintanilla-Dieck et al. 2013). This is significant, as higher serum levels of individual cytokines (e.g., IL-1β; IL-10) was not replicated in cochlear tissues, suggestive of a general paucity of paracellular flux between the tight junction-coupled endothelial cells comprising the blood-labyrinth barrier of the cochlear lateral wall (Koo et al. 2015). The cochlear expression of specific cytokines after LPS challenge was further emphasized at later time points, when cochlear tissues continued to express higher levels of individual cytokines (e.g., IL-6, IL-8; MIP-1α) while serum levels returned to very low baseline levels, suggestive of local, cochlear (parenchymal) production of cytokines. This was confirmed by upregulation of mRNA for these cytokines in cochlear tissues (Koo et al. 2015). Thus, the cochlea contributes to inflammatory responses induced by systemic, as well as cochlear, immunogenic bacterial ligands, or other sources of inflammation, e.g., cellular debris.

2   Aminoglycoside Antibiotics

Severe bacterial infections, such as bacteremia, meningitis or sepsis, are often treated with aminoglycosides antibiotics, despite their well-known ototoxic effects, due to their broad-spectrum bactericidal activity. Figure 7.1 shows a classic cross-sectional diagram of the cochlear duct, with the sensory hair cells within the organ of Corti, and the lateral wall containing the highly-vascularized stria vascularis and adjacent spiral ligament. The cochlear duct is divided into three fluid-filled compartments, two with perilymph, containing a typical composition of extracellular fluids, akin to cerebrospinal fluid (CSF) that is high in sodium and low in potassium. In contrast, endolymph is high in potassium and low in sodium, and is generated by the stria vascularis (Nin et al. 2008). The distinct electrochemical composition of

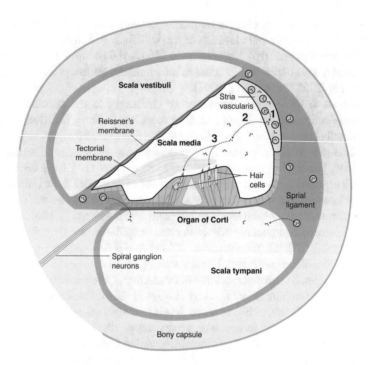

Fig. 7.1 A cross-section of the cochlear duct, with sensory hair cells (light blue) in the organ of Corti, and the lateral wall containing the stria vascularis and adjacent spiral ligament, within which are blood vessels (pink). There are three fluid-filled compartments, the perilymphatic scala vestibuli and scala tympani with CSF-like extracellular fluid, and the scala media with endolymph. The two fluid compartments are separated by tight junctions between all cells lining the scala media. Tight junctions are also present between adjoining cells enclosing the stria vascularis to form a fourth distinct intra-cochlear compartment. Tight junctions also couple endothelial cells lining cochlear blood vessels. Schematic diagram not to relative scale

endolymph and perilymph is maintained by tight junctions between all adjacent epithelial cells lining the scala media. Tight junctions are also present between adjoining basal cells of the stria vascularis, which together with the marginal cells, enclose a fourth distinct intra-cochlear compartment – the stria vascularis with intra-strial fluid in the very small volume of space between strial cells. In addition, endothelial cells lining cochlear capillaries are coupled by tight junctions to form the primary blood-labyrinth barrier. In contrast, endothelial cells in capillary beds of most other tissues are not conjoined by tight junctions, allowing paracellular flux (extravasation) between endothelial cells, especially when capillaries are dilated during inflammation.

In the 1980s, Tran Ba Huy et al. (1986) intravenously infused rats with aminoglycosides and readily detected these drugs in perilymph, not in endolymph. Aminoglycosides were also readily detected in hair cells in vivo prior to loss of hair cell function and hair cell death (Hiel et al. 1993). In vitro studies by Kroese et al. (1989) reported that aminoglycosides blocked the mechanoelectrical transduction

(MET) channel located at the tips of stereocilia projecting from hair cell apices, which in vivo are in endolymph. Marcotti et al. (2005) demonstrated that aminoglycosides could also permeate through MET channels into hair cells in vitro. Furthermore, hair cell uptake of aminoglycosides can be blocked by MET channel blockers, such as curare, or high levels of extracellular calcium ions (Alharazneh et al. 2011, Coffin et al. 2009), and that hair cell death was dependent on aminoglycosides entering via the MET channel (Vu et al. 2013, Coffin et al. 2009). However, these in vitro studies begged the question: were intravascular aminoglycosides trafficked directly into endolymph, and if so, how could these drugs cross the blood-labyrinth barrier?

3 Trafficking across the Blood-Labyrinth Barrier

Numerous prior studies have localized aminoglycosides within cochlear cells, especially in hair cells, but also transiently in other cochlear tissues, especially within the lateral wall (Balogh et al. 1970, Imamura and Adams 2003, Hiel et al. 1993, Yamane et al. 1988). When we systemically administered fluorescently-tagged gentamicin, we found the fluorescent conjugate preferentially localized within the stria vascularis, and near the endolymphatic surface of hair cells and their stereocilia, as well as their adjacent supporting cells within 30 min (Wang and Steyger 2009). Therefore, we tested the hypothesis that intravascular aminoglycosides are primarily trafficked across the strial BLB into endolymph and hair cells. We developed two in vivo cochlear perfusion paradigms (Li and Steyger 2011). In the first, we systemically administered fluorescent gentamicin, and simultaneously perfused the scala tympani with artificial perilymph to remove any aminoglycosides from perilymph bathing the basolateral membranes of hair cells. Within 30 min, we found gentamicin prominently localized in the stria vascularis, and strikingly, readily localized within hair cells. In the reverse experiment, by perfusing the scala tympani with artificial perilymph containing gentamicin at a generous 20% dose (Desjardins-Giasson and Beaubien 1984) of serum levels obtained in the first experiment for 30 min. Gentamicin was not detected in the stria vascularis, and only weak localization of gentamicin in hair cells. Cochlear sensitivity was verified as auditory compound action potentials (CAP) were retained, demonstrating that the cochlea was still physiologically active. Thus, aminoglycosides preferentially traffic across the BLB into the interstitial fluid of the stria vascularis, prior to clearance into endolymph and enter hair cells (Li and Steyger 2011).

Several physiological mechanisms could facilitate aminoglycoside trafficking across endothelial cells (Fig. 7.2), collectively categorized as transcellular trafficking, including; (i) transcytosis, which can be one or more of the following: receptor-mediated endocytosis, non-specific endocytosis, pinocytosis, and/or exocytosis. Aminoglycosides can also permeate through non-selective cation channels, e.g., TRPV4 (Karasawa et al. 2008), and potentially clear into the interstitial fluids via the same mechanism, since such channels are often permeant in both directions, if

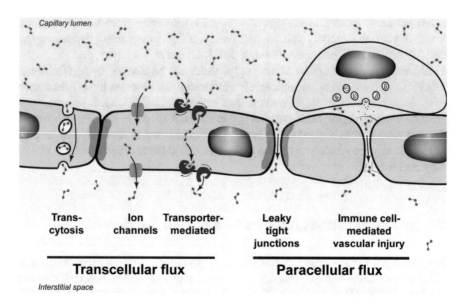

Fig. 7.2 Aminoglycosides could traffic across the BLB by trans-cellular trafficking, including; (i) transcytosis, e.g., receptor-mediated endocytosis, non-specific endocytosis, pinocytosis, and/or exocytosis; (ii) permeate through non-selective cation channels, e.g., TRPV4; and/or (iii) electrogenic sodium transporters. Alternatively, aminoglycosides could take a para-cellular route if (iv) tight junctions between endothelial cells become permeable, or (v) the BLB is compromised by immune cells releasing cytotoxic compounds to enable diapedesis between endothelial cells into the interstitial space. Schematic diagram not to relative scale

expressed on luminal and ablumenal membranes (Huang et al. 2000, Karasawa et al. 2008). Also, electrogenic sodium transporters, similar to the sodium-glucose transporter-2 that can traffic aminoglycosides into renal cells (Jiang et al. 2014), may also exist in cochlear endothelial cells.

Alternatively, aminoglycosides could cross the BLB via paracellular flux if the tight junctions between adjacent activated endothelial cells become increasingly permeable during inflammation (Pober and Sessa 2007). In addition, the integrity of the endothelial barrier can be directly compromised by immune cells, like neutrophils or macrophages, releasing cytotoxic compounds that enable them to pass through the endothelial cell layer (Stearns et al. 1993, Stamatovic et al. 2016), potentially allowing aminoglycosides to inadvertently escape the capillary lumen into the interstitial space.

For aminoglycosides trafficking across the stria vascularis into endolymph and thence into hair cells (Li and Steyger 2011), there are additional barriers, and permutations to reach endolymph (Fig. 7.3). Strial endothelial cells readily take up aminoglycosides (Dai and Steyger 2008, Koo et al. 2015), and these capillaries are largely surrounded by basal cells and intermediate cells, each connected to the other by gap junctions (Takeuchi and Ando 1998, Takeuchi et al. 2001), however the presence of gap junctions between endothelial cells and strial cells is disputed

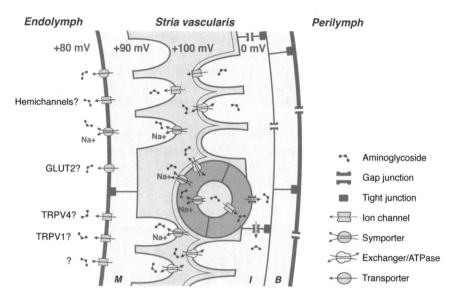

Fig. 7.3 To cross the strial BLB, aminoglycosides must first enter endothelial cells (dark grey) as described in Fig. 7.2. Gap junctions could allow aminoglycosides to permeate into intermediate cells (*I*) and basal cells (*B*). Aminoglycosides clear endothelial, intermediate and basal cells via transporters, exchangers, and/or cation channels, or if within endosomes by exocytosis, into the intra-strial space. Aminoglycosides are taken up by marginal cells (*M*) across their basolateral membranes, presumptively by ATPases, exchangers, and transporters (and ion channels?). Once in marginal cells, aminoglycosides can clear into endolymph down the electrochemical gradient, presumptively via permeation of hemi-channels, facilitated glucose transporters (GLUT), and at least two TRP channels, TRPV1 and TRPV4. Schematic diagram not to relative scale

(Cohen-Salmon et al. 2007). Hemi-channels (akin to gap junctions, allowing permeation between the cell cytoplasm and interstitial fluid) are electrophysiologically blocked by aminoglycosides (Figueroa et al. 2014), and gap junctions (composed of two hemi-channels directly connecting the cytoplasm of two cells) are permeant to fluorescently-tagged gentamicin (unpublished studies). Aminoglycosides could also enter endothelial cells by permeating non-selective cation channels (like TRPV4, and the candidate TRPC3) or electrogenic transporters, akin to SGLT2, expressed by vascular cells (Fian et al. 2007, Karasawa et al. 2008, Jiang et al. 2014). Once in endothelial cells, aminoglycosides could permeate into adjacent pericytes, intermediate cells and basal cells (Koo et al. 2015, Takeuchi and Ando 1998).

Aminoglycosides are cleared from endothelial cells (or intermediate and basal cells) directly into the intra-strial space, presumptively via transporters, exchangers, and/or cation channels (Karasawa et al. 2008, Takeuchi and Ando 1997), or if within endosomes by exocytosis. Aminoglycosides are readily localized within marginal cells, presumably following uptake across their basolateral membranes that are highly populated with a wide variety of ATPases, exchangers, and transporters (Koo et al. 2015, Yoshihara et al. 1999, Takeuchi and Ando 1997, Iwano et al. 1989,

Crouch et al. 1997). It is not known if the basolateral membranes of marginal cells possess non-selective, aminoglycoside-permeant cation channels, as this could potentially ground the elegant electrophysiological environment of the stria vascularis, where the intra-strial space has a potential of +100 mv relative to perilymph (Nin et al. 2008). Once in marginal cells with a potential of +90 mV, aminoglycosides could passively be cleared into endolymph (+80 mV) down the electrochemical gradient.

Aminoglycosides are cleared from marginal cells into endolymph by one or more hypothesized mechanisms, including permeation of hemi-channels (Zhao et al. 2005), facilitated glucose transporters (Takeuchi and Ando 1997, Yoshihara et al. 1999), and by at least two TRP channels that are permeant to aminoglycosides and expressed near the apical surface of marginal cells, TRPV4 (Karasawa et al. 2008), and TRPV1 (Jiang et al. 2015). Strial endothelial cells have the highest intensity of fluorescent gentamicin (Koo et al. 2015) that, presumptively, overcomes the intra-strial electrical barrier of +100 mV barrier, potentially in a similar manner as potassium (Nin et al. 2008, Marcus et al. 2002).

Once aminoglycosides have traversed the strial blood-labyrinth barrier into endolymph, they preferentially enter hair cells via hair cell stereociliary MET channels (Li and Steyger 2011, Alharazneh et al. 2011, Marcotti et al. 2005). As aminoglycosides are cationic molecules, and the potential of endolymph is +80 mV, there is a tremendous driving force acting on these drugs to enter the negatively-polarized hair cells via their MET channels, and also supporting cells (Dai et al. 2006, Li and Steyger 2011), via other activated non-selective, aminoglycoside-permeant cation channels, such as TRPV1 and TRPV4 (Karasawa et al. 2008, Jiang et al. 2015), or other potential candidate aminoglycoside-permeant cation channels yet to be identified. Furthermore, although aminoglycosides do readily enter the perilymphatic scala tympani (Tran Ba Huy et al. 1986), they are not readily taken up by hair cells (Li and Steyger 2011), unless the basolaterally-expressed TRPA1 channels are activated by reactive oxygen species (due to noise or drug intoxication) or other endogenous intracellular signaling molecules (Stepanyan et al. 2011). Once in hair cells, aminoglycosides can then induce hair cell death pathways, including intracellular inflammatory pathways that are discussed in more detail elsewhere in this book.

4   The Impact of Inflammation on Ototoxicity

Most prior studies of aminoglycoside-induced ototoxicity have been conducted in *healthy* preclinical models. Since systemic inflammation readily modulates the permeability of the blood-brain barrier (Abbott et al. 2006), we examined whether inflammation induced physiological changes in the blood-labyrinth barrier to modulate cochlear uptake of aminoglycosides and subsequent ototoxicity. Low systemic dosing with LPS to induce inflammation did not alter serum levels of aminoglycosides (relative to control subjects without LPS administration), yet there was increased cochlear uptake of aminoglycosides, particularly in the stria vascularis

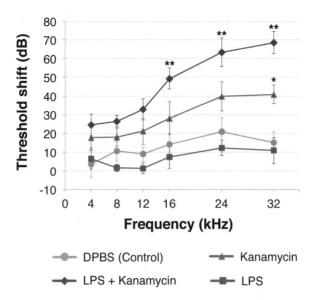

Fig. 7.4 Three weeks after chronic systemic exposure to low doses of LPS, ABR threshold shifts for LPS-only mice were similar to DPBS-treated mice. Kanamycin (700 mg/kg, twice daily) alone induced a significant permanent threshold shifts (PTS) at only 32 kHz (*P < 0.05) compared to DPBS-treated mice. LPS + kanamycin-treated mice had significant PTS at 16, 24 (**P < 0.01) and 32 kHz (P < 0.05) compared to kanamycin-treated mice, and also DPBS- and LPS-only mice at 32 kHz (**P < 0.01). Error bars = s.d. From: Koo JW, Quintanilla-Dieck L, Jiang M, Liu J et al (2015) Endotoxemia-mediated inflammation potentiates aminoglycoside-induced ototoxicity. Sci Transl Med 7:298ra118. Reprinted with permission from AAAS

(Koo et al. 2015). There was also a lack of evidence for paracellular flux between the endothelial cells of dilated cochlear capillaries in the stria vascularis and spiral ligament, suggestive of *transcellular* trafficking of aminoglycosides (Koo et al. 2015).

Interestingly, significant auditory threshold shifts occur in cochlear regions where hair cells appear morphologically intact following ototoxic drug administration (Koo et al. 2015, Nicol et al. 1992). Recent studies now show that aminoglycosides can, in specific situations, induce cochlear synaptopathy, disrupting the synapses between inner hair cells and their innervating afferent nerve fibers as well as decreased neuronal density in the spiral ganglion of the cochlea (Oishi et al. 2015). Remarkably, experimental meningitis and the consequent inflammatory response, also induced cochlear synaptopathy and significantly decreased spiral ganglion density (Perny et al. 2016). Thus, cochlear synaptopathy may contribute to the greater degree of cochlear dysfunction observed relative to that suggested by actual hair cell loss (Steyger 2017) (Fig. 7.4).

Crucially, chronic low-dose LPS induction of systemic inflammation synergistically potentiated aminoglycoside-induced ototoxicity. The frequency range of auditory threshold shifts and the degree of hair cell death induced by kanamycin was significantly greater than in preclinical models without inflammation (Koo et al.

2015). Inflammation also potentiated cisplatin-induced ototoxicity (Oh et al. 2011). This raises an important point for the development of otoprotective agents, such as bisbenzylisoquinoline derivatives (Kruger et al. 2016) or D-methionine (Campbell et al. 2016, Fox et al. 2016), to reduce drug-induced ototoxicity in preclinical models: are otoprotective compounds against ototoxicity effective in subjects *with* inflammation during treatment, as shown for otoprotectants used for ameliorating NIHL (Tieu and Campbell 2012)?

5 Human Studies

If systemic inflammation synergistically potentiates aminoglycoside-induced ototoxicity, then the very patients treated with aminoglycosides for severe and life-threatening infections are potentially most at risk from infection- and inflammation-potentiation of aminoglycoside-induced hearing loss. Prior observations that bacteremia and hyperthermia (an experimental model for sepsis-induced fever) also enhanced aminoglycoside-induced ototoxicity in humans and mice, respectively (Moore et al. 1984, Henry et al. 1983). Endotoxemia (akin to intravenous LPS induction of inflammation) also potentiated aminoglycoside-induced nephrotoxicity (Zager 1992), and heightened the degree of cisplatin-induced cochleotoxicity (Oh et al. 2011).

Routine intravenous use of aminoglycosides remains common in neonatal intensive care units (NICUs) and cystic fibrosis clinics in the US (Garinis et al. 2017a, Cross et al. 2015) for both prophylactically preventing sepsis, as well as eradicating systemic or exacerbated respiratory infections, respectively. Yet, strikingly, there is a lack of clinical data showing the dose-dependency of aminoglycoside-induced ototoxicity in humans (Ahmed et al. 2012, Kushner et al. 2015, Rizzi and Hirose 2007), although several risk factors, e.g., renal insufficiency or age, can predispose patients to drug-induced ototoxicity (Rizzi and Hirose 2007, Forge and Schacht 2000). Furthermore, the risk of aminoglycoside-induced hearing loss is underestimated as the total dose or duration dosing is rarely stratified, resulting in heterogeneous study populations (EBSR 2010).

Fortunately, cumulative intravenous gentamicin dosing can be more accurately monitored in hospital, and recent pilot data suggest that the risk of hearing loss is associated with greater cumulative aminoglycoside dosing in NICU infants with 4 or more days of gentamicin dosing (Cross et al. 2015, Garinis et al. 2017b), and also in subjects with cystic fibrosis with an increasing number of intravenous (14-day) courses of tobramycin dosing (Garinis et al. 2017a). When NICU subjects with (suspected) sepsis or clinical signs of inflammation were treated with gentamicin for ≥5 days, these subjects were twice as likely to be referred on a *higher* frequency distortion product otoaccoustic emissions assessment compared to all other subjects in this pilot study (Cross et al. 2015). Thus, it appears that patients with (suspected) sepsis are at an even greater risk of aminoglycoside-induced hearing loss than previously recognized and a larger observational trial is now needed. If the above data is validated in human studies, there will be a need to implement ototoxicity monitoring protocols in NICUs (Garinis et al. 2018), and extend the provision of ototoxicity monitoring in cystic fibrosis clinics worldwide (Al-Malky et al. 2015, Garinis et al. 2017a).

6 Summary

Cochlear tissues are capable of sustained cellular inflammatory responses to both local and systemic immunogenic stimuli, despite the blood-labyrinth barrier excluding the rapid entry of immunogens, immune cells and antibodies. Typically, the BLB and immune responses would be evolutionarily beneficial for preserving inner ear function, yet modulation of the BLB physiology by inflammatory responses has unintended consequences, such as potentiating the uptake and ototoxicity of specific medications developed only in recent decades. Thus, further work is required to unravel the implications of local and systemic inflammation on cochlear physiology as well as cochlear immune responses to acoustic trauma; infections of the middle ear, meninges or CSF, in addition to ototoxic medications.

Acknowledgements Figures drafted by Karen Thiebes, of Simplified Science Publishing, LLC. I thank lab members for discussion on the manuscript. This research was supported by R01 awards (DC004555, DC12588) from the National Institute of Deafness and Other Communication Disorders. The content is solely the responsibility of the author and does not represent the official views of the NIH, Oregon Health & Science University or the VA Portland Health Care System. The author declares no existing or potential conflict of interest.

This work was supported by NIDCD R01 awards DC04555 and DC012588.

References

Abbott NJ, Ronnback L, Hansson E. Astrocyte-endothelial interactions at the blood-brain barrier. Nat Rev Neurosci. 2006;7:41–53.

Ahmed RM, Hannigan IP, Macdougall HG, Chan RC, et al. Gentamicin ototoxicity: a 23-year selected case series of 103 patients. Med J Aust. 2012;196:701–4.

Alharazneh A, Luk L, Huth M, Monfared A, et al. Functional hair cell mechanotransducer channels are required for aminoglycoside ototoxicity. PLoS One. 2011;6:e22347.

Al-Malky G, Dawson SJ, Sirimanna T, Bagkeris E, et al. High-frequency audiometry reveals high prevalence of aminoglycoside ototoxicity in children with cystic fibrosis. J Cyst Fibros. 2015;14:248–54.

Balogh K Jr, Hiraide F, Ishii D. Distribution of radioactive dihydrostreptomycin in the cochlea. An autoradiographic study. Ann Otol Rhinol Laryngol. 1970;79:641–52.

Bhatt SM, Lauretano A, Cabellos C, Halpin C, et al. Progression of hearing loss in experimental pneumococcal meningitis: correlation with cerebrospinal fluid cytochemistry. J Infect Dis. 1993;167:675–83.

Campbell KC, Martin SM, Meech RP, Hargrove TL, et al. D-methionine (d-met) significantly reduces kanamycin-induced ototoxicity in pigmented Guinea pigs. Int J Audiol. 2016;55:273–8.

Caye-Thomasen P, Dam MS, Omland SH, Mantoni M. Cochlear ossification in patients with profound hearing loss following bacterial meningitis. Acta Otolaryngol. 2012;132:720–5.

Coffin AB, Reinhart KE, Owens KN, Raible DW, et al. Extracellular divalent cations modulate aminoglycoside-induced hair cell death in the zebrafish lateral line. Hear Res. 2009;253:42–51.

Cohen-Salmon M, Regnault B, Cayet N, Caille D, et al. Connexin30 deficiency causes instrastrial fluid-blood barrier disruption within the cochlear stria vascularis. Proc Natl Acad Sci U S A. 2007;104:6229–34.

Cross CP, Liao S, Urdang ZD, Srikanth P, et al. Effect of sepsis and systemic inflammatory response syndrome on neonatal hearing screening outcomes following gentamicin exposure. Int J Pediatr Otorhinolaryngol. 2015;79:1915–9.

Crouch JJ, Sakaguchi N, Lytle C, Schulte BA. Immunohistochemical localization of the na-k-cl co-transporter (nkcc1) in the gerbil inner ear. J Histochem Cytochem. 1997;45:773–8.

Dai CF, Steyger PS. A systemic gentamicin pathway across the stria vascularis. Hear Res. 2008;235:114–24.

Dai CF, Mangiardi D, Cotanche DA, Steyger PS. Uptake of fluorescent gentamicin by vertebrate sensory cells in vivo. Hear Res. 2006;213:64–78.

Darrow DH, Keithley EM, Harris JP. Effects of bacterial endotoxin applied to the Guinea pig cochlea. Laryngoscope. 1992;102:683–8.

Desjardins-Giasson S, Beaubien AR. Correlation of amikacin concentrations in perilymph and plasma of continuously infused Guinea pigs. Antimicrob Agents Chemother. 1984;26:87–90.

Ebsr A. Evidence-based systematic review: drug-induced hearing loss—gentamicin [Online]. 2010. Available: www.asha.org/uploadedFiles/EBSRGentamicin.pdf [Accessed].

Fian R, Grasser E, Treiber F, Schmidt R, et al. The contribution of trpv4-mediated calcium signaling to calcium homeostasis in endothelial cells. J Recept Signal Transduct Res. 2007;27:113–24.

Figueroa VA, Retamal MA, Cea LA, Salas JD, et al. Extracellular gentamicin reduces the activity of connexin hemichannels and interferes with purinergic ca(2+) signaling in hela cells. Front Cell Neurosci. 2014;8:265.

Forge A, Schacht J. Aminoglycoside antibiotics. Audiol Neurootol. 2000;5:3–22.

Fox DJ, Cooper MD, Speil CA, Roberts MH, et al. D-methionine reduces tobramycin-induced ototoxicity without antimicrobial interference in animal models. J Cyst Fibros. 2016;15:518–30.

Francis SP, Cunningham LL. Non-autonomous cellular responses to ototoxic drug-induced stress and death. Front Cell Neurosci. 2017;11:252.

Fujioka M, Okano H, Ogawa K. Inflammatory and immune responses in the cochlea: potential therapeutic targets for sensorineural hearing loss. Front Pharmacol. 2014;5:287.

Garinis AC, Cross CP, Srikanth P, Carroll K, et al. The cumulative effects of intravenous antibiotic treatments on hearing in patients with cystic fibrosis. J Cyst Fibros. 2017a;16:401–9.

Garinis AC, Liao S, Cross CP, Galati J, et al. Effect of gentamicin and levels of ambient sound on hearing screening outcomes in the neonatal intensive care unit: a pilot study. Int J Pediatr Otorhinolaryngol. 2017b;97:42–50.

Garinis AC, Kemph A, Tharpe AM, Weitkamp JH, et al. Monitoring neonates for ototoxicity. Int J Audiol. 2018. (in press). https://doi.org/10.1080/14992027.2017.1339130. PMID: 28949262 and PMCID: PMC5741535 [Available on 2018-12-22].

Henry KR, Guess MB, Chole RA. Hyperthermia increases aminoglycoside ototoxicity. Acta Otolaryngol. 1983;95:323–7.

Hiel H, Erre JP, Aurousseau C, Bouali R, et al. Gentamicin uptake by cochlear hair cells precedes hearing impairment during chronic treatment. Audiology. 1993;32:78–87.

Hirose K, Discolo CM, Keasler JR, Ransohoff R. Mononuclear phagocytes migrate into the murine cochlea after acoustic trauma. J Comp Neurol. 2005;489:180–94.

Hirose K, Hartsock JJ, Johnson S, Santi P, et al. Systemic lipopolysaccharide compromises the blood-labyrinth barrier and increases entry of serum fluorescein into the perilymph. J Assoc Res Otolaryngol. 2014a;15:707–19.

Hirose K, Li SZ, Ohlemiller KK, Ransohoff RM. Systemic lipopolysaccharide induces cochlear inflammation and exacerbates the synergistic ototoxicity of kanamycin and furosemide. J Assoc Res Otolaryngol. 2014b;15:555–70.

Huang CJ, Favre I, Moczydlowski E. Permeation of large tetra-alkylammonium cations through mutant and wild-type voltage-gated sodium channels as revealed by relief of block at high voltage. J Gen Physiol. 2000;115:435–54.

Ikeda K, Sakagami M, Morizono T, Juhn SK. Permeability of the round window membrane to middle-sized molecules in purulent otitis media. Arch Otolaryngol Head Neck Surg. 1990;116:57–60.

Imamura S, Adams JC. Distribution of gentamicin in the Guinea pig inner ear after local or systemic application. J Assoc Res Otolaryngol. 2003;4:176–95.

Iwano T, Yamamoto A, Omori K, Akayama M, et al. Quantitative immunocytochemical localization of na+,k+−atpase alpha-subunit in the lateral wall of rat cochlear duct. J Histochem Cytochem. 1989;37:353–63.

Jiang M, Wang Q, Karasawa T, Koo JW, et al. Sodium-glucose transporter-2 (sglt2; slc5a2) enhances cellular uptake of aminoglycosides. PLoS One. 2014;9:e108941.

Jiang M, Johnson A, Karasawa T, Kachelmeier A et al. Role of transient receptor potential vanilloid 1 (trpv1) in the cellular uptake of aminoglycosides. ARO Midwinter Meeting Abstracts. 2015;38:PS-582.

Karasawa T, Wang Q, Fu Y, Cohen DM, et al. Trpv4 enhances the cellular uptake of aminoglycoside antibiotics. J Cell Sci. 2008;121:2871–9.

Kaur T, Hirose K, Rubel EW, Warchol ME. Macrophage recruitment and epithelial repair following hair cell injury in the mouse utricle. Front Cell Neurosci. 2015a;9:150.

Kaur T, Zamani D, Tong L, Rubel EW, et al. Fractalkine signaling regulates macrophage recruitment into the cochlea and promotes the survival of spiral ganglion neurons after selective hair cell lesion. J Neurosci. 2015b;35:15050–61.

Kawauchi H, Demaria TF, Lim DJ. Endotoxin permeability through the round window. Acta Otolaryngol Suppl. 1989;457:100–15.

Koo JW, Quintanilla-Dieck L, Jiang M, Liu J, et al. Endotoxemia-mediated inflammation potentiates aminoglycoside-induced ototoxicity. Sci Transl Med. 2015;7:298ra118.

Kroese AB, Das A, Hudspeth AJ. Blockage of the transduction channels of hair cells in the bullfrog's sacculus by aminoglycoside antibiotics. Hear Res. 1989;37:203–17.

Kruger M, Boney R, Ordoobadi AJ, Sommers TF, et al. Natural bizbenzoquinoline derivatives protect zebrafish lateral line sensory hair cells from aminoglycoside toxicity. Front Cell Neurosci. 2016;10:83.

Kushner B, Allen PD, Crane BT. Frequency and demographics of gentamicin use. Otol Neurotol. 2015;37:190–5.

Li H, Steyger PS. Systemic aminoglycosides are trafficked via endolymph into cochlear hair cells. Sci Rep. 2011;1:159.

Macarthur CJ, Hausman F, Kempton JB, Trune DR. Murine middle ear inflammation and ion homeostasis gene expression. Otol Neurotol. 2011;32:508–15.

Macarthur C, Hausman F, Kempton B, Trune DR. Intratympanic steroid treatments may improve hearing via ion homeostasis alterations and not immune suppression. Otol Neurotol. 2015;36(6):1089–95.

Marcotti W, Van Netten SM, Kros CJ. The aminoglycoside antibiotic dihydrostreptomycin rapidly enters mouse outer hair cells through the mechano-electrical transducer channels. J Physiol. 2005;567:505–21.

Marcus DC, Wu T, Wangemann P, Kofuji P. Kcnj10 (kir4.1) potassium channel knockout abolishes endocochlear potential. Am J Physiol Cell Physiol. 2002;282:C403–7.

Monzack EL, May LA, Roy S, Gale JE, et al. Live imaging the phagocytic activity of inner ear supporting cells in response to hair cell death. Cell Death Differ. 2015;22:1995–2005.

Moore RD, Smith CR, Lietman PS. Risk factors for the development of auditory toxicity in patients receiving aminoglycosides. J Infect Dis. 1984;149:23–30.

Nicol KM, Hackney CM, Evans EF, Pratt SR. Behavioural evidence for recovery of auditory function in Guinea pigs following kanamycin administration. Hear Res. 1992;61:117–31.

Nin F, Hibino H, Doi K, Suzuki T, et al. The endocochlear potential depends on two k+ diffusion potentials and an electrical barrier in the stria vascularis of the inner ear. Proc Natl Acad Sci U S A. 2008;105:1751–6.

Oh GS, Kim HJ, Choi JH, Shen A, et al. Activation of lipopolysaccharide-tlr4 signaling accelerates the ototoxic potential of cisplatin in mice. J Immunol. 2011;186:1140–50.

Oh S, Woo JI, Lim DJ, Moon SK. Erk2-dependent activation of c-Jun is required for nontypeable haemophilus influenzae-induced cxcl2 upregulation in inner ear fibrocytes. J Immunol. 2012;188:3496–505.

Oishi N, Duscha S, Boukari H, Meyer M, et al. Xbp1 mitigates aminoglycoside-induced endoplasmic reticulum stress and neuronal cell death. Cell Death Dis. 2015;6:e1763.

Perny M, Roccio M, Grandgirard D, Solyga M, et al. The severity of infection determines the localization of damage and extent of sensorineural hearing loss in experimental pneumococcal meningitis. J Neurosci. 2016;36:7740–9.

Pober JS, Sessa WC. Evolving functions of endothelial cells in inflammation. Nat Rev Immunol. 2007;7:803–15.

Quintanilla-Dieck L, Larrain B, Trune D, Steyger PS. Effect of systemic lipopolysaccharide-induced inflammation on cytokine levels in the murine cochlea: a pilot study. Otolaryngol Head Neck Surg. 2013;149:301–3.

Richardson MP, Reid A, Tarlow MJ, Rudd PT. Hearing loss during bacterial meningitis. Arch Dis Child. 1997;76:134–8.

Rizzi MD, Hirose K. Aminoglycoside ototoxicity. Curr Opin Otolaryngol Head Neck Surg. 2007;15:352–7.

Rock KL, Latz E, Ontiveros F, Kono H. The sterile inflammatory response. Annu Rev Immunol. 2010;28:321–42.

Stamatovic SM, Johnson AM, Keep RF, Andjelkovic AV. Junctional proteins of the blood-brain barrier: new insights into function and dysfunction. Tissue Barriers. 2016;4:e1154641.

Stearns GS, Keithley EM, Harris JP. Development of high endothelial venule-like characteristics in the spiral modiolar vein induced by viral labyrinthitis. Laryngoscope. 1993;103:890–8.

Stepanyan RS, Indzhykulian AA, Velez-Ortega AC, Boger ET, et al. Trpa1-mediated accumulation of aminoglycosides in mouse cochlear outer hair cells. J Assoc Res Otolaryngol. 2011;12:729–40.

Steyger PS. Is auditory synaptopathy a result of drug-induced hearing loss? Hearing J. 2017;70:8–9.

Takeuchi S, Ando M. Marginal cells of the stria vascularis of gerbils take up glucose via the facilitated transporter glut: application of autofluorescence. Hear Res. 1997;114:69–74.

Takeuchi S, Ando M. Dye-coupling of melanocytes with endothelial cells and pericytes in the cochlea of gerbils. Cell Tissue Res. 1998;293:271–5.

Takeuchi S, Ando M, Sato T, Kakigi A. Three-dimensional and ultrastructural relationships between intermediate cells and capillaries in the gerbil stria vascularis. Hear Res. 2001;155:103–12.

Takumida M, Anniko M. Localization of endotoxin in the inner ear following inoculation into the middle ear. Acta Otolaryngol. 2004;124(7):772.

Tarlow MJ, Comis SD, Osborne MP. Endotoxin induced damage to the cochlea in Guinea pigs. Arch Dis Child. 1991;66:181–4.

Tieu C, Campbell KC. Current pharmacologic otoprotective agents in or approaching clinical trials: how they elucidate mechanisms of noise-induced hearing loss. Otolaryngology. 2012;3:130.

Tran Ba Huy P, Bernard P, Schacht J. Kinetics of gentamicin uptake and release in the rat. Comparison of inner ear tissues and fluids with other organs. J Clin Invest. 1986;77:1492–500.

Trune DR, Kempton B, Hausman FA, Larrain BE, et al. Correlative mrna and protein expression of middle and inner ear inflammatory cytokines during mouse acute otitis media. Hear Res. 2015;326:49–58.

Vu AA, Nadaraja GS, Huth ME, Luk L, et al. Integrity and regeneration of mechanotransduction machinery regulate aminoglycoside entry and sensory cell death. PLoS One. 2013;8:e54794.

Wang Q, Steyger PS. Trafficking of systemic fluorescent gentamicin into the cochlea and hair cells. J Assoc Res Otolaryngol. 2009;10:205–19.

Yamane H, Nakai Y, Konishi K. Furosemide-induced alteration of drug pathway to cochlea. Acta Otolaryngol Suppl. 1988;447:28–35.

Yoshihara T, Satoh M, Yamamura Y, Itoh H, et al. Ultrastructural localization of glucose transporter 1 (glut1) in Guinea pig stria vascularis and vestibular dark cell areas: an immunogold study. Acta Otolaryngol. 1999;119:336–40.

Zager RA. Endotoxemia, renal hypoperfusion, and fever: interactive risk factors for aminoglycoside and sepsis-associated acute renal failure. Am J Kidney Dis. 1992;20:223–30.

Zhang W, Dai M, Fridberger A, Hassan A, et al. Perivascular-resident macrophage-like melanocytes in the inner ear are essential for the integrity of the intrastrial fluid-blood barrier. Proc Natl Acad Sci U S A. 2012;109:10388–93.

Zhao HB, Yu N, Fleming CR. Gap junctional hemichannel-mediated atp release and hearing controls in the inner ear. Proc Natl Acad Sci U S A. 2005;102:18724–9.

Peter Steyger, PhD, is Professor of Otolaryngology—Head & Neck Surgery at Oregon Health & Science University, and an affiliate investigator at the National Center for Rehabilitative Auditory Research, at the VA Portland Heath Care Center. Over the last 25 years, Peter has investigated cellular mechanisms of ototoxicity and more recently trafficking of ototoxins into the cochlea. His long-term goal is to improve clinical awareness and identification of ototoxicity.

Chapter 8
The Contribution of Anti-oxidant and Anti-inflammatory Functions of Adenosine A$_1$ Receptor in Mediating Otoprotection

Sandeep Sheth, Debashree Mukherjea, Leonard P. Rybak, and Vickram Ramkumar

Abstract The production of high levels of adenosine into the extracellular fluid during enhanced metabolic activity or ischemic conditions confers cytoprotection to the affected tissue. This action is mediated by adenosine receptors (ARs) which are ubiquitously expressed on the surface of cells which respond to the elevation in levels of adenosine in the extracellular fluid. While endogenous adenosine released to the extracellular fluid could confer protection under normal physiological condition, exogenously administered adenosine analogs are required to boost the protective capacity of these receptors under stress conditions. This chapter provides a summary of the adenosine/AR system in the cochlea and shows that the adenosine A$_1$ receptor (A$_1$AR) could protect against hearing loss by inhibiting cochlear oxidative stress, inflammation and apoptosis of cochlear cells.

Keywords Adenosine A$_1$ receptor · Hearing loss · Oxidative stress · Cochlear inflammation · Otoprotection

S. Sheth (✉) · V. Ramkumar
Department of Pharmacology, Southern Illinois University School of Medicine, Springfield, IL, USA
e-mail: ssheth@siumed.edu

D. Mukherjea · L. P. Rybak
Department of Surgery (Otolaryngology), Southern Illinois University School of Medicine, Springfield, IL, USA

1   Adenosine and Adenosine Receptors

Adenosine Production and Metabolism

Adenosine is a purine nucleoside composed of adenine base attached to a ribose molecule which is ubiquitously present in the body. The primary source of adenosine, extracellularly and intracellularly, is adenosine triphosphate (ATP). Tissues with high metabolic rates, such as the cochlea exposed to noise or hypoxia (Muñoz et al. 2001), have high turnover of ATP and accumulate high concentrations of adenosine. During noise exposures, ATP is released from cells of the organ of Corti and marginal cells of the stria vascularis. The passage of ATP to the extracellular fluid space is mediated by connexin/pannexin hemichannels (Zhao et al. 2005). The extracellular metabolism of ATP to adenosine is tightly regulated by a group of cell surface-located enzymes called as ectonucleotidases. Specifically, members of the ecto-nucleoside triphosphate diphosphohydrolase (E-NTPDase/CD39) family, NTPDase 1, 2, 3, and 8, hydrolyze extracellular nucleoside tri-and diphosphates (ATP and ADP) into AMP. It was reported that exposure to noise up-regulates the expression and activity of NTPDase 1, 2 and 3 enzymes in the cochlea which results in increased hydrolysis of extracellular ATP and accumulation of ADP and AMP (Vlajkovic et al. 2004, 2006). The final breakdown of AMP to adenosine is catalyzed by ecto-5′-nucleotidase (CD73) (Robson et al. 2006). Intracellularly, ATP is broken down into AMP by ATPase and adenylate kinase. AMP is then metabolized by cytosolic-5′-nucleotidase to produce adenosine. Adenosine is also produced intracellularly from the hydrolysis of S-adenosylhomocysteine (SAH) by SAH hydrolase. However, adenosine generated from SAH contributes minimally towards total adenosine produced intracellularly. The metabolism of adenosine is regulated by two enzymes: (a) adenosine deaminase which deaminates adenosine to form inosine and hypoxanthine and (b) adenosine kinase which phosphorylates adenosine to AMP. AMP is subsequently converted back to ADP and ATP (for review see Adair 2005). These enzymes are essential for maintaining adenosine homeostasis by lowering its concentration. Inhibition of either of these enzymes would increase the availability of adenosine in the extracellular space. A study investigating the role of adenosine kinase in the cochlea reported that inhibition of this enzyme enhanced adenosine signaling and delayed the onset of age-related hearing loss (Vlajkovic et al. 2011).

Adenosine Transporters

Extracellular adenosine concentration is dependent on nucleoside transporters which facilitate its release and reuptake across the cell membrane. There are two main classes of nucleoside transporters, namely equilibrative nucleoside transporters (ENT) and concentrative nucleoside transporters (CNT). The ENT family (ENT1–4) is bidirectional and passive transporters which can transport adenosine through facilitated diffusion in either direction depending on its concentration gradient and does not require any energy. On the other hand, CNTs are unidirectional,

Na^+-dependent transporters which require energy from Na^+/K^+-ATPase to transport adenosine against its concentration gradient. Investigation into the expression of nucleoside transporters and adenosine uptake in rat cochlea revealed the presence of mRNA transcripts for 2 equilibrative transporters (ENT1 and ENT2) and two concentrative transporters (CNT1 and CNT2) (Khan et al. 2007). Under physiological conditions the extracellular levels of adenosine are relatively low (in the nanomolar range). However, when cells are under stress, adenosine levels can increase significantly and reach as high as 100 μm (Fredholm 2010). Therefore, stress signals differentially modulates the expression and activity of enzymes and transporters that are involved in the production and metabolism of adenosine in order to increase the availability of adenosine in the extracellular milieu (Kobayashi et al. 2000). As such, drugs which target these transporters to enhance extracellular adenosine could provide otoprotection.

Adenosine Receptors

Higher adenosine concentrations in the extracellular fluids are important to activate the adenosine receptors (ARs), which are G protein-coupled receptors. Four distinct subtypes of these receptors, namely the A_1, A_{2A}, A_{2B} and A_3ARs, have been cloned and characterized. The A_1 and A_3ARs are coupled to $G_{i/o}$ proteins that inhibit adenylyl cyclase, leading to reduction in the intracellular levels of cyclic adenosine monophosphate (cAMP). In contrast, the A_{2A} and A_{2B}ARs couple to G_s proteins which stimulate adenylyl cyclase and increase cAMP levels. ARs can also initiate other signaling pathways such as phospholipase C (PLC), protein kinase B (PKB), intracellular Ca^{2+} release and mitogen-activated protein kinases (MAPK) (Jacobson and Gao 2006; Merighi et al. 2003). Although ARs are ubiquitously present, each AR subtype show tissue specific distribution and function. With respect to the auditory system, ARs has been implicated in the regulation of cochlear blood flow (Muñoz et al. 1999) and protection of cochlear cells against oxidative damage (Hu et al. 1997; Ramkumar et al. 2004). All four subtypes of ARs are differentially expressed in the outer and inner hair cells of the organ of Corti, spiral ganglion neurons, lateral wall tissues and cochlear blood vessels of the rat cochlea (Vlajkovic et al. 2007; Kaur et al. 2016).

2   Hearing Loss Induced by Oxidative Stress

ROS Production as a Primary Mechanism for Ototoxicity

Reactive oxygen species (ROS) generation in the cochlea has been considered a critical event that initiates damage to the outer hair cells (OHC), spiral ganglion neurons and cells of stria vascularis and spiral ligament, leading to hearing loss (Rybak et al. 2009; Henderson et al. 2006). It is well established that hair cells exposed to noise (Yamane et al. 1995) or ototoxic drugs, such as cisplatin (Ford et al. 1997b) and aminoglycosides (Choung et al. 2009), generate excessive amounts of ROS as a result of increased metabolic activity (Kopke et al. 1999). ROS also play a major role in the degeneration of these cochlear cells during aging (Someya et al. 2009). NOX3, a specific isoform of NADPH oxidase, was found to be responsible for ROS generation in the cochlea (Bánfi et al. 2004). This enzyme is activated and induced by cisplatin in the rat cochlea, as knockdown of NOX3 by pretreatment with NOX3 siRNA abrogated cisplatin ototoxicity (Mukherjea et al. 2010). Interestingly, NOX3 is also activated by cisplatin in ROS-dependent manner, as scavenging of ROS by antioxidants reduced cisplatin-induced expression of NOX3 (Mukherjea et al. 2008).

Increased ROS interferes with the physiological function of the cell by reacting with DNA, proteins, membrane lipids, cytosolic molecules, cell surface receptors and antioxidant enzymes resulting in the initiation of mitochondrial apoptotic process (Kopke et al. 1999; Wong and Ryan 2015). ROS induce lipid peroxidation in the cell membrane of cochlear tissue through formation of malondialdehyde (MDA) (Rybak et al. 2000) and 4-hydroxynonenal (4-HNE) (Yamashita et al. 2004). Increased levels of 4-HNE is associated with excessive Ca^{2+} influx and apoptosis in OHCs (Ikeda et al. 1993; Clerici et al. 1995). Moreover, superoxide ions (O_2^-) interacts with nitric oxide to form peroxynitrite radical, which reacts with and inactivates cell membrane proteins to form 3-nitrotyrosine (3-NT) (Lee et al. 2004a, b). MDA (Kaygusuz et al. 2001), 4-HNE and 3-NT (Yamashita et al. 2004, 2005; Jiang et al. 2007) are frequently used as markers for oxidative stress and free radical damage in the cochlea. Overall, there is overwhelming evidence suggesting that ROS production and accumulation in the cochlear tissue is the key pathological mechanism responsible for hearing impairment (Kopke et al. 1999; Rybak et al. 2009).

Cochlear Antioxidant Defense System

The harmful oxidative effects of ROS are neutralized by an endogenous antioxidant defense system present in the cochlea. This defense system includes antioxidant molecules, such as reduced glutathione (GSH) and enzymatic scavengers, such as superoxide dismutase (SOD), catalase, glutathione peroxidase (GSHPx), glutathione reductase (GSHRed) and glutathione S-transferase (GST). SOD converts superoxide (O_2^-) into hydrogen peroxide (H_2O_2) and oxygen (O_2), while catalase and GSHPx converts hydrogen peroxide into water (H_2O) and oxygen. GSHRed reduces the oxidized GSH (GSSG) to its reduced form (GSH) which acts as free-radical scavenger. GST catalyzes the conjugation of GSH to xenobiotics (e.g., cisplatin and aminoglycosides) to detoxify them, thus playing an important role defending the cochlea from damage caused by ototoxic agents (Kopke et al. 1999). Despite having an elaborate antioxidant defense system, the cochlea is sensitive to excessive ROS production which could overwhelms its antioxidant defenses, resulting in redox imbalance which can trigger cell death. For instance, depletion of GSH levels makes the cochlea more susceptible to noise- (Yamasoba et al. 1998), cisplatin- (Kopke et al. 1997) and aminoglyco-side-induced hearing loss (Lautermann et al. 1995), and supplementing GSH led to significantly reduced severity of hearing loss. Thus, intervention to either prevent excessive production of ROS or initiate a repair mechanism may inhibit the oxidative damage to the sensory cells of the cochlea and restore their function.

3 Role of A_1AR in Otoprotection

Cochlear Expression and Regulation of A_1AR

Immunohistochemical studies in the rat cochlea suggest that A_1ARs are localized in the spiral ganglion neurons and the cells of the organ of Corti, namely the inner hair cells (IHCs), OHCs and the supporting Deiter's cells (Vlajkovic et al. 2007; Kaur et al. 2016). In the organ of Corti, A_1AR immunoreactivity is higher in IHCs than OHCs (Fig. 8.1) (Kaur et al. 2016). Interestingly, the IHCs are more resistant than the OHCs to ototoxic damage caused by noise (Chen and Fechter 2003) or cisplatin (Nakai et al. 1982; Kaur et al. 2016). Therefore, it could be speculated that the over-expression of A_1AR in IHCs contributes to their resistance against ototoxic insults. However, a recent study suggests that this is not the case, as inhibition of these receptors by a selective antagonist did not lead to killing of IHCs (Kaur et al. 2016).

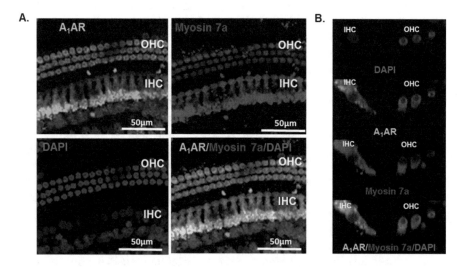

Fig. 8.1 A_1AR is expressed in the rat hair cells. (**a**) Whole-mount preparation obtained from adult rat shows A_1AR immunoreactivity (green) on OHCs and IHCs. A_1AR staining is co-localized with myosin 7a (red), a marker of hair cells. Cell nuclei is stained with DAPI (blue). (**b**) Orthogonal sections of images in (**a**) shows A_1AR immunoreactivity in IHCs and OHCs with a significant amount of co-localization with myosin 7a. This figure is adapted with permission from Kaur et al. (2016)

Radioligand binding studies performed in chinchilla cochlea showed ~5 fold increase in A_1AR levels within 24 h of cisplatin treatment (Ford et al. 1997a). In another study, cisplatin was shown to induce ~2 fold increase in the mRNA levels of A_1AR in rat cochlea (Kaur et al. 2016). Moderate noise exposure (10 dB SPL) also increased A_1AR transcript levels by two folds in rat cochlea (Wong et al. 2010), which is consistent with a previous study demonstrating increased A_1AR radioligand binding in the chinchilla cochlea after exposure to moderate narrow band noise (Ramkumar et al. 2004). This increase in A_1AR expression may represent a unique compensatory cytoprotective mechanism by the cochlea to counter the toxic effects of ROS produced as a result of ototoxic insults. ROS can regulate the transcription of A_1AR gene through activation of transcription factor such as nuclear factor (NF)-κB (Nie et al. 1998). Increased oxidative stress induced by noise exposure in chinchilla cochlea upregulated A_1AR through activation of NF-κB (Ramkumar et al. 2004). Another mechanism for A_1AR regulation involves nitric oxide, which is found to endogenously regulate A_1AR expression also through the activation of NF-κB (Jhaveri et al. 2006). Such feedback regulation of A_1AR could enhance its cytoprotective activity in response to oxidative stress.

A_1AR Stimulates Cochlear Antioxidant Defense System

Activation of A_1AR has been reported to stimulate antioxidant defenses by increasing the activity of antioxidant enzymes in the cochlea. Local application of A_1AR specific agonist, R-phenylisopropyladenosine (R-PIA), to the round window of chinchilla cochlea exhibited significant increase in the activities of antioxidant enzymes such as SOD and GSHPx and reduced the levels of MDA, a marker of lipid peroxidation and cell damage (Ford et al. 1997b). In organotypic explants from P-3 rat organ of Corti, R-PIA significantly increased the levels of catalase, gamma glutamylcysteine synthetase (an enzyme involved in GSH production) and GST (Kopke et al. 1997), confirming the antioxidant function of A_1AR in the cochlea. A_1AR agonists, R-PIA and 2-chloro-N^6-cyclopentyl-adenosine (CCPA), protected against cisplatin-induced hair cell damage (Whitworth et al. 2004). This effect of A_1AR was associated with a reduction in lipid peroxidation induced by cisplatin. Additionally, co-administration of the A_1AR antagonist, 8-cyclopentyl-1,3-dipropylxanthine (DPCPX), completely reversed the protective effects of R-PIA, which strongly supports the otoprotective role of A_1AR. By contrast, pre-treatment with $A_{2A}AR$ agonist significantly increased cisplatin-induced toxicity (Whitworth et al. 2004). In animal model of noise-induced hearing loss, round window application of R-PIA prior to noise exposure resulted in less hearing and hair cell loss in treated ears than in the untreated ears (Hu et al. 1997). In another study, application of CCPA to the round window membrane of the rat cochlea 6 h after noise exposure provided significant recovery in hearing thresholds, which was associated with reduced free radical generation particularly in Dieter's cells and inner sulcus cells of the

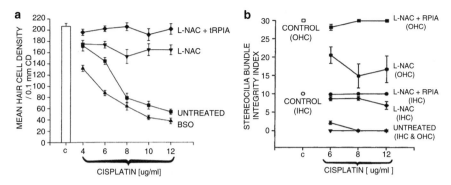

Fig. 8.2 Otoprotective effect of L-NAC and L-NAC + *R*-PIA in cisplatin exposed P-3 rat organ of Corti explants. (**a**) Mean hair cell density per 0.1 mm of cochlear duct length (**a**) and stereocilia bundle integrity index (**b**) were determined after explants were exposed to either increasing doses of cisplatin only (untreated), cisplatin + buthionine sulfoximine (BSO) (5 μmol/L), cisplatin + L-NAC (10^{-2} mol/L) or Cisplatin + L-NAC + *R*-PIA (100 μmol/L). The explants were pre-treated with BSO, L-NAC and L-NAC + *R*-PIA for 24 h after which there were exposed to cisplatin for another 48 h. Control explants (C) are explants not exposed to cisplatin. (**a**) BSO treatment increased the damage to hair cells caused by cisplatin. In contrast, L-NAC protected the hair cells from cisplatin toxicity and its combination with *R*-PIA resulted in total protection of hair cells at all csplatin concentrations tested. (**b**) Cisplatin eliminated almost all IHC and OHC stereocilia bundle integrity. Treatment with L-NAC and L-NAC + *R*-PIA equally preserved IHCs stereocilia integrity. L-NAC alone significantly enhanced OHC bundle integrity, but its combination with *R*-PIA brought the OHC stereocilia bundle integrity index to control levels. Error bars indicate SEM ± mean. This figure is adapted with permission form Kopke et al. (1997)

cochlea (Wong et al. 2010). These studies have identified A_1AR as a potential target for pharmacological interventions to reduce ototoxicity resulting from oxidative stress.

Treatment of A_1AR agonist in combination with different antioxidants has also shown promising results in ameliorating hearing loss. The combination of *R*-PIA with an esterified GSH analogue, GSH monoethylester, provided protection against both impulse and continuous noise-induced hearing loss in chinchilla cochlea (Hight et al. 2003). The otoprotection afforded by this combination therapy was due to increased levels of GSH in the perilymph, offsetting the oxidative stress induced by noise. Kopke et al. (1997) studied the combination of L-N-acetyl cysteine (L-NAC), a potential source of cellular glutathione, with *R*-PIA in cisplatin-treated organotypic explants from P-3 rat organ of Corti. The mean hair cell density of explants exposed to L-NAC + *R*-PIA was preserved at the concentrations of cisplatin tested (4–12 μg/mL) (Fig 8.2a). The protection produced by this combination was significantly greater than L-NAC alone, suggesting that *R-PIA* augments the otoprotective actions of antioxidant molecules presumably by increasing GSH levels. Interestingly, depleting GSH levels using buthionine sulfoximine (BSO), an inhibitor of GSH synthesis, sensitized hair cells to the damage induced by cisplatin, which confirms the importance of antioxidants in protecting the cochlea against oxidative damage. The combination of L-NAC + *R*-PIA was also effective in maintaining the integrity of IHC and OHC stereocilia, which were severely damaged by cisplatin (Fig. 8.2b). Addition of R-PIA amplified the protective effect of L-NAC so that complete preservation of IHC and OHC stereocilia was achieved even at the highest concentration of cisplatin (Kopke et al. 1997). These findings support the otoprotective role of A_1AR mediated in part by enhancing the cochlear antioxidant defense mechanisms.

Inhibition of A_1AR Sensitizes the Cochlea to Hearing Loss

Caffeine, the active ingredient in coffee, is an antagonist of ARs, including the A_1AR. Based on this information, we predicted that caffeine contribute to auditory deficits when its consumption is paired with loud sounds or other forms of acoustic trauma. Caffeine administration to guinea pig (120 mg/kg/day) by the intraperitoneal route aggravated noise-induced (120 dB for 1 h at 6 kHz) hearing loss by day 10. This was associated with a significant loss of spiral ganglion neurons in the middle and apical turns of the cochlea, but had no effect on OHC morphology (Mujica-Mota et al. 2014). In another study, it was shown that patients diagnosed with tinnitus reported reduced perception of tinnitus when they reduced their consumption of coffee over a 30 day period. Greater reductions in tinnitus perception were observed in patients who were previously consuming higher amounts of caffeine (Figueiredo et al. 2014). However, an earlier study failed to observe a beneficial effect of caffeine abstinence as a therapy for tinnitus (Claire et al. 2010). A recent study in rats showed that caffeine consumption exacerbated cisplatin-induced hearing loss, without significant altering loss of OHCs. However, these investigators demonstrated a significant loss of IHC synapses greater that that observed for cisplatin alone (Sheth et al. 2017). Based on these data, it appears that caffeine consumption should be reduced or avoided in patients who have some pre-existing hearing deficits or who are being treated with ototoxic drugs.

As mentioned above, otoprotection was mediated primarily by activation of A_1AR, as $A_{2A}AR$ activation exacerbated cisplatin ototoxicity (Whitworth et al. 2004). An interesting finding of this study was that inhibition of the A_1AR by a selective antagonist enhanced cisplatin-induced ototoxicity (Whitworth et al. 2004), suggesting an endogenous protective role of adenosine. However, this endogenous protective system could be overwhelmed by the intense toxicity produced by cisplatin. In such a case exogenously administered drug is needed to supplement the endogenous system. A recent study indicated that mice which are deficient in A_1AR demonstrate high frequency hearing loss (as determined by lower amplitudes of waves I and II) under ambient noise condition and are more sensitive to noise than their wild type or $A_{2A}AR^{-/-}$ counterparts (Vlajkovic et al. 2017). Moreover, these mice demonstrate reduced basal levels of ribbon synapses. These data support the conclusion that the A_1AR confer a tonic protective role against cochlear trauma.

4   Adenosine and Cochlear Inflammation

Adenosine Decreases Inflammatory Markers in Resident Cells of the Cochlea

Recent studies have indicated a role ROS in mediating hearing loss through the induction of cochlear inflammation (Mukherjea et al. 2011). The role of inflammation in sensorineural hearing loss is supported by the fact that middle ear infection (otitis media) (Paparella et al. 1972) and labyrinthitis (Merchant and Gopen 1996) are usually associated with hearing loss. Noise trauma induces an inflammatory response in the inner ear (Fujioka et al. 2006). Cisplatin-induced ROS generation is reported to be a main contributor to cochlear inflammation and apoptosis of cells in the cochlea (Mukherjea et al. 2011; Kaur et al. 2011). Cisplatin-induced ROS initiate the inflammatory process and hearing loss by activating the signal transducer and activator of transcription 1 (STAT1) transcription factor (Kaur et al. 2011) and possibly NF-κB (Ramkumar et al. 2004). STAT1 is shown to couple the activation of transient receptor potential vanilloid receptor (TRPV)-1 via NOX3 NADPH oxidase to the induction of inflammation in the cochlea (Mukherjea et al. 2011). Accordingly, localized knockdown of STAT1 by siRNA protects against cisplatin-induced hearing loss and damage to OHCs in rats by reducing inflammatory mediators, such as tumor necrosis factor-α (TNF-α) (Kaur et al. 2011). Thus, targeting STAT1-dependent inflammation could serve as a useful approach to treat ototoxicity. To this end, recent finding from our laboratory has demonstrated that transtympanic administration of the A_1AR agonist, *R*-PIA, protected from cisplatin ototoxicity by suppressing an inflammatory response initiated by ROS generation via NOX3 NADPH oxidase, leading to inhibition of STAT1 (Fig. 8.3). *R*-PIA also decreased the expression of STAT1 target genes, such as TNF-α, inducible nitric oxide synthase (iNOS) and cyclooxygenase-2 (COX-2) and reduced cisplatin-mediated apoptosis (Kaur et al. 2016).

It is not clear how activation of A_1AR mediates inhibition of ROS and inflammation. One possibility is that A_1AR activation contributes to activation of antioxidant enzymes as described previously (Ford et al. 1997a). In addition, we show that more prolonged exposure to A_1AR agonist could also suppress the expression of *NOX3* (Kaur et al. 2016), a major source of ROS generation in the cochlea. Thus, activation of the A_1AR could produce an acute reduction in ROS by stimulating ROS scavenging and long term reduction in ROS by suppressing NOX3 expression. An important insight gleaned from this study is that ROS could serve as proinflammatory molecules in the cochlea through their activation of STAT1 (Fig. 8.4).

A clearer link between oxidative stress, inflammation and hearing loss has recently been described (Kaur et al. 2011, 2016). These investigators showed that cisplatin-induced activation of STAT1 is dependent on activation of the mitogen activated protein kinase (MAPK) pathway. This pathway represents the first step in the cascade which culminates into death of cochlear cells and hearing loss. We propose that activation of the A_1AR inhibits the MAPK activation thereby suppressing

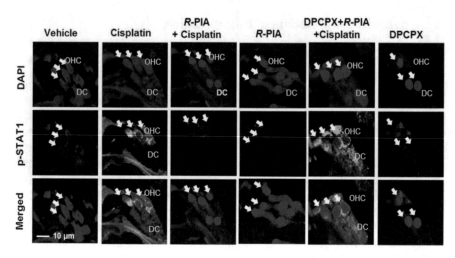

Fig. 8.3 *R*-PIA inhibits cisplatin-induced activation of STAT1 in the rat cochlea. Immunolabeling studies for Ser[727] p-STAT1 in OHCs was performed on the cochlear sections isolated from the rats treated with vehicle or cisplatin (11 mg/kg i.p.) for 72 h following trans-tympanic administration of *R*-PIA (1 μM). Cisplatin increased Ser[727] p-STAT1 immunoreactivity in the OHCs, which was reduced by pretreatment with *R*-PIA. Vehicle, *R*-PIA, and DPCPX alone showed low baseline Ser[727] p-STAT1 immunoreactivity. The arrow indicates three rows of OHCs and DC represents Deiters cells. This figure is adapted with permission from Kaur et al. (2016)

STAT1 activation and inhibition of downstream events. STAT1 appears to play a crucial role in promoting p53 activation following DNA damage (Townsend et al. 2004) and we have shown that cochlear-derived cells in which STAT1 was inhibited were relatively resistant to cisplatin (Kaur et al. 2011). Thus, STAT1 represents an ideal target of drugs, such as A_1AR agonists (Kaur et al. 2016) and epigallocatechin gallate (Borse et al. 2017), which provide otoprotection.

Does Adenosine Blocks Activation and Recruitment of Circulating Immune Cells to the Cochlea?

The above discussion focuses on examining the expression of molecules in the inflammatory pathways and cytokines in cells which are resident to the cochlea. However, the impact of these cells to the overall inflammatory response in the cochlea is unclear at present. Based on studies which showed that round window or trans-tympanic administration of protective agents provide sufficient protection against hearing loss, we speculate that modulating the local immune response in the cochlea by drugs is an adequate otoprotective strategy. This observation begs to question the role of the circulating inflammatory cells in mediating cochlear damage, resolution of cochlear inflammation and hearing loss. As indicated below,

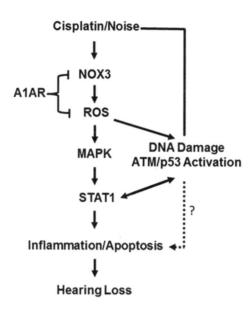

Fig. 8.4 Proposed mechanism underlying the otoprotective action of A_1AR agonists. Oxidative stress produced by cisplatin or noise increases the activity and expression of NOX3 which promotes ROS generation and oxidative stress. ROS stimulate MAPK activation, followed by STAT1 activation. ROS could also increase DNA strand breaks and activation of the ATM/p53 cascade. Full activation of STAT1, apparently through a second signal mediated by DNA damage (activation of ATM and p53), promotes inflammation and apoptosis of cochlear cells and hearing loss. It is also possible that DNA damage response could affect the inflammatory and apoptosis pathway independent of STAT1. Activation of A_1AR can inhibit NOX3 expression to reduce ROS production. Additionally, A_1AR can directly reduce ROS by activating the antioxidant defence system in the cochlea

circulating immune cells demonstrate high expression of different ARs which suppress inflammatory processes.

The anti-inflammatory role of adenosine on established immune cells in the circulation has been well described. Adenosine can significantly modulate the extent of the initial inflammatory response. For example, adenosine inhibits the recruitment and activation of neutrophils to the site of injury through activation of the $A_{2A}AR$. This is mediated through inhibition of leukocyte recruitment, neutrophil adhesion and damage to the vascular endothelium (Cronstein et al. 1986). Adenosine can also reduce adhesion of neutrophils to the endothelium mediated by both selectin and integrin (Cronstein et al. 1992). The source of adenosine is likely ATP released by neutrophils and metabolized by ectonucleotidases present on these cells. This could inhibit further recruitment of leukocytes by activation of the A_3AR (Chen et al. 2006). Adenosine can also decrease the production of oxygen free radicals and other mediators in neutrophils by activating the $A_{2A}AR$ subtype (Taylor et al. 2005).

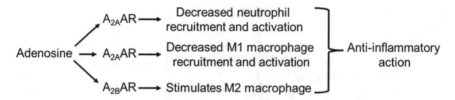

Fig. 8.5 Potential targets for mediating the anti-inflammatory actions of adenosine in the cochlea. Adenosine acts on $A_{2A}AR$ on neutrophils to decrease its recruitment and activation. Its actions on this receptor on M1 macrophages decrease recruitment and activation of these immune cells. Furthermore, activation of $A_{2B}AR$ stimulates M2 macrophages and promotes resolution of the inflammation

Macrophages represent another potential target for mediating the anti-inflammatory action of adenosine in the cochlea. For example, activation of the $A_{2A}ARs$ on pro-inflammatory M1 macrophages suppresses the synthesis of chemokines and cytokines, such as TNF-α, interleukin (IL)-1β, IL-6, IL-12 and macrophage inhibitory peptide (MIP)-1α (Mosser and Edwards 2008). Activation of the $A_{2A}AR$ could also suppress the release of cytotoxic mediators such as oxygen and nitrogen free radicals from activated M1 macrophages (Haskó and Cronstein 2013) and increase the release of the anti-inflammatory cytokine, IL-10 (Haskó et al. 2007). M2 macrophages mediate anti-inflammatory functions and are activated by Th2 cytokines, such as IL-4 and IL-13. These cells are characterized by markers including arginase-1, tissue inhibitor of metalloproteinases (TIMP-1), macrophage-galactose type C-type lectin (mgl-1), IL-4 receptor and resistin-like molecules. Adenosine promotes IL-4 and IL-13-dependent M2 macrophage activation, as indicated by upregulation of its characteristic markers (Ferrante et al. 2013). This action of adenosine is mediated via the $A_{2B}AR$. Overall, these studies suggest that multiple ARs present on immune cells contribute to the anti-inflammatory role of adenosine (Fig. 8.5). The existence of these receptors on circulating immune cells would suggest that they would contribute this anti-inflammatory function if and when these cells are recruited to the cochlea in response to cochlear damage. Nevertheless, the contribution of these mechanisms regulated by the A_{2A} and $A_{2B}R$ in otoprotection is unclear at present, given the observation that activation of A_1AR alone could protect against cisplatin and noise-induced hearing loss.

5   Conclusion

This study highlights the protective role of activation of the A_1AR in the cochlea against cochlear trauma which produces hearing loss. It also highlights the role of exogenously applied adenosine analogs in mediating otoprotection, as the endogenous A_1AR signaling system could be overwhelmed under certain circumstances. Therefore, localized administration of selective A_1AR agonists by the trans-tympanic or round window routes could serve as useful strategy to treat hearing loss, especially that produced by drugs such as cisplatin. Our current understanding is that these drugs target A_1AR on resident cochlear cells where they modulate oxidative stress and/or inflammatory pathways, leading to a reduction in cell apoptosis. Less clear is the contribution of circulating immune cells to the hearing loss induced by cochlear trauma and the role(s) the A_1AR and other ARs subtypes (such as A_2AR) play in modulating the recruitment of these cells to the cochlea. Future studies in this area should shed more light as to the relative roles of the different AR subtypes in otoprotection.

Acknowledgement The authors would like to acknowledge NIH grant support: NCI RO1 CA166907 to VR, NIDCD RO1-DC 002396 to LPR and RO3 DC011621 to DM, and a grant from the American Hearing Research Foundation to SS.

References

Adair TH. Growth regulation of the vascular system: an emerging role for adenosine. Am J Physiol Regul Integr Comp Physiol. 2005;289(2):R283–96.

Bánfi B, Malgrange B, Knisz J, Steger K, Dubois-Dauphin M, Krause KH. NOX3, a superoxide-generating NADPH oxidase of the inner ear. J Biol Chem. 2004;279(44):46065–72.

Borse V, Al Aameri RFH, Sheehan K, Sheth S, Kaur T, Mukherjea D, Tupal S, Lowy M, Ghosh S, Dhukhwa A, Bhatta P, Rybak LP, Ramkumar V. Epigallocatechin-3-gallate, a prototypic chemopreventative agent for protection against cisplatin-based ototoxicity. Cell Death Dis. 2017;8(7):e2921.

Chen GD, Fechter LD. The relationship between noise-induced hearing loss and hair cell loss in rats. Hear Res. 2003;177(1–2):81–90.

Chen Y, Corriden R, Inoue Y, Yip L, Hashiguchi N, Zinkernagel A, Nizet V, Insel PA, Junger WG. ATP release guides neutrophil chemotaxis via P2Y2 and A3 receptors. Science. 2006;314(5806):1792–5.

Choung YH, Taura A, Pak K, Choi SJ, Masuda M, Ryan AF. Generation of highly-reactive oxygen species is closely related to hair cell damage in rat organ of Corti treated with gentamicin. Neuroscience. 2009;161(1):214–26.

Claire LS, Stothart G, McKenna L, Rogers PJ. Caffeine abstinence: an ineffective and potentially distressing tinnitus therapy. Int J Audiol. 2010;49(1):24–9.

Clerici WJ, DiMartino DL, Prasad MR. Direct effects of reactive oxygen species on cochlear outer hair cell shape in vitro. Hear Res. 1995;84(1–2):30–40.

Cronstein BN, Levin RI, Belanoff J, Weissmann G, Hirschhorn R. Adenosine: an endogenous inhibitor of neutrophil-mediated injury to endothelial cells. J Clin Invest. 1986;78(3):760–70.

Cronstein BN, Levin RI, Philips M, Hirschhorn R, Abramson SB, Weissmann G. Neutrophil adherence to endothelium is enhanced via adenosine A1 receptors and inhibited via adenosine A2 receptors. J Immunol. 1992;148(7):2201–6.

Ferrante CJ, Pinhal-Enfield G, Elson G, Cronstein BN, Hasko G, Outram S, Leibovich SJ. The adenosine-dependent angiogenic switch of macrophages to an M2-like phenotype is independent of interleukin-4 receptor alpha (IL-4Rα) signaling. Inflammation. 2013;36(4):921–31.

Figueiredo RR, Rates MJ, Azevedo AA, Moreira RK, Penido Nde O. Effects of the reduction of caffeine consumption on tinnitus perception. Braz J Otorhinolaryngol. 2014;80(5):416–21.

Ford MS, Maggirwar SB, Rybak LP, Whitworth C, Ramkumar V. Expression and function of adenosine receptors in the chinchilla cochlea. Hear Res. 1997a;105(1–2):130–40.

Ford MS, Nie Z, Whitworth C, Rybak LP, Ramkumar V. Up-regulation of adenosine receptors in the cochlea by cisplatin. Hear Res. 1997b;111(1–2):143–52.

Fredholm BB. Adenosine receptors as drug targets. Exp Cell Res. 2010;316(8):1284–8.

Fujioka M, Kanzaki S, Okano HJ, Masuda M, Ogawa K, Okano H. Proinflammatory cytokines expression in noise-induced damaged cochlea. J Neurosci Res. 2006;83(4):575–83.

Haskó G, Cronstein B. Regulation of inflammation by adenosine. Front Immunol 2013;4

Haskó G, Pacher P, Deitch EA, Vizi ES. Shaping of monocyte and macrophage function by adenosine receptors. Pharmacol Ther. 2007;113(2):264–75.

Henderson D, Bielefeld EC, Harris KC, Hu BH. The role of oxidative stress in noise-induced hearing loss. Ear Hear. 2006;27(1):1–19.

Hight NG, McFadden SL, Henderson D, Burkard RF, Nicotera T. Noise-induced hearing loss in chinchillas pre-treated with glutathione monoethylester and R-PIA. Hear Res. 2003;179(1–2):21–32.

Hu BH, Zheng XY, McFadden SL, Kopke RD, Henderson D. R-phenylisopropyladenosine attenuates noise-induced hearing loss in the chinchilla. Hear Res. 1997;113(1–2):198–206.

Ikeda K, Sunose H, Takasaka T. Effects of free radicals on the intracellular calcium concentration in the isolated outer hair cell of the guinea pig cochlea. Acta Otolaryngol. 1993;113(2):137–41.

Jacobson KA, Gao ZG. Adenosine receptors as therapeutic targets. Nat Rev Drug Discov. 2006;5(3):247–64.

Jhaveri KA, Toth LA, Sekino Y, Ramkumar V. Nitric oxide serves as an endogenous regulator of neuronal adenosine A1 receptor expression. J Neurochem. 2006;99(1):42–53.

Jiang H, Talaska AE, Schacht J, Sha SH. Oxidative imbalance in the aging inner ear. Neurobiol Aging. 2007;28(10):1605–12.

Kaur T, Borse V, Sheth S, Sheehan K, Ghosh S, Tupal S, Jajoo S, Mukherjea D, Rybak LP, Ramkumar V. Adenosine A1 receptor protects against cisplatin ototoxicity by suppressing the NOX3/STAT1 inflammatory pathway in the cochlea. J Neurosci. 2016;36(14):3962–77.

Kaur T, Mukherjea D, Sheehan K, Jajoo S, Rybak LP, Ramkumar V. Short interfering RNA against STAT1 attenuates cisplatin-induced ototoxicity in the rat by suppressing inflammation. Cell Death Dis. 2011;2:e180.

Kaygusuz I, Oztürk A, Ustündağ B, Yalçin S. Role of free oxygen radicals in noise-related hearing impairment. Hear Res. 2001;162(1–2):43–7.

Khan AF, Thorne PR, Muñoz DJ, Wang CJ, Housley GD, Vlajkovic SM. Nucleoside transporter expression and adenosine uptake in the rat cochlea. Neuroreport. 2007;18(3):235–9.

Kobayashi S, Zimmermann H, Millhorn DE. Chronic hypoxia enhances adenosine release in rat PC12 cells by altering adenosine metabolism and membrane transport. J Neurochem. 2000;74(2):621–32.

Kopke R, Allen KA, Henderson D, Hoffer M, Frenz D, Van de Water T. A radical demise. Toxins and trauma share common pathways in hair cell death. Ann N Y Acad Sci. 1999;884:171–91.

Kopke RD, Liu W, Gabaizadeh R, Jacono A, Feghali J, Spray D, Garcia P, Steinman H, Malgrange B, Ruben RJ, Rybak L, Van de Water TR. Use of organotypic cultures of Corti's organ to study the protective effects of antioxidant molecules on cisplatin-induced damage of auditory hair cells. Am J Otol. 1997;18(5):559–71.

Lautermann J, McLaren J, Schacht J. Glutathione protection against gentamicin ototoxicity depends on nutritional status. Hear Res. 1995;86:15–24.

Lee JE, Nakagawa T, Kita T, Kim TS, Iguchi F, Endo T, Shiga A, Lee SH, Ito J. Mechanisms of apoptosis induced by cisplatin in marginal cells in mouse stria vascularis. ORL J Otorhinolaryngol Relat Spec. 2004a;66(3):111–8.

Lee JE, Nakagawa T, Kim TS, Endo T, Shiga A, Iguchi F, Lee SH, Ito J. Role of reactive radicals in degeneration of the auditory system of mice following cisplatin treatment. Acta Otolaryngol. 2004b;124(10):1131–5.

Merchant SN, Gopen Q. A human temporal bone study of acute bacterial meningogenic labyrinthitis. Am J Otol. 1996;17(3):375–85.

Merighi S, Mirandola P, Varani K, Gessi S, Leung E, Baraldi PG, Tabrizi MA, Borea PA. A glance at adenosine receptors: novel target for antitumor therapy. Pharmacol Ther. 2003;100(1):31–48.

Mosser DM, Edwards JP. Exploring the full spectrum of macrophage activation. Nat Rev Immunol. 2008;8(12):958–69.

Mujica-Mota MA, Gasbarrino K, Rappaport JM, Shapiro RS, Daniel SJ. The effect of caffeine on hearing in a guinea pig model of acoustic trauma. Am J Otolaryngol. 2014;35(2):99–105.

Mukherjea D, Jajoo S, Kaur T, Sheehan KE, Ramkumar V, Rybak LP. Transtympanic administration of short interfering (si)RNA for the NOX3 isoform of NADPH oxidase protects against cisplatin-induced hearing loss in the rat. Antioxid Redox Signal. 2010;13(5):589–98.

Mukherjea D, Jajoo S, Sheehan K, Kaur T, Sheth S, Bunch J, Perro C, Rybak LP, Ramkumar V. NOX3 NADPH oxidase couples transient receptor potential vanilloid 1 to signal transducer and activator of transcription 1-mediated inflammation and hearing loss. Antioxid Redox Signal. 2011;14:999–1010.

Mukherjea D, Jajoo S, Whitworth C, Bunch JR, Turner JG, Rybak LP, Ramkumar V. Short interfering RNA against transient receptor potential vanilloid 1 attenuates cisplatin-induced hearing loss in the rat. J Neurosci. 2008;28(49):13056–65.

Muñoz DJ, Kendrick IS, Rassam M, Thorne PR. Vesicular storage of adenosine triphosphate in the guinea-pig cochlear lateral wall and concentrations of ATP in the endolymph during sound exposure and hypoxia. Acta Otolaryngol. 2001;121(1):10–5.

Muñoz DJ, McFie C, Thorne PR. Modulation of cochlear blood flow by extracellular purines. Hear Res. 1999;127(1–2):55–61.

Nakai Y, Konishi K, Chang KC, Ohashi K, Morisaki N, Minowa Y, Morimoto A. Ototoxicity of the anticancer drug cisplatin. An experimental study. Acta Otolaryngol. 1982;93(1–6):227–32.

Nie Z, Mei Y, Ford M, Rybak L, Marcuzzi A, Ren H, Stiles GL, Ramkumar V. Oxidative stress increases A1 adenosine receptor expression by activating nuclear factor kappa B. Mol Pharmacol. 1998;53(4):663–9.

Paparella MM, Oda M, Hiraide F, Brady D. Pathology of sensorineural hearing loss in otitis media. Ann Otol Rhinol Laryngol. 1972;81(5):632–47.

Ramkumar V, Whitworth CA, Pingle SC, Hughes LF, Rybak LP. Noise induces A1 adenosine receptor expression in the chinchilla cochlea. Hear Res. 2004;188(1–2):47–56.

Robson SC, Sévigny J, Zimmermann H. The E-NTPDase family of ectonucleotidases: Structure function relationships and pathophysiological significance. Purinergic Signal. 2006;2(2):409–30.

Rybak LP, Husain K, Morris C, Whitworth C, Somani S. Effect of protective agents against cisplatin ototoxicity. Am J Otol. 2000;21(4):513–20.

Rybak LP, Mukherjea D, Jajoo S, Ramkumar V. Cisplatin ototoxicity and protection: clinical and experimental studies. Tohoku J Exp Med. 2009;219(3):177–86.

Sheth S, Sheehan K, Lowy M, Borse B, Dhukhwa A, Al-aameri R, Mukherjea D, Rybak LP, Ramkumar V. The detrimental effects of caffeine consumption on hearing in the rat model of cisplatin ototoxicity. Assoc Res Otolaryngol (Abstract 597) 2017:398. Abstract retrieved from Abstract Archives in Association for Research in Otolaryngology database.

Someya S, Xu J, Kondo K, Ding D, Salvi RJ, Yamasoba T, Rabinovitch PS, Weindruch R, Leeuwenburgh C, Tanokura M, Prolla TA. Age-related hearing loss in C57BL/6J mice is mediated by Bak-dependent mitochondrial apoptosis. Proc Natl Acad Sci U S A. 2009;106(46):19432–7.

Taylor PR, Martinez-Pomares L, Stacey M, Lin HH, Brown GD, Gordon S. Macrophage receptors and immune recognition. Annu Rev Immunol. 2005;23:901–44.

Townsend PA, Scarabelli TM, Davidson SM, Knight RA, Latchman DS, Stephanou A. STAT-1 interacts with p53 to enhance DNA damage-induced apoptosis. J Biol Chem. 2004;279(7):5811–20.

Vlajkovic SM, Abi S, Wang CJ, Housley GD, Thorne PR. Differential distribution of adenosine receptors in rat cochlea. Cell Tissue Res. 2007;328(3):461–71.

Vlajkovic SM, Ambepitiya K, Barclay M, Boison D, Housley GD, Thorne PR. Adenosine receptors regulate susceptibility to noise-induced neural injury in the mouse cochlea and hearing loss. Hear Res. 2017;345:43–51.

Vlajkovic SM, Guo CX, Telang R, Wong AC, Paramananthasivam V, Boison D, Housley GD, Thorne PR. Adenosine kinase inhibition in the cochlea delays the onset of age-related hearing loss. Exp Gerontol. 2011;46(11):905–14.

Vlajkovic SM, Housley GD, Muñoz DJ, Robson SC, Sévigny J, Wang CJ, Thorne PR. Noise exposure induces up-regulation of ecto-nucleoside triphosphate diphosphohydrolases 1 and 2 in rat cochlea. Neuroscience. 2004;126(3):763–73.

Vlajkovic SM, Vinayagamoorthy A, Thorne PR, Robson SC, Wang CJ, Housley GD. Noise-induced up-regulation of NTPDase3 expression in the rat cochlea: Implications for auditory transmission and cochlear protection. Brain Res. 2006;1104(1):55–63.

Whitworth CA, Ramkumar V, Jones B, Tsukasaki N, Rybak LP. Protection against cisplatin ototoxicity by adenosine agonists. Biochem Pharmacol. 2004;67(9):1801–7.

Wong AC, Guo CX, Gupta R, Housley GD, Thorne PR, Vlajkovic SM. Post exposure administration of A(1) adenosine receptor agonists attenuates noise-induced hearing loss. Hear Res. 2010;260(1–2):81–8.

Wong AC, Ryan AF. Mechanisms of sensorineural cell damage, death and survival in the cochlea. Front Aging Neurosci. 2015;21(7):58.

Yamane H, Nakai Y, Takayama M, Iguchi H, Nakagawa T, Kojima A. Appearance of free radicals in the guinea pig inner ear after noise-induced acoustic trauma. Eur Arch Otorhinolaryngol. 1995;252:504–8.

Yamashita D, Jiang HY, Le Prell CG, Schacht J, Miller JM. Post-exposure treatment attenuates noise-induced hearing loss. Neuroscience. 2005;134(2):633–42.

Yamashita D, Jiang HY, Schacht J, Miller JM. Delayed production of free radicals following noise exposure. Brain Res. 2004;1019(1–2):201–9.

Yamasoba T, Nuttall AL, Harris C, Raphael Y, Miller JM. Role of glutathione in protection against noise-induced hearing loss. Brain Res. 1998;784:82–90.

Zhao HB, Yu N, Fleming CR. Gap junctional hemichannel-mediated ATP release and hearing controls in the inner ear. Proc Natl Acad Sci U S A. 2005;102(51):18724–9.

Chapter 9
Trauma, Inflammation, Cochlear Implantation Induced Hearing Loss and Otoprotective Strategies to Limit Hair Cell Death and Hearing Loss

Stefania Goncalves, Enrique Perez, Esperanza Bas, Christine T. Dinh, and Thomas R. Van De Water

Abstract Hair cells are highly sensitive units that in response to trauma, inflammation, and cochlear implantation activate different signaling pathways leading to hair cell death and hearing impairment. In this chapter we discuss the most recent literature regarding signaling pathways of hair cell loss, mechanisms and inflammatory responses after noise exposure and electrode insertion, and otoprotective strategies that can limit hair cell death and hearing loss.

Keywords Electrode insertion trauma · Apoptosis · Inflammatory process signaling molecules · Foreign body reaction · Pharmacological treatment · Cochlear drug delivery · Protective hypothermia

1 Mechanisms of Inflammatory Process Initiated Cell Death in the Cochlea

Hearing is a complex function relying on a coordinated effort by a multitude of specialized cells. As the principal organ involved in the conversion of acoustic energy into electrical stimuli in the inner ear, the cochlea houses some of the most important cells comprising the auditory system. Like other tissues in the body, these inner ear structures are subject to inflammatory insults resulting in either transient or permanent injury.

Sensorineural hearing loss (SNHL) is a type of hearing impairment that may develop as a result of injury to cochlear or retrocochlear structures along the auditory pathway. This form of hearing loss is highly prevalent, affecting over 48 million

S. Goncalves · E. Perez · E. Bas · C. T. Dinh · T. R. Van De Water (✉)
Department of Otolaryngology, University of Miami Ear Institute, University of Miami Miller School of Medicine, Miami, FL, USA
e-mail: tvandewater@med.miami.edu

© Springer International Publishing AG, part of Springer Nature 2018
V. Ramkumar, L. P. Rybak (eds.), *Inflammatory Mechanisms in Mediating Hearing Loss*, https://doi.org/10.1007/978-3-319-92507-3_9

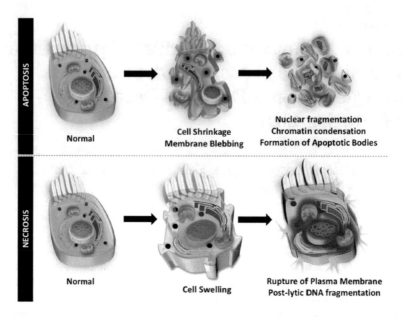

Fig. 9.1 Schematic representation of apoptotic and necrotic cell death (Reproduced from Dinh et al. 2015)

Americans and resembling the most common sensory deficit in both the pediatric and adult populations (Lin et al. 2011; Deltenre and Van Maldergem 2013; Li-Korotky 2012). Hair cells located inside the cochlea are key to auditory function and their loss is responsible for the most common cause of SNHL. Unlike in birds and other selected vertebrates, mammals do not possess the ability to regenerate nor replace hair cells and therefore any postnatal loss of these cells is associated with permanent damage and hearing loss (Wan et al. 2013).

There are a variety of insults that may lead to hair cell loss including but not limited to: noise-induced trauma, ototoxic drugs, infections, cochlear hypoxia, radiation exposure, and cochlear implant electrode insertion trauma (Dinh et al. 2015). Although these resemble environmental insults, a host of genetic factors have also been linked with hereditary hearing loss or a higher propensity to cochlear injury in many cases of acquired hearing loss (Walsh et al. 2010; Ahmed et al. 2011). To explore potential therapeutic options for SNHL it is essential to understand the underlying molecular mechanisms that govern cochlear inflammation and ultimately cell death, following an insult. The two most widely studied pathways to cell death as a result of cochlear injury are apoptosis and necrosis (Fig. 9.1). These, as well as other forms of non-apoptotic cell death affecting cochlear cell populations will be discussed in the following sections. It is important to note that some of the known mechanisms of cell death in cochlear injury are partially extrapolated from work done in other cell types (Dinh et al. 2015).

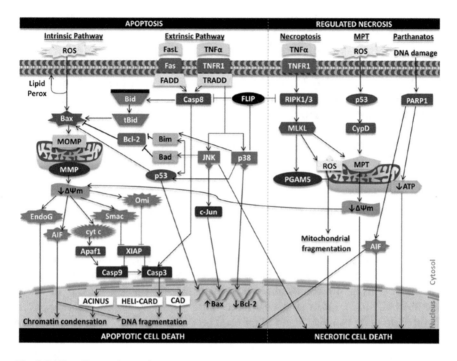

Fig. 9.2 Signaling pathways in apoptosis and necrotic cell death (Reproduced from Dinh et al. 2015)

Apoptosis

Apoptosis refers to the active orchestration of specific energy-requiring molecular pathways resulting in cell death (Kerr et al. 1972). This form of programmed cell death can affect all cells in the body and is governed by the balance between pro-inflammatory, pro-death, and pro-survival molecular pathways often involving caspases. Morphologically, hair cells undergoing apoptosis may demonstrate specific features such as the disruption or loss of stereocilia, shrinkage of cuticular plates, mitochondrial swelling, nuclear chromatin condensation, and disruption of junctional complexes, which results in extrusion of the cell from the sensory epithelium (Hirose et al. 2004). Apoptotic cells are eventually broken down into apoptotic bodies, which are then removed by phagocytosis. There are two distinct pathways that may lead to apoptosis: the Intrinsic and Extrinsic cell death pathways (Fig. 9.2).

Intrinsic Pathway

The intrinsic cell death pathway is most often governed by shifts in the balance of pro- and anti-apoptotic proteins of the Bcl-2 family, ultimately resulting in mitochondrial leakage of pro-death proteins into the cytosol of the affected cell. These molecules include cytochrome c (cyt c), apoptosis inducing factor (AIF), endonuclease G (EndoG), second mitochondria-derived activator of caspases/direct inhibitor of apoptosis protein binding protein with low pI (Smac/DIABLO), and mammalian homolog of bacterial high temperature requirement protein A2 (Omi/HtrA2) (Dinh et al. 2015). Upon release into the cytosol some of these proteins such as cyt c contribute to apoptotic cell death through direct or indirect activation of caspase proteins 3, 7, and 9. Others such as nuclease EndoG lead to caspase-independent apoptotic pathways via nucleosomal fragmentation of DNA or chromatin condensation, once translocated into the nucleus.

Of particular importance to cochlear function, the intrinsic pathway is often triggered by oxidative stress. Reactive oxygen species (ROS) are derived predominantly from enzymatic processes (most frequently associated with the host-defense function of inflammatory cells such as neutrophils and macrophages) and non-enzymatic processes such as those carried out by the electron transport chain (ETC) of a cell's mitochondria (Dworakowski et al. 2006). In a delicate system such as the cochlea, stresses such as acoustic trauma can lead to direct mechanical injury to the organ of Corti, followed by diminished blood supply secondary to edema of the stria vascularis. This in turn results in cochlear hypoxia and oxidative stress to hair cells (Smith et al. 1985). Other stresses such as those caused by aminoglycoside or cisplatin ototoxicity can also lead to increased production of ROS in cochlear tissues (Rybak et al. 2007; Lesniak et al. 2005). Accumulation of ROS overwhelms the anti-oxidant protective mechanisms of hair cells and promotes Bax (pro-apoptotic Bcl-2) protein activation, resulting in a shift towards the intrinsic pathway with translocation of mitochondrial proteins as previously described. Interestingly, this mechanism of inflammation followed by oxidative stress and programmed cell death can partially explain a tonotopic pattern of hair cell loss. It is widely known that basal outer hair cells are more vulnerable to various cochlear insults such as ototoxic drugs and acoustic trauma, resulting in reproducible patterns of high-frequency SNHL (Jensen-Smith et al. 2012). Prior reports suggest that this may be secondary to an inherent susceptibility to oxidative stress in these cells secondary to lower production of glutathione, a potent anti-oxidant (Sha et al. 2001).

Extrinsic Pathway

The extrinsic cell death pathway is characterized by a complex interaction of various molecular cascades that ultimately occur as a result of the binding of a death ligand to its complimentary death ligand receptor. Amongst the most widely studied death ligand-to-receptor interactions are the TNFα-TNFR1 and FasL-FasR ligand-receptor interactions (Dinh et al. 2015). The extrinsic pathway often involves downstream activation of initiator caspases 8 and 10, which in turn can cleave and activate effector caspase-3. There is likely crossover with the intrinsic pathway as both caspase-8 and -10 are also capable of activating pro-apoptotic members of the Bcl-2 protein family, Bax and Bak, which in turn can propagate mitochondrial release of death proteins as previously described (Chandler et al. 1998). Insults to the cochlea such as acoustic trauma can also result in the local production of cytokines and death ligands such as TNFα. Prior studies have demonstrated increased expression of this molecule by the stria vascularis and spiral ligament, which may in turn serve to recruit inflammatory cells into the cochlea (Keithley et al. 2008). In a form of positive feedback, these leukocytes may then release further cytokines and ROS via enzymatic processes. TNFα signaling in the cochlea can lead to activation of various members of the mitogen-activated protein kinases (MAPK) as part of the extrinsic pathway. Two distinct MAPK pathways studied in hair cells, the p38 and JNK kinase cascades, have been implicated in noise-, drug-, radiation-, and direct TNFα-induced loss of hair cells (Dinh et al. 2015). Considering this intricate and multi-interactive form of programmed cell death it is no surprise various therapeutic agents have been proposed to target specific parts of these cascades such as CEP-1347, a JNK inhibitor molecule shown to protect against hair cell loss and hearing loss in various forms of cochlear injury (Ylikoski et al. 2002; Wang et al. 2003).

Necrosis

Morphological analysis still remains the gold standard to differentiate between apoptosis and necrosis (Op de Beeck et al. 2011). Unlike the characteristic cell shrinkage (pyknosis) observed in apoptosis, necrosis is characterized by advance swelling (oncosis) and early cell membrane disruption (Fig. 9.1). A compromised cell membrane results in spillage of intracellular contents to the surrounding extracellular space (Nagańska and Matyja 2001). These events are thought to propagate further inflammation and initiation of cell death within surrounding cells. Although initially thought to only occur passively and unchecked, several different mechanisms of regulated necrosis have now been elucidated. These include receptor interacting protein kinase (RIPK)-dependent necroptosis and mitochondrial permeability transition (PMT)-dependent necrosis (Galluzzi et al. 2010; Tsujimoto and Shimizu 2007). Although still not fully understood, these mechanisms of "programmed" necrosis appear to share common signaling elements with extrinsic and intrinsic apoptotic pathways (Fig. 9.2). Various morphological and molecular studies have demonstrated a combination of both necrotic and apoptotic cell death following common cochlear insults such as ototoxicity and acoustic trauma (Zheng et al. 2014). Many of these studies have consistently demonstrated the greater vulnerability of outer hair cells of the basal cochlea to these mechanisms of cell death. Ototoxicity, such as that seen with kanamycin and acoustic trauma has been shown to induce hair cell death through apoptosis more frequently than necrosis (Taylor et al. 2008; Yang et al. 2004).

Other Forms of Non-apoptotic Cell Death

Several less-studied forms of cell death exist, which may resemble biochemically, genetically, and morphologically distinct pathways from those seen in necrosis and apoptosis. Ferroptosis is defined as a non-apoptotic iron-dependent oxidative form of cell death distinct from necrosis and autophagy (Dixon et al. 2012). The hallmark morphologic features of this type of cell death include small mitochondria with increased membrane density. Although still controversial Autosis refers to autophagy induced cell death (Liu et al. 2013). Autophagy has been traditionally understood as a pro-survival mechanism for cells, morphologically characterized by the formation of cytoplasmic vacuoles (Levine and Kroemer 2008). Recent investigations suggest that at high levels, autophagy itself may function as a driver of cell death (Liu et al. 2013). Yet another potential distinct pathway for cell death is the caspase-1 dependent Pyroptosis. This form of cell death appears to share morphologic features resembling both apoptosis and necrosis with caspase-independent DNA fragmentation and early disruption of plasma membrane, respectively (Kepp et al. 2010). Finally, if determined to be distinct from regulated necrosis, Parthanatos may also resemble another form of non-necrotic, non-apoptotic cell death. Morphologically, these cells appear shrunken with condensed nuclei and undergo membrane disintegration. The key molecular mechanisms of Parthanatos cell death involve caspase-independent rapid activation of poly (ADP-ribose) polymerase-1 (PARP-1) and accumulation of poly (ADP-ribose) (PAR), with subsequent nuclear translocation of apoptosis inducing factor (AIF) from the mitochondria (Andrabi et al. 2008). Limited research exists into the contribution of any one of these pathways in cochlear tissue cell death. One prior study reported on autophagic stress as a possible mechanism for spiral ganglion neuronal death in a model of age-related hearing loss (ARHL). This pathway for cell death did not seem to affect other populations of cells in the organ of Corti or the stria vascularis (Menardo et al. 2012). Indeed, like the simultaneous activation of both apoptotic and necrotic cell death pathways observed with certain cochlear insults, it is likely that these other forms of cell deaths may also play a simultaneous role in the cellular demise of cochlear tissues. This may present new opportunities for targeted therapies.

2   Electrode Insertion Trauma (EIT) Initiation of the Inflammatory Process

Some authors (Olivetto et al. 2015) described three phases of cochlea injury as a result of cochlea implantation: (1) EIT within the first 2 days after implantation; (2) inflammatory response within 14 days; and (3) development of fibrosis over the long term. Trauma associated to the insertion of an electrode array into the scala tympani of the cochlea initiates a long-term process from the first second after the

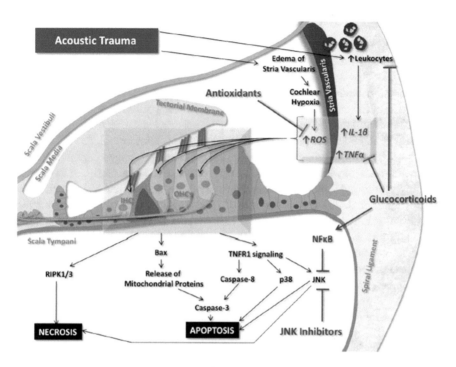

Fig. 9.3 Noise-induced cell death pathway leading to hearing loss (Reproduced from Dinh et al. 2015)

insertion of the electrode that triggers the release of different pro-inflammatory molecules, chemokines, and cytokines that change the environment of the auditory organ (Fig. 9.3). This new biochemical environment compromises the normal function of hair cells (HC) and cochlear neural elements impairing residual hearing putting at higher risk the structures located within the apical and middle turns of the cochlea responsible for the transduction of low and mid frequencies (Bas et al. 2016). Some of the trauma-associated stimuli during cochlear implantation that leads to oxidative stress and inflammation within the cochlea affecting HC and spiral ganglion neuron (SGN) survival are: (1) acoustic and vibrational trauma associated to surgical drilling; (2) insertional trauma to the basilar membrane, spiral ligament, and/or osseous spiral lamina; (3) displacement of blood and bone particles into the scala tympani; and (4) bacterial infection of the cochlea (Lehnhardt 1993; Cohen 1997; Kiefer et al. 2004; Eshraghi 2006; Postelmans et al. 2011; Bas et al. 2012). Therefore, non-traumatic cochlear implantation techniques (soft surgery) have been proposed. These techniques have been found to help preserve residual hearing resulting in patients with improvement in music perception as well as significant gain in hearing in quiet and in noisy environment.

Mechanisms of Hearing and Hair Cell Losses

Although, the mechanisms behind EIT remain unclear some hypothesis have been proposed. During the insertion of an electrode array, injury to the spiral ligament, stria vascularis, and/or trauma to the boundaries that separate the different compartments within the cochlea may result in mixing of the perilymphatic and endolymphatic fluids compromising the endolymphatic potential and affects residual hearing (Teubner et al. 2003; Wangemann et al. 2004). Cohen-Salmon et al. (2002), Teubner et al. (2003) and collaborators have also suggested that prolonged depolarization of HC membranes may lead to apoptosis. Furthermore, EIT can also trigger the expression of different cytokines and inflammatory molecules that affect hearing preservation (Takumi et al. 2014) as is the case of TNF alpha, which is a pro-inflammatory cytokine that has been used in multiple experiments as an ototoxic agent for studying the protective effect of different drugs (Bas et al. 2012; Dinh et al. 2008a, b, 2011; Dinh and Van De Water 2009; Haake et al. 2009, Hoang et al. 2009). 4-hydroxynonenal (HNE, a product of lipid peroxidation), interleukin-1 beta (IL-1B), intercellular cell adhesion mediator-1 (ICAM-1), inducible nitric oxide synthetase (iNOS) and cyclooxygenase-2 (COX-2), and reactive oxygen and nitrogen species are other inflammatory molecules that are found in the cochlea after EIT (Bas et al. 2012; Eshraghi et al. 2013) leading to HC death and loss of residual hearing (Fig. 9.3).

In vitro and *in vivo* studies have shown the suppressive effects of dexamethasone on the expression of these cytokines and inflammatory molecules (Guzman et al. 2006; Moriyama et al. 2007, Ichimiya et al. 2000; Maeda et al. 2005; Bas et al. 2012) helping in preservation of the residual hearing after cochlea implantation (Fig. 9.4).

Mechanisms of Damage to Cochlear Neural Elements

The loss of auditory HCs and Schawnn cells (SC) due to EIT during cochlear implantation, such as direct trauma to the SGNs through the osseous spiral lamina, can lower the number and activity of afferent SGNs, which are necessary for hearing (Roehm and Hansen 2005). Even isolated damage to the SC can also compromise SGNs viability considering that these SCs express neurotrophins and their respective receptors to cross-communicate with SGNs promoting homeostasis within the ganglion (Hansen et al. 2001), besides assuring proper insulation of the spiral ganglia axons by myelination of type 1 SGNs (Romand and Romand 1990). After injury, SCs release pro-inflammatory cytokines, chemokines and adhesion molecules that result in recruitment of inflammatory cells to the injured area (Shamash et al. 2002; Tofaris et al. 2002).

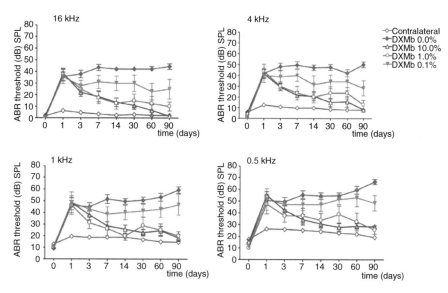

Fig. 9.4 Dexamethasone protects against electrode insertion trauma induced hearing loss in guinea pigs. ABR thresholds prominently increased after electrode insertion across all frequencies in non-drug-eluted electrodes. Dexamethasone eluted electrodes at a concentration of 1–10% significantly reduced the negative effects of electrode insertion trauma in hearing. *ABR* auditory brainstem response, *DXMb* dexamethasone base (Reproduced from Bas et al. 2016)

3 Inflammatory Process Initiated Cellular and Molecular Signaling That Occurs Post-EIT. The Role of Cytokines, Chemokines, and Cell Adhesion Molecules in the Inflammatory Process

Cochlear implantation can restore hearing perception by bypassing the injured HCs; however, electrode insertion can initiate a robust inflammatory reaction with the recruitment of inflammatory cells and deposition of fibrotic tissue surrounding the electrode that can progressively affect the survival of SGN and residual HCs compromising hearing outcome and enhancing the inflammatory environment within the cochlea (Hughes et al. 2001; Choi and Oghalai 2005; Jia et al. 2011; Wolfe et al. 2013; Mosca et al. 2014). In the nervous system SCs (peripheral nervous system) and astrocytes (central nervous system) are similar to macrophages and display polarized phenotypes induced by classical or alternative activation in response to nerve injury (Reichert et al. 1994; de Waele et al. 1996). Bas and collaborators characterized the action of SGNs and macrophages in neuro-inflammation and scarring following cochlear implantation using adult mouse *in vivo* and neonatal rat *in vitro* models of electrode analog insertion trauma. Leukocytes were isolated from the spleen of neonatal rats and co-cultured with the neonatal organotypic organ of Corti cultures. This study reported an overexpression of chemokines (Ccl2, Ccl3, Ccl12,

TGF-B) in both damaged cochlear tissues and leukocytes, and cells adhesion molecules (Vcam, Icam) in damaged cochlear tissues explants that was reduced to levels comparable to undamaged tissue when treating insulted cultures with dexamethasone (Bas et al. 2016). Therefore, damage tissue in response to EIT induces an inflammatory phase characterized by the expression and release of chemokines for cells recruitment, adhesion molecules to increase cell-cell and cell-tissue interaction in response to the injury, and cytokines inducing the activation of inflammatory cells, the release of more pro-inflammatory molecules, and the overall enhancement of the inflammatory environment. Following the initial inflammatory phase, a proliferative phase is initiated that is characterized by the release of matrix proteins, mainly fibrins and collagens, by recruited fibroblasts and activated macrophages that will constitute the extracellular matrix that will promote migration, growth, and cellular differentiation (de Waele et al. 1996; Stout 2010; Ydens et al. 2012). After peripheral nerve injury, diverse cell adhesion molecules are released such as integrin 4-alpha, which is a receptor that binds fibronectin, playing an important role in neuron regeneration and cell-matrix interactions during leukocyte recruitment and can also mediate the recruitment of immune cells propagating the inflammatory response (Vogelezang et al. 2001; Bas et al. 2016). Bas and colleagues showed an increased expression of Integrin 4-alpha at 1 and 3 days after implantation that later dropped at 7 days post-implantation in lateral wall tissues, organ of Corti, SGNs, and site of electrode insertion. However, 14 days post-implantation the expression of this molecule returned to baseline in the lateral wall and organ of Corti, decreased in the site of electrode insertion, but increased SGNs. Other integrin molecules (Itga1, Itga2, Itga3, Itgav, Itga6) were also described to be upregulated in EIT cochlear tissues (Bas et al. 2016). After the initial inflammatory and proliferative phases, remodeling of the injury site occurs followed by a reduction in the inflammatory environment and angiogenesis, together with the maturation of scar tissue.

Leukocyte Recruitment to Trauma Site and Cell-Cell Interaction

After the traumatic insertion of an electrode into the cochlea, resident SCs release pro-inflammatory cytokines, chemokines, and cell adhesion molecules for the recruitment of inflammatory cells to the injured area (Shamash et al. 2002; Tofaris et al. 2002), where monocytes are transformed into M1 and M2 type macrophages. The M2 macrophages secrete anti-inflammatory molecules and promote tissue regeneration and remodeling, in contrast, the M1 macrophages degrades cellular debris, including bacteria, and release reactive oxygen species and anti-inflammatory cytokines (IL-1a, IL-1b, IL-4, IL-13) that can be noxious to the cellular environment (Martinez et al. 2008; Stout 2010).

Bas and collaborators described the behavior of leukocytes exposed to organotypic cultures of the organ of Corti after an EIT event showing a reduction in the trajectory traveled by leukocytes within the culture dish while increasing their contact interaction with other inflammatory and injured cells. These findings were significantly different from leukocyte response seen in response to undamaged tissue and dexamethasone treated EIT damaged cochlear tissue explants, where the trajectory traveled by the leukocytes was increased and the interaction between these and normal or dexamethasone treated damaged tissues was decreased (Bas et al. 2016).

Role of Foreign Body Reaction Post-implantation

The robust inflammatory response following EIT also induces the release of different growth factors, especially Tgfb1, 3 days after EIT *in vitro* together with the downregulation of Tgfb3 and its receptor Tgfbr2, which are associated with reduction of scarring during wound healing (Bas et al. 2016). In the literature, TFGb1 has been involved in acute, chronic and malignant environments. In acute settings, this molecule is involved in damage and regeneration of tissues typical of physiological wound healing. However, in chronic environments, TFGb1 has negative effects as consequence of the long-term continuous perpetuation of cell proliferation and fibrinogenesis leading to fibrosis through the persistent induction of fibronectin production, hypertrophy, and hyperplasia.

After the acute phases of inflammation, the receptors of fibronectin (ITGA 4alpha) persist in the organ of Corti, SGNs, and wound bed even 30 days after EIT *in vivo* (Bas et al. 2016). Simultaneously, fibroblasts and differentiated myofibroblasts proliferate and new collagen is produced and deposited around the electrode (Bas et al. 2016) together with the upregulation of F-actin suggesting the presence of reorganization of the actin skeleton and extracellular matrix remodeling that leads to scar formation and maturation around the electro array where the presence of myofibroblasts collagen type 1A has been reported (Bas et al. 2016). These events perpetuate within the cochlea as an unbreakable cycle suggesting an atypical chronic inflammatory response characterized by the presence of activated inflammatory cells even 30 days after the injury, which some authors have also described as a possible immune response against the electrode (foreign body reaction) (Bas et al. 2016).

Role of Macrophages and Schwann Cells in Early and Chronic Inflammatory Responses Post-implantation

SCs are a type of glial cells that are found in the peripheral nervous system. These cells can be activated or inactivated and their responsibilities differ (cytotoxic versus cytoprotective) depending on the environment they are interacting with. In their inactivated state they are found surrounding neurons aiding in axon myelination,

Table 9.1 Pattern of expression of F4/80, IL-1B, and Arginase in an *in vivo* model of EIT

Tissue	Inflammatory environment
Lateral Wall	Progressive increase in F4/80 and IL-1B (peak 14 and 30 days post-implantation)
Organ of Corti	IL-1B increase by day 14 and remained stable until day 30. Biphasic pattern of F4/80 and Arginase 1 with peaks at day 3 and day 14 post-implantation. After 4 weeks, levels of Arginase 1 remained higher than IL-1B and F4/80 was reduced
Spiral Ganglion	Arginase 1 levels and F4/80 started rising at day 1 showing a maximum level at day 7 post-implantation. Arginase 1 levels predominate over IL-1B indicating involvement of M2 macrophages
Wound site	IL-1B and Arginase 1 expression overlapped at all time
Cochlear nerve	F4/80 and IL-1B expression progressively increase beginning at post-implantation day 3

phagocytosis of cellular debris, and promotion of nerve regeneration. However, after these cells are exposed to an inflammatory environment become active developing actions on neurons that can be detrimental due to the release of different inflammatory cytokines (such as IL-1B) (Shamash et al. 2002; Tofaris et al. 2002). In a similar way monocytes can have either similar or opposite actions depending on the phenotype they acquired after they are activated. While M1 macrophages are also pro-inflammatory and participate in the production and release of different cytokines (such as IL-1B), the actions of M2 macrophages are anti-inflammatory promoting cell survival and tissue regeneration (Martinez et al. 2008; Stout 2010; Ydens et al. 2012). Bas and colleagues, described monocytes/macrophage infiltration (F4/80), IL-1B, and Arginase 1 (indirect marker for the alternate activated M2 macrophages) levels in an *in vivo* model of EIT. The pattern of expression of these molecules varies based on the tissue under observation (see Table 9.1 for details).

4   Mechanisms of Pharmacological Treatment to Prevent Post-EIT Initiated Inflammatory Process Damage to the Cochlea and Loss of Hearing

The most extensively studied pharmacologic treatment of electrode insertion trauma (EIT) revolves around the use of corticosteroids, partially owed to the abundance of glucocorticoid receptors located in cochlear tissues, which offer a direct target for treatment (Bas et al. 2016). Thorough *in vivo* and *in vitro* molecular studies on the signaling pathways of corticosteroids has revealed that medications such as dexamethasone can inhibit the ototoxic inflammatory effects of the TNFα signaling cascade, a known mechanism of injury in EIT. This inhibition appears to be modulated by activation of NFκB signaling pathways, which promote upregulation of anti-apoptotic genes *Bcl-2* and *Bcl-xl* as well as decrease in expression of pro-apoptotic genes *Bax* and *TNFR1* (Dinh et al. 2008a, b; Dinh and Van De Water 2009).

The JNK kinase cascade has also served as a target of pharmacotherapy in the setting of EIT. Through the use of a mini-osmotic pump as well as implantation of a drug eluting hyaluronate-gel at the round window niche, a JNK inhibitor AM-111 (i.e. dJNKI peptide 1) has been shown to effectively block upstream steps along the extrinsic apoptotic pathway, which may offer protection against EIT-induced inflammation cell death (Eshraghi et al. 2006, 2013). This in turn has been shown to preserve residual hearing (Eshraghi et al. 2006, 2013).

Antioxidants have long been the subject of investigation for the treatment noise- and drug-induced hearing loss showing positive results at hearing preservation with a variety of compounds such as sodium thiosulfate (STS), D-methionine, and N-acetylcysteine (NAC) (Dinh and Van De Water 2009). In a model of EIT pre-implantation application of NAC-impregnated pledgets to the round window niche demonstrated protection against residual hearing loss (Eastwood et al. 2010). It is believed that its ability to dramatically increase levels of intrinsic antioxidant gluta-thione can make it a potent anti-inflammatory/otoprotective agent inside the cochlea (Zou et al. 2003). Nevertheless, this therapeutic agent is also associated with a tran-sient increase in hearing thresholds as well as possible promotion of unwanted intra-cochlear osteoneogenesis following electrode insertion, likely limiting its clinical utility (Eastwood et al. 2010).

Other less studied pharmacologic therapies in the setting of EIT include the use of TNFα-inhibitor etanercept, insulin-like growth factor (IGF), and hepatocyte growth factor (Ihler et al. 2014; Kikkawa et al. 2014). These compounds have proven to protect against hearing loss when delivered to the cochlea but require further research to elucidate their clinical potential (Ihler et al. 2014, Kikkawa et al. 2014).

Soft Surgery

With a better understanding of the mechanisms behind EIT, various recommenda-tions for soft surgical techniques in cochlear implantation have been proposed to preserve residual hearing (Bas et al. 2012). Research on vibratory trauma secondary to drilling in the temporal bone has revealed that high frequency vibrations can lead to inner ear trauma and significant threshold shifts (Sutinen et al. 2007). Direct vibrational forces as well as high intensity sounds may mediate this type of injury. At a molecular level this stimulus has been linked to pro-inflammatory TNFα sig-naling with downstream apoptotic pathway activation (Zou et al. 2005). Delicate dissection near the facial recess and the use of lower frequency drilling has been suggested as a means to reduce this unwanted effect.

Vibration injury also plays an important factor when deciding between inser-tion approaches to the cochlea. Prior studies have demonstrated that average drilling time is significantly longer when using a traditional cochleostomy versus a round window approach to prepare a route for insertion, since the latter often involves merely removal of small bony over-hangs at the round window niche (Usami et al. 2011). Success with preserving residual low-frequency hearing by using a round window approach for electrode insertion has also been shown (Skarzynski et al. 2007a, b). Nevertheless, the trajectory of electrode insertion using this approach may limit its use in cases of perimodiolar implants (Souter et al. 2011). Furthermore, careful drilling of anterior-inferior cochleostomies may allow for reasonable residual hearing preservation (Garcia-Ibanez et al. 2009; Kiefer et al. 2004). To balance the potential for injury secondary to vibratory forces with those of direct intracochlear trauma from inappropriate angle of insertion, soft surgery calls for appropriate selection of insertion technique in each individual case. This may depend on the electrode to be used and the cochlear anatomy of the patient. In either case, care should be taken to avoid a drill injury to the endostium of the scala tympani by using a 1-mm burr at low speeds and with adequate irrigation (Bas et al. 2012).

Extensive irrigation, hemostasis, and wiping gloved hands in order to prevent any bone dust or blood products from entering the cochlea should be perform prior to a cochleostomy (Bas et al. 2012). Intracochlear osteoneogenesis potentially arising from bone particles displaced into the scala tympani as well as alterations in the endocochlear potential secondary to lysis of displaced erythrocytes can be detrimental to residual hearing (Clark et al. 1995; Radeloff et al. 2007).

Another soft surgery technique for reducing the risk of contaminating intracochlear contents with bone dust or blood is the application of sodium hyaluronate gel over the cochleostomy or round window site prior to penetration of endostium with a micro-lancet and then the electrode array. This technique can prevent unwanted particles from entering the scala tympani and also avoid excessive perilymph leakage (Laszig et al. 2002). The gel can also serve as a lubricant to reduce intracochlear friction forces when inserting the electrode (Kontorinis et al. 2011).

Finally, electrode parameters may also play an important roll in soft surgery technique. The use of short electrode arrays for shallow insertions has proven efficacious at preserving residual low-frequency hearing in appropriately selected cases (Helbig et al. 2011). Nevertheless, deep insertion can also lead to adequate hearing preservation with the use of soft surgery techniques (Bas et al. 2012).

Transtympanic and Transcochlear Drug Delivery

Although proven to be otoprotective in models of EIT, soft surgery alone does not solve the dilemma of intracochlear injury during cochlear implant surgery. With the expanding knowledge on pharmacological agents capable of offering otoprotection as previously mentioned in this chapter, significant research efforts have been made to find the ultimate delivery mechanism for these compounds to reach the inner ear. Like the blood-brain barrier, the blood-labyrinth barrier offers a significant obstacle to delivery of drugs into the cochlea via a systemic route (Inamura and Salt 1992).

Otoprotective drugs such as corticosteroids can also carry significant undesirable side effects when administered systemically, further encouraging researchers to develop local inner ear delivery mechanisms, which can obviate the need for systemic therapy.

The literature on intratympanic steroid treatment of sudden sensorineural hearing loss has demonstrated significant benefits of using a middle ear drug delivery strategy to treat an inner ear process (Wei et al. 2013). Adopting from this knowledge, researchers have demonstrated hearing preservation in partially deaf patients undergoing cochlear implantation with the use of immediate preoperative intratympanic steroid injections (Rajan et al. 2012). An inherent problem with this technique is the suboptimal delivery of therapeutic agent to the round window niche, with unpredictable levels of medication reaching intracochlear tissues. The loss of medication through the Eustachian tube can partially explain the limitations of an uncontrolled middle-ear delivery strategy (Silverstein et al. 2004). Researchers looking at the treatment of other inner ear pathologies such as Meniere's have further refined transtympanic treatment strategies with the use of micro-wicks or micro-catheters inserted through the tympanic membrane into the vicinity of the round window niche (Silverstein et al. 2004; Marks et al. 2000). Although these drug delivery methods may supply more predictable levels of medication to the inner ear in repeated doses, their safety in the setting of an implanted ear has not been studied and the possibility of introducing a route for developing middle ear infections cannot be ignored.

Sizable research efforts into the treatment of EIT have revolved around developing intracochlear drug-delivery techniques that can achieve predictable diffusion of medication throughout the cochlea. Using animal models of this form of hearing loss, researchers have demonstrated several techniques for intracochlear delivery of therapeutic agents with promising results. These include the pre-implantation use of drug-eluting polymers placed in the round window niche, direct injection into the scala tympani with a microsyringe, and the temporary insertion of drug-eluting silicon catheters (Chang et al. 2009; Paasche et al. 2009; Jolly et al. 2010). Studies employing these delivery mechanisms have demonstrated otoprotective effects including preservation of low frequency hearing using steroid formulations (Chang et al. 2009; Jolly et al. 2010). Nevertheless, most of these treatment techniques only provide a short-term perfusion of medication throughout the cochlea, placing in question their efficacy at ameliorating delayed injury from continued inflammation.

To study the prolonged treatment of electrode insertion trauma and combat delayed pathways of intracochlear inflammation, cell death, and eventual fibrosis, several authors have used novel osmotic minipump or reciprocating perfusion system devices, which can provide continuous controlled delivery of drugs into the inner ear (Brown et al. 1993; Eshraghi et al. 2007; Chen et al. 2005). Ruling out a purely washout effect that may occur from the removal of cytokines in the perilymph, these devices have demonstrated significant short- and long-term otoprotection with the administration of steroid solutions (Eshraghi et al. 2007).

Drug Elution from the Electrode Array

Evolving biomedical nanotechnology has allowed for the development of drug-eluting cochlear electrode arrays. Most of the research behind this technology involves the use of silicon-based electrode array coatings, modified to contain mixtures of micronized dexamethasone (mDXM). The steroid compound can be embedded in the silicon itself or be used as a coating. Embedding allows for a slow sustained release while coating allows for a faster more short-term release of an incorporated drug (Bohl et al. 2012). Pharmacokinetic studies have also demonstrated that this form of intracochlear drug delivery can produce sustained release of medication for up to 2 years, which may provide hearing protection for as long as 1 year post-implantation (Farahmand-Ghavi et al. 2010). Furthermore, *in vivo* studies comparing different steroid concentrations have shown that compared to a 2% mixture, a 10% dexamethasone eluting silicon rod can provide higher levels of steroid in the perilymphatic space up to 24 h post-implantation (Liu et al. 2015). This may allow for a greater total cumulative dose as well as a stronger early burst release, which may provide a therapeutic benefit for the EIT pattern of hearing loss (Liu et al. 2015; Bas et al. 2016). Most importantly, a recent *in vivo* dose-response study using electrode arrays containing different concentrations of silicon-micronized dexamethasone mix showed significant otoprotection against electrode insertion trauma in a guinea pig model of scala tympani electrode insertion (Bas et al. 2016). The authors of this work demonstrated significant protection of hair cells and intracochlear neural elements, as well as a reduction in implant impedance and intra-scalar fibrosis. Ultimately, this was associated with reductions in the initial EIT-induced elevations in auditory brainstem response (ABR)- and cochlear action potential (CAP)- hearing thresholds. Electrode arrays containing the highest steroid concentration (10% dexamethasone base) provided the best otoprotection (Bas et al. 2016) (Figs. 9.4 and 9.5).

Other studies addressing the relationship between surviving spiral ganglion neurons and hearing outcomes following cochlear implant insertion have also explored the use of neurotrophin-eluting polymer electrode coatings with promising results (Richardson et al. 2009; Staecker and Garnham 2010). With a similar interest some have also studied the use of growth factor-eluting electrode arrays, which has demonstrated favorable hearing outcomes, although no significant improvement in hair cell or spiral ganglion cell survival was observed (Kikkawa et al. 2014).

Mild Protective Hypothermia

Therapeutic hypothermia has been established as an important strategy for the treatment of neurological injuries (Polderman 2008). Hypothermia has also been previously shown to improve hearing outcomes in models of noise-induced hearing loss (Drescher 1976). Although not extensively studied, recent research has demonstrated that mild hypothermia (34 °C), provided peri- and intra-operatively in a rat

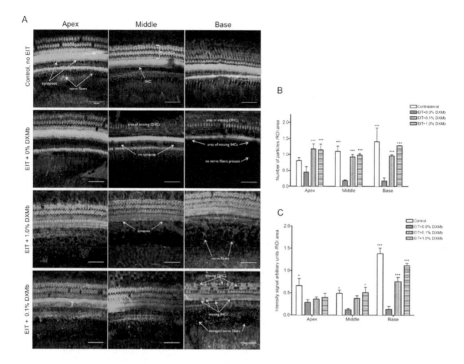

Fig. 9.5 Dexamethasone protects against electrode insertion trauma induced Hair cell loss in guinea pigs 90 days post electrode insertion eluted with different concentrations of dexamethasone (0, 0.1, and 1%). (a) Whole mounts of organ of Corti specimens immunestained for synaptic contacts (*magenta*), nerve fibers (*orange*), hair cells (*green*), and cell nuclei (*blue*). (b and c) Dexamethasone eluting electrodes significantly protected against electrode insertion trauma measured by the presence of nerve fibers (c, analysis of neurofilament-H immunostaining) and neural synapsis (b, analysis of synapsis immunostaining). *DXMb* dexamethasone base, *IHCs* inner hair cells, *OHCs* outer hair cells. In (b) and (c) mean values are plotted as bars depicting +/- SEM. *p> 0.05, **p> 0,01, ***p> 0.001. n=3 OC whole mounts/group. (Reproduced from Bas et al. 2016)

model of cochlear insertion trauma, can significantly improve hearing outcomes as measured by post-operative ABR threshold shifts and distortion product otoacoustic emission (DPOE) amplitudes (Balkany et al. 2005). Both an immediate as well as a delayed protective effect was noted in this study. Borrowing from the traumatic brain injury (TBI) literature, it is likely that the therapeutic properties of hypothermia, which have been shown to modulate inflammatory pathways like the TNFα, a receptor cascade, may protect against damage by similar inflammatory molecular pathways in the cochlea (Lotocki et al. 2006). Furthermore, alterations in the blood-brain barrier permeability, which favor accumulation of cytokines and leukocytes following TBI, have been ameliorated with the use of therapeutic hypothermia (Lotocki et al. 2009). It is possible that similar effects on the blood-labyrinth barrier may occur with the use of mild hypothermia, partially explaining its otoprotective properties in a model of EIT. Further research into the efficacy of mild hypothermia as a therapeutic agent for preservation of hearing in EIT is needed.

References

Ahmed ZM, Yousaf R, Lee BC, et al. Functional null mutations of MSRB3 encoding methionine sulfoxide reductase are associated with human deafness DFNB74. Am J Hum Genet. 2011;88(1):19–29. https://doi.org/10.1016/j.ajhg.2010.11.010; Epub 2010 Dec 23.

Andrabi SA, Dawson TM, Dawson VL. Mitochondrial and nuclear cross talk in cell death: parthanatos. Ann N Y Acad Sci. 2008;1147:233–41. https://doi.org/10.1196/annals.1427.014.

Balkany TJ, Eshraghi AA, Jiao H, et al. Mild hypothermia protects auditory function during cochlear implant surgery. Laryngoscope. 2005;115(9):1543–7.

Bas E, Bohorquez J, Goncalves S, et al. Electrode array-eluted dexamethasone protects against electrode insertion trauma induced hearing and hair cell losses, damage to neural elements, increases in impedance and fibrosis: a dose response study. Hear Res. 2016;337:12–24. https://doi.org/10.1016/j.heares.2016.02.003.

Bas E, Dinh CT, Garnham C, Polak M, Van de Water TR. Conservation of hearing and protection of hair cells in cochlear implant patients' with residual hearing. Anat Rec (Hoboken). 2012;295(11):1909–27. https://doi.org/10.1002/ar.22574; Epub 2012 Oct 8.

Bohl A, Rohm HW, Ceschi P, et al. Development of a specially tailored local drug delivery system for the prevention of fibrosis after insertion of cochlear implants into the inner ear. J Mater Sci Mater Med. 2012;23(9):2151–62. https://doi.org/10.1007/s10856-012-4698-z.

Brown JN, Miller JM, Altschuler RA, Nuttall AL. Osmotic pump implant for chronic infusion of drugs into the inner ear. Hear Res. 1993;70(2):167–72.

Chandler JM, Cohen GM, MacFarlane M. Different subcellular distribution of caspase-3 and caspase-7 following Fas-induced apoptosis in mouse liver. J Biol Chem. 1998;273(18):10815–8.

Chang A, Eastwood H, Sly D, James D, Richardson R, O'Leary S. Factors influencing the efficacy of round window dexamethasone protection of residual hearing post-cochlear implant surgery. Hear Res. 2009;255(1–2):67–72. https://doi.org/10.1016/j.heares.2009.05.010; Epub 2009 Jun 17.

Chen Z, Kujawa SG, McKenna MJ, et al. Inner ear drug delivery via a reciprocating perfusion system in the guinea pig. J Control Release. 2005;110(1):1–19.

Choi CH, Oghalai JS. Predicting the effect of post-implant cochlear fibrosis on residual hearing. Hear Res. 2005;205(1–2):193–200.

Clark GM, Shute SA, Shepherd RK, Carter TD. Cochlear implantation: osteoneogenesis, electrode-tissue impedance, and residual hearing. Ann Otol Rhinol Laryngol Suppl. 1995;166:40–2.

Cohen-Salmon M, Ott T, Michel V, et al. Targeted ablation of connexin 26 in the inner ear epithelial gap junction netwoek causes hearing impairment and cell death. Curr Biol. 2002;12(13):1106–11.

Cohen NL. Cochlear implant soft surgery: fact or fantasy? Otolaryngol Head Neck Surg. 1997;117(3 Pt 1):214–6.

de Waele C, Campos Torres A, Josset P, Vidal PP. Evidence for reactive astrocytes in rat vestibular and cochlear nuclei following unilateral inner ear lesion. Eur J Neurosci. 1996;8(9):2006–18.

Deltenre P, Van Maldergem L. Hearing loss and deafness in the pediatric population: causes, diagnosis, and rehabilitation. Handb Clin Neurol. 2013;113:1527–38. https://doi.org/10.1016/B978-0-444-59565-2.00023-X.

Dinh C, Hoang K, Haake S, et al. Biopolymer-released dexamethasone prevents tumor necrosis factor alpha-induced loss of auditory hair cells in vitro: implications toward the development of a drug-eluting cochlear implant electrode array. Otol Neurotol. 2008b;29(7):1012–9.

Dinh CT, Bas E, Chan SS, Dinh JN, Vu L, Van De Water TR. Dexamethasone treatment of tumor necrosis factor-alpha challenged organ of Corti explants activates nuclear factor kappa B signaling that induces changes in gene expression that favor hairs cell survival. Neuroscience. 2011;188:157–67. https://doi.org/10.1016/j.neuroscience.2011.04.061.

Dinh CT, Goncalves S, Bas E, Van De Water TR, Zine A. Molecular regulation of auditory hair cell death and approaches to protect sensory receptor cells and/or stimulate repair following acoustic trauma. Front Cell Neurosci. 2015;31(9):96. https://doi.org/10.3389/fncel.2015.00096. eCollection 2015.

Dinh CT, Haake S, Chen S, et al. Dexamethasone protects organ of corti explants against tumor necrosis factor-alpha-induced loss of auditory hair cells and alters the expression levels of apoptosis-related genes. Neuroscience. 2008a;157(2):405–13.

Dinh CT, Van De Water TR. Blocking pro-cell-death signal pathways to conserve hearing. Audiol Neurootol. 2009;14(6):383–92. https://doi.org/10.1159/000241895.

Dixon SJ, Lemberg KM, Lamprecht MR, et al. Ferroptosis: an iron-dependent form of nonapoptotic cell death. Cell. 2012;149(5):1060–72. https://doi.org/10.1016/j.cell.2012.03.042.

Drescher DG. Effect of temperature on cochlear responses during and after exposure to noise. J Acoust Soc Am. 1976;59(2):401–7.

Dworakowski R, Anilkumar N, Zhang M, Shah AM. Redox signalling involving NADPH oxidase-derived reactive oxygen species. Biochem Soc Trans. 2006;34(Pt 5):960–4.

Eastwood H, Pinder D, James D, et al. Permanent and transient effects of locally delivered n-acetyl cysteine in a guinea pig model of cochlear implantation. Hear Res. 2010;259(1–2):24–30. https://doi.org/10.1016/j.heares.2009.08.010; Epub 2009 Sep 2.

Eshraghi AA, Adil E, He J, Graves R, Balkany TJ, Van De Water TR. Local dexamethasone therapy conserves hearing in an animal model of electrode insertion trauma-induced hearing loss. Otol Neurotol. 2007;28(6):842–9.

Eshraghi AA, Gupta C, Van De Water TR, et al. Molecular mechanisms involved in cochlear implantation trauma and the protection of hearing and auditory sensory cells by inhibition of c-Jun-N-terminal kinase signaling. Laryngoscope. 2013;123(Suppl 1):S1–14. https://doi.org/10.1002/lary.23902.

Eshraghi AA, He J, Mou CH, et al. D-JNKI-1 treatment prevents the progression of hearing loss in a model of cochlear implantation trauma. Otol Neurotol. 2006;27(4):504–11.

Eshraghi AA. Prevention of cochlear implant electrode damage. Curr Opin Otolaryngol Head Neck Surg. 2006;14(5):323–8.

Farahmand-Ghavi F, Mirzadeh H, Imani M, et al. Corticosteroid-releasing cochlear implant: a novel hybrid of biomaterial and drug delivery system. J Biomed Mater Res B Appl Biomater. 2010;94(2):388–98. https://doi.org/10.1002/jbm.b.31666.

Galluzzi L, Vitale I, Abrams JM, et al. Molecular definitions of cell death subroutines: recommendations of the Nomenclature Committee on Cell Death 2012. Cell Death Differ. 2010;19(1):107–20. https://doi.org/10.1038/cdd.2011.96.

Garcia-Ibanez L, Macias AR, Morera C, et al. An evaluation of the preservation of residual hearing with the Nucleus Contour Advance electrode. Acta Otolaryngol. 2009;129(6):651–64. https://doi.org/10.1080/00016480802369278.

Guzman J, Ruiz J, Eshraghi AA. Triamcinolone acetonide protects auditory hair cells from 4-hydroxy-2,3-nonenal (HNE) toxicity in vitro. Acta Otolaryngol. 2006;126(7):685–90.

Haake SM, Dinh CT, Chen S, Eshraghi AA, van de Water TR. Dexamethasone protects auditory HCs against TNF alpha-initiated apoptosis via activation of PI3K/Akt and NF-kappa B signaling. Hear Res. 2009;255(1–2):22–32. https://doi.org/10.1016/j.heares.2009.05.003.

Hansen MR, Vijapurkar U, Koland JG, Gree SH. Reciprocal signaling between spiral ganglion neurons and schwann cells involves neuregulin and nuerotrophins. Hear Res. 2001;161:87–98. https://doi.org/10.1016/S0378-5955(01)00360-4.

Helbig S, Van de Heyning P, Kiefer J, et al. Combined electric acoustic stimulation with the PULSARCI(100) implant system using the FLEX(EAS) electrode array. Acta Otolaryngol. 2011;131(6):585–95. https://doi.org/10.3109/00016489.2010.544327.

Hirose K, Westrum LE, Cunningham DE, Rubel EW. Electron microscopy of degenerative changes in the chick basilar papilla after gentamicin exposure. J Comp Neurol. 2004;470(2):164–80.

Hoang KN, Dinh CT, Bas E, Chen S, Eshraghi AA, van de Water TR. Dexamethasone treatment of naïve organ of corti explants alters the expression pattern of apoptosis-related genes. Brain Res. 2009;1301:1–8. https://doi.org/10.1016/j.brainres.2009.08.097.

Hughes ML, Vander Weff KR, Brown CJ, et al. A longitudinal study of electrode impedance, the electrically evoked compound action potential, and behavioral measures in nucleus 24 cochlear implant users. Ear Hear. 2001;22:471–86. https://doi.org/10.1097/00003446-2001120000-00004.

Ichimiya I, Yoshida K, Kirano T, Suzuki M, Mogi G. Significance of spiral ligament fibrocytes with cochlear inflammation. Int J Pediatr Otorhinolaryngol. 2000;56(1):45–51.

Ihler F, Pelz S, Coors M, Matthias C, Canis M. Application of a TNF-alpha-inhibitor into the scala tympany after cochlear electrode insertion trauma in guinea pigs: preliminary audiologic results. Int J Audiol. 2014;53(11):810–6.

Inamura N, Salt AN. Permeability changes of the blood-labyrinth barrier measured in vivo during experimental treatments. Hear Res. 1992;61(1–2):12–8.

Jensen-Smith HC, Hallworth R, Nichols MG. Gentamicin rapidly inhibits mitochondrial metabolism in high-frequency cochlear outer hair cells. PLoS One. 2012;7(6):e38471. https://doi.org/10.1371/journal.pone.0038471; Epub 2012 Jun 8.

Jia S, Zhao Y, Law M, Galiano R, Mustoe TA. The effects of collagenase ointment on the prevention of hypertrophic scarring in a rabbit ear scarring model: a pilot study. Wounds. 2011;23(6):160–5.

Jolly C, Garnham C, Mirzadeh H, et al. Electrode features for hearing preservation and drug delivery strategies. Adv Otorhinolaryngol. 2010;67:28–42. https://doi.org/10.1159/000262594.

Keithley EM, Wang X, Barkdull GC. Tumor necrosis factor alpha can induce recruitment of inflammatory cells to the cochlea. Otol Neurotol. 2008;29(6):854–9. https://doi.org/10.1097/MAO.0b013e31818256a9.

Kepp O, Galluzzi L, Zitvogel L, Kroemer G. Pyroptosis—a cell death modality of its kind? Eur J Immunol. 2010;40(3):627–30. https://doi.org/10.1002/eji.200940160.

Kerr JF, Wyllie AH, Currie AR. Apoptosis: a basic biological phenomenon with wide-ranging implications in tissue kinetics. Br J Cancer. 1972;26(4):239–57.

Kiefer J, Gstoettner W, Baumgartner W, et al. Conservation of low-frequency hearing in cochlear implantation. Acta Otolaryngol. 2004;124(3):272–80.

Kikkawa YS, Nakagawa T, Ying L, et al. Growth factor-eluting cochlear implant electrode: impact on residual auditory function, insertional trauma, and fibrosis. J Transl Med. 2014;12:280. https://doi.org/10.1186/s12967-014-0280-4.

Kontorinis G, Paasche G, Lenarz T, Stöver T. The effect of different lubricants on cochlear implant electrode insertion forces. Otol Neurotol. 2011;32(7):1050–6. https://doi.org/10.1097/MAO.0b013e31821b3c88.

Laszig R, Ridder GJ, Fradis M. Intracochlear insertion of electrodes using hyaluronic acid in cochlear implant surgery. J Laryngol Otol. 2002;116(5):371–2.

Lehnhardt E. Intracochlear placement of cochlear implant electrodes in soft surgery technique. HNO. 1993;41(7):356–9.

Lesniak W, Pecoraro VL, Schacht J. Ternary complexes of gentamicin with iron and lipid catalyze formation of reactive oxygen species. Chem Res Toxicol. 2005;18(2):357–64.

Levine B, Kroemer G. Autophagy in the pathogenesis of disease. Cell. 2008;132(1):27–42. https://doi.org/10.1016/j.cell.2007.12.018.

Li-Korotky HS. Age-related hearing loss: quality of care for quality of life. Gerontologist. 2012;52(2):265–71. https://doi.org/10.1093/geront/gnr159.

Lin FR, Niparko JK, Ferrucci L. Hearing loss prevalence in the United States. Arch Intern Med. 2011;171(20):1851–2. https://doi.org/10.1001/archinternmed.2011.506.

Liu Y, Shoji-Kawata S, Sumpter RM Jr, et al. Autosis is a Na+,K+-ATPase-regulated form of cell death triggered by autophagy-inducing peptides, starvation, and hypoxia-ischemia. Proc Natl Acad Sci U S A. 2013;110(51):20364–71. https://doi.org/10.1073/pnas.1319661110.

Liu Y, Jolly C, Braun S, et al. Effects of dexamethasone-releasing implant on cochleae: a functional, morphological, and pharmacokinetic study. Hear Res. 2015;327:89–101. https://doi.org/10.1016/j.heares.2015.04.019.

Lotocki G, de Rivero Vaccari JP, Perez ER, et al. Therapeutic hypothermia modulates TNFR1 signaling in the traumatized brain via early transient activation of the JNK pathway and suppression of XIAP cleavage. Eur J Neurosci. 2006;24(8):2283–90.

Lotocki G, de Rivero Vaccari JP, Perez ER. Alterations in blood-brain barrier permeability to large and small molecules and leukocyte accumulation after traumatic brain injury: effects of post-traumatic hypothermia. J Neurotrauma. 2009;26(7):1123–34. https://doi.org/10.1089/neu.2008.0802.

Maeda K, Yoshida K, Ichimiya I, Suzuki M. Dexamethasone inhibits tumor necrosis factor-alpha-induced cytokine secretion from spiral ligament fibrocytes. Hear Res. 2005;202(1–2):154–60.

Marks S, Arenberg IK, Hoffer ME. Round window microcatheter administered microdose of gentamycin: an alternative in the treatment of tinnitus in patients with Menière's disease. Laryngorhinootologie. 2000;79(6):327–31.

Martinez FO, Sica A, Mantovani A, Locati M. Macrophage activation and polarization. Front Biosci. 2008;13:453–61. https://doi.org/10.2741/2692.

Menardo J, Tang Y, Ladrech S, et al. Oxidative stress, inflammation, and autophagic stress as the key mechanisms of premature age-related hearing loss in SAMP8 mouse Cochlea. Antioxid Redox Signal. 2012;16(3):263–74. https://doi.org/10.1089/ars.2011.4037.

Moriyama M, Yoshida K, Ichimiya I, Suzuki M. Nitric Oxide production from cultures spiral ligament fibrocytes: effects of corticosteroids. Acta Otolaryngol. 2007;127(7):676–81.

Mosca F, Grassia R, Leone CA. Longitudinal variations in fitting parameters for adult cochlear implant recipients. Acta Otorhinolaryngol Ital. 2014;34:111–6.

Nagańska E, Matyja E. Ultrastructural characteristics of necrotic and apoptotic mode of neuronal cell death in a model of anoxia in vitro. Folia Neuropathol. 2001;39(3):129–39.

Olivetto E, Simoni E, Guaran V, Astolfi L, Martini A. Sensorineural hearing loss and ischemic injury: Development of animal models to assess vascular and oxidative effects. Hear Res. 2015;327:58–68. https://doi.org/10.1016/j.heares.2015.05.004.

Op de Beeck K, Schacht J, Van Camp G. Apoptosis in acquired and genetic hearing impairment: the programmed death of the hair cell. Hear Res. 2011;281(1–2):18–27. https://doi.org/10.1016/j.heares.2011.07.002.

Paasche G, Tasche C, Stöver T, Lesinski-Schiedat A, Lenarz T. The long-term effects of modified electrode surfaces and intracochlear corticosteroids on postoperative impedances in cochlear implant patients. Otol Neurotol. 2009;30(5):592–8. https://doi.org/10.1097/MAO.0b013e3181ab8fba.

Polderman KH. Induced hypothermia and fever control for prevention and treatment of neurological injuries. Lancet. 2008;371(9628):1955–69. https://doi.org/10.1016/S0140-6736(08)60837-5.

Postelmans JT, van Spronsen E, Grolman W, et al. An evaluation of preservation of residual hearing using the suprameatal approach for cochlear implantation: can this implantation technique be used for preservation of residual hearing? Laryngoscope. 2011;121(8):1794–9. https://doi.org/10.1002/lary.21866.

Radeloff A, Unkelbach MH, Tillein J, et al. Impact of intrascalar blood on hearing. Laryngoscope. 2007;117(1):58–62.

Rajan GP, Kuthubutheen J, Hedne N, Krishnaswamy J. The role of preoperative, intratympanic glucocorticoids for hearing preservation in cochlear implantation: a prospective clinical study. Laryngoscope. 2012;122(1):190–5. https://doi.org/10.1002/lary.22142.

Reichert F, Saada A, Rotshenker S. Peripheral nerve injury induces schwann cells to express two macrophage phenotypes: phagocytosis and the galactose-specific lectin MAC-2. J Neurosci. 1994;14:3231–45.

Richardson RT, Wise AK, Thompson BC, et al. Polypyrrole-coated electrodes for the delivery of charge and neurotrophins to cochlear neurons. Biomaterials. 2009;30(13):2414–24. https://doi.org/10.1016/j.biomaterials.2009.01.015.

Roehm PC, Hansen MR. Strategies to preserve or regenerate spiral ganglion neurons. Curr Opin Otolaryngol Head Neck Surg. 2005;13:294–300. https://doi.org/10.1097/01.moo.0000180919.68812.b9.

Romand MR, Romand R. Development of spiral ganglion cells in mammalian cochlea. J Electron Microsc Tech. 1990;15:144–54.

Rybak LP, Whitworth CA, Mukherjea D, Ramkumar V. Mechanisms of cisplatin-induced ototoxicity and prevention. Hear Res. 2007;226(1–2):157–67.

Sha SH, Taylor R, Forge A, Schacht J. Differential vulnerability of basal and apical hair cells is based on intrinsic susceptibility to free radicals. Hear Res. 2001;155:1–8.

Shamash S, Reichert F, Rotshenker S. The cytokine nerwork of wallerian degeneration: turmor necrosis factor alpha, interleukin-1 alpha, and interleukin-1 beta. J Neuroscie. 2002;22:3052–60.

Silverstein H, Thompson J, Rosenberg SI, Brown N, Light J. Silverstein MicroWick. Otolaryngol Clin N Am. 2004;37(5):1019–34.

Skarzynski H, Lorens A, Piotrowska A, Anderson I. Preservation of low frequency hearing in partial deafness cochlear implantation (PDCI) using the round window surgical approach. Acta Otolaryngol. 2007a;127(1):41–8.

Skarzynski H, Lorens A, Piotrowska A, Anderson I. Partial deafness cochlear implantation in children. Int J Pediatr Otorhinolaryngol. 2007b;71(9):1407–13.

Smith DI, Lawrence M, Hawkins JE Jr. Effects of noise and quinine on the vessels of the stria vascularis: an image analysis study. Am J Otolaryngol. 1985;6(4):280–9.

Souter MA, Briggs RJ, Wright CG, Roland PS. Round window insertion of precurved perimodiolar electrode arrays: how successful is it? Otol Neurotol. 2011;32(1):58–63. https://doi.org/10.1097/MAO.0b013e3182009f52.

Staecker H, Garnham C. Neurotrophin therapy and cochlear implantation: translating animal models to human therapy. Exp Neurol. 2010;226(1):1–5. https://doi.org/10.1016/j.expneurol.2010.07.012.

Stout RD. Editorial: macrophage functional phenotypes: no alternatives in dermal wound healing? J Leukoc Biol. 2010;87:19–21. https://doi.org/10.1189/jlb.0509311.

Sutinen P, Zou J, Hunter LL, Toppila E, Pyykkö I. Vibration-induced hearing loss: mechanical and physiological aspects. Otol Neurotol. 2007;28(2):171–7.

Takumi Y, Nishio SY, Mugridge K, et al. Gene expression pattern after insertion of dexamethasone-eluting electrode into the guinea pig cochlea. PLoS One. 2014;9(10):e110238. https://doi.org/10.1371/journal.pone.0110238.

Taylor RR, Nevill G, Forge A. Rapid hair cell loss: a mouse model for cochlear lesions. J Assoc Res Otolaryngol. 2008;9(1):44–64.

Teubner B, Michel V, Pech J, et al. Connexin 30 (Gjb6)-deficiency causes severe hearing impairment and lack of endocochlear potential. Hum Mol Genet. 2003;12(1):13–21.

Tofaris GK, Patterson PH, Jessen KR, Mirsky R. Denervated Schwann cells attract macrophages by secretion of leukemia inhibitory factor (LIF) and monocyte chemoattractant protein-1 in a process regulated by interleukin-6 and LIF. J Neurosci. 2002;22:6696–703.

Tsujimoto Y, Shimizu S. Role of the mitochondrial membrane permeability transition in cell death. Apoptosis. 2007;12(5):835–40.

Usami S, Moteki H, Suzuki N, et al. Achievement of hearing preservation in the presence of an electrode covering the residual hearing region. Acta Otolaryngol. 2011;131(4):405–12. https://doi.org/10.3109/00016489.2010.539266; Epub 2011 Jan 5.

Vogelezang MG, Liu Z, Relvas JB, Raivich G, Scherer SS, French-Constant C. Alpha4 integrin is expressed during peripheral nerve regeneration and enhances neurite outgrowth. J Neurosci. 2001;21:6732–44.

Walsh T, Pierce SB, Lenz DR, et al. Genomic duplication and overexpression of TJP2/ZO-2 leads to altered expression of apoptosis genes in progressive nonsyndromic hearing loss DFNA51. Am J Hum Genet. 2010;87(1):101–9. https://doi.org/10.1016/j.ajhg.2010.05.011; Epub 2010 Jun 17.

Wan G, Corfas G, Stone JS. Inner ear supporting cells: rethinking the silent majority. Semin Cell Dev Biol. 2013;24(5):448–59. https://doi.org/10.1016/j.semcdb.2013.03.009; Epub 2013 Mar 29.

Wang J, Van De Water TR, Bonny C, de Ribaupierre F, Puel JL, Zine A. A peptide inhibitor of c-Jun N-terminal kinase protects against both aminoglycoside and acoustic trauma-induced auditory hair cell death and hearing loss. J Neurosci. 2003;23(24):8596–607.

Wangemann P, Itza EM, Albrecht B, et al. Loss of KCNJ10 protein expression abolishes endocochlear potential and causes deafness in Pendred syndrome mouse model. BMC Med. 2004;2:30.

Wei BP, Stathopoulos D, O'Leary S. Steroids for idiopathic sudden sensorineural hearing loss. Cochrane Database Syst Rev. 2013;7:CD003998. https://doi.org/10.1002/14651858. CD003998.pub3.

Wolfe J, Baker RS, Wood M. Clinical case study review: steroid-responsive change in electrode impedance. Otol Neurotol. 2013;34:227–32. https://doi.org/10.1097/MAO.0B013e31827b4bba.

Yang WP, Henderson D, Hu BH, Nicotera TM. Quantitative analysis of apoptotic and necrotic outer hair cells after exposure to different levels of continuous noise. Hear Res. 2004;196(1–2):69–76.

Ydens E, Cauwels A, Asselbergh B, et al. Acute injury in the peripheral nervous system triggers an alternative macrophage response. J Neuroinflammation. 2012;9:176. https://doi.org/10.1186/1742-2094-9-176.

Ylikoski J, Xing-Qun L, Virkkala J, Pirvola U. Blockade of c-Jun N-terminal kinase pathway attenuates gentamicin-induced cochlear and vestibular hair cell death. Hear Res. 2002;163(1–2):71–81.

Zheng HW, Chen J, Sha SH. Receptor-interacting protein kinases modulate noise-induced sensory hair cell death. Cell Death Dis. 2014;5:e1262. https://doi.org/10.1038/cddis.2014.177.

Zou J, Bretlau P, Pyykkö I, et al. Comparison of the protective efficacy of neurotrophins and antioxidants for vibration-induced trauma. ORL J Otorhinolaryngol Relat Spec. 2003;65(3):155–61.

Zou J, Pyykkö I, Sutinen P, Toppila E. Vibration induced hearing loss in guinea pig cochlea: expression of TNF-alpha and VEGF. Hear Res. 2005;202(1–2):13–20.

Chapter 10
Anti-inflammatory Therapies for Sensorineural Hearing Loss

Alanna M. Windsor and Michael J. Ruckenstein

Abstract Anti-inflammatory and immunosuppressive therapies have been widely employed in the treatment of sensorineural hearing loss in the context of autoimmune inner ear disease (AIED) and idiopathic sudden sensorineural hearing loss (ISSHL). While steroids are the mainstay of treatment for these disorders, numerous other therapies, including cyclophosphamide, methotrexate, azathioprine, rituximab, anakinra, anti- TNF-α agents, and plasmapharesis have been investigated. Here we will describe the most commonly-studied of these immune-modulating therapies and review the evidence for their efficacy in the treatment of inner ear disorders, focusing on AIED and ISSHL. Further investigation of the potential inflammatory mechanisms mediating these forms of sensorineural hearing loss may ultimately identify targets for future treatments.

Keywords Sensorineural hearing loss · Autoimmune inner ear disease · Corticosteroids · Immunomodulation · Sudden hearing loss

1 Introduction

The role of inflammatory mechanisms in the pathogenesis of sensorineural hearing loss (SNHL) holds great interest for researchers as it suggests the possibility of reversing hearing loss through the use of anti-inflammatory or immunosuppressive medical therapies. Corticosteroids, in particular, have been employed in the treatment of hearing loss since the 1950s with varying degrees of success, and their effectiveness in certain cases has been used as evidence of an underlying immune-mediated mechanism (Trune and Canlon 2012). One entity, autoimmune inner ear disease (AIED), has been partially defined by its response to immunosuppressive medications (McCabe 1979). However, immunosuppressive therapies have also been used in other conditions including idiopathic sudden sensorineural hearing

A. M. Windsor (✉) · M. J. Ruckenstein
Department of Otorhinolaryngology-Head and Neck Surgery,
University of Pennsylvania, Philadelphia, PA, USA
e-mail: alanna.windsor@uphs.upenn.edu; michael.ruckenstein@uphs.upenn.edu

© Springer International Publishing AG, part of Springer Nature 2018
V. Ramkumar, L. P. Rybak (eds.), *Inflammatory Mechanisms in Mediating Hearing Loss*, https://doi.org/10.1007/978-3-319-92507-3_10

loss (ISSHL), Meniere's disease, and SNHL related to systemic autoimmune diseases such as granulomatosis with polyangiitis, systemic lupus erythematous, and Cogan syndrome. Each of these conditions can have considerable overlap in their presentation, may be difficult to distinguish on initial presentation, and may in fact encompass many different disorders with heterogeneous pathologic mechanisms.

This lack of clarity in the underlying inner ear pathophysiology of these diseases makes the directed study of therapeutic options challenging. Indeed, many treatments have been tested empirically on the basis of their efficacy in systemic autoimmune diseases, under the presumption that the SNHL seen in AIED and a least a subset of patients with ISSHL and Meniere's disease is related to an underlying inflammatory or immune-mediated process. In several studies examining the effects of various immunosuppressive medications on hearing loss, patients considered to have Meniere's disease have been included under the umbrella of 'AIED' or been labeled as having 'immune-mediated Meniere's disease' (Matsuoka and Harris 2013; Matteson et al. 2000, 2005). The absence of a definitive diagnostic test, variable presentation, fluctuating course, low incidence, and often spontaneous improvement of hearing in these diseases create additional challenges for study design in this population.

Despite an extensive body of literature examining the use of steroids in hearing loss, our understanding of the primary mechanisms through which steroids act in the inner ear is limited. Steroids have become first-line therapy for both AIED and ISSHL although the optimal choice of drug, dose, and route are debated. Many other immunosuppressive therapies have been studied in the treatment of AIED, though these studies tend to be retrospective or observational in nature and are limited by small sample sizes. The clinician must therefore balance the potential benefits against the considerable risk of side effects from these therapies. This chapter will review the role of anti-inflammatory and immunosuppressive therapies in various inner ear disorders, with a particular focus on AIED and ISSHL.

2 Autoimmune Inner Ear Disease

Steroids in Autoimmune Inner Ear Disease

Corticosteroids are a class of molecule with a wide array of effects in nearly every organ system. Upon binding to the glucocorticoid receptor, a member of the nuclear receptor family, they allow translocation of the receptor to the cell nucleus, where they regulate transcription of corticosteroid-responsive genes. Signaling through this pathway leads to apoptosis of inflammatory cells and suppression of expression of the proinflammatory cytokine tumor necrosis factor-α (TNF-α) and interleukin-1β (IL-1β) expression, among other effects (Flammer and Rogatsky 2011). Steroids have been the primary therapy for AIED ever since McCabe first described, in 1979, a series of patients with an unusual form progressive, bilateral sensorineural hearing loss which he postulated was autoimmune in etiology and which responded to treatment with dexamethasone and cyclophosphamide (McCabe 1979). Indeed, response to steroids has been used as a diagnostic criterion for AIED, as no single definitive diagnostic test exists (García-Berrocal et al. 2003). Since McCabe's study, numerous case series and animal studies have emerged to evaluate the efficacy of steroids in AIED, develop treatment algorithms, and elucidate the mechanisms underlying the steroid response in AIED.

Animal Studies

Despite their effectiveness in reversing hearing loss related to AIED, the actions of steroids in the inner ear are unclear. Animal models have therefore proven useful not only in investigating the pathogenesis of AIED but also in revealing potential pathways through which steroids may exert their effects. One such model is the MRL-Fas[lpr] mouse, which carries a mutation in the *Fas* gene that prevents apoptosis of self-recognizing T lymphocytes. These mice develop a systemic autoimmune disease similar to systemic lupus erythematous as well as elevated auditory brainstem response (ABR) thresholds and pathologic changes within the stria vascularis, which is responsible for maintaining the endocochlear potential; these changes include intracellular edema, cellular degeneration and intra-capillary antibody deposition (Ruckenstein et al. 2009; Ruckenstein and Hu 1999).

Trune et al. demonstrated that administration of oral prednisolone in MRL-Fas[lpr] mice prior to the onset of systemic autoimmune disease and hearing loss could prevent hearing loss in treated animals compared to untreated controls (Trune et al. 1999a). Moreover, a companion study showed that when prednisolone was administered after the onset of clinical disease, ABR thresholds stabilized or improved in 53% of mice as compared to 25% in untreated controls (Trune et al. 1999b). Ruckenstein et al. investigated the pathogenesis of strial disease in AIED by treating MRL-Fas[lpr] mice with dexamethasone beginning at 6 weeks of age, before autoimmune disease onset, and examined inner ear histology in animals sacrificed at 20 weeks (Ruckenstein et al. 1999). The authors found that dexamethasone administration reduced serum immunoglobulin levels, decreased lymphoid hyperplasia, improved renal function, and prevented antibody deposition in the stria vascularis of treated mice. However, steroid-treated mice still developed strial cellular edema and degeneration similar to mice that were untreated, suggesting that, while steroids were able to eliminate antibody deposition, strial degeneration was mediated through another process.

Though steroids' improvement of cochlear dysfunction in AIED has often been ascribed to inhibition of the immune-mediated inflammatory response, glucocorticoids can also bind to mineralocorticoid receptors expressed in the inner ear and thereby influence ion transport (Trune and Canlon 2012). Trune et al. have hypothesized that corticosteroids, by acting on mineralocorticoid receptors, may reverse hearing loss in AIED by restoring ion homeostasis in the stria vascularis (Trune et al. 2006). Using the MRL-Fas[lpr] mouse model for autoimmune SNHL, the authors tested hearing in mice treated with either aldosterone, a mineralocorticoid, or prednisolone (Ruckenstein 2004). Mice treated with aldosterone experienced similar hearing improvement to those given prednisolone. Examination of stria vascularis morphology of mice in the aldosterone treatment group revealed a reversal of the edema and degeneration seen in the untreated mice. Mice in the prednisolone group showed some improvement in the appearance of the stria, though not to the same degree as in the aldosterone group. A follow-up study demonstrated that mice treated with prednisolone and spironolactone, a mineralocorticoid receptor antagonist, had a hearing decline similar to mice who were not treated at all, suggesting that prednisolone's hearing effects in this mouse model were mediated through its action on the mineralocorticoid receptor (Trune et al. 2006). While these studies point to an interesting means by which steroids can reverse inner ear damage, whether or not the mouse model accurately reflects the true pathogenic events of AIED in human populations is unknown.

Human Studies

An early report of AIED treatment in human subjects emerged in 1984, when Hughes et al. reviewed the clinical experience with AIED at their institution (Hughes et al. 1984). The authors advocated initiating treatment with high-dose, short-term prednisone, followed by a lower maintenance dose over a subsequent period of weeks to months, reserving cytotoxic medications for those patients who did not respond to steroids. They noted that response time to treatment was variable, ranging from rapid recovery within weeks to a delayed recovery of hearing over the course of months. While no prospective, randomized clinical trials have compared the efficacy of various steroid doses, routes of administration, and duration of treatment, initial treatment with oral prednisone 1 mg/kg/day or 60 mg for a period of 4 weeks has come to be most commonly used (Broughton et al. 2004; Niparko et al. 2005; Ryan et al. 2009; Ruckenstein 2004). Hearing is tested at the start of treatment and at the end of 4 weeks. Various criteria have been used to define who is a steroid-responder, for example: if pure-tone thresholds improve by at least 15 dB at one frequency or 10 dB at two or more consecutive frequencies; if speech discrimination scores improve by 12%; if the average (PTA) threshold improves by 10 dB (Broughton et al. 2004; Niparko et al. 2005). Steroid-responders are then tapered over a variable length of time. If steroid-responders experience a deterioration in hearing after tapering of steroids, they are then restarted on high-dose steroids. Patients who show no response after the initial treatment period, however, are rapidly weaned off and considered for alternative therapies.

Clinical response to steroids is variable. Rauch et al. reported an overall steroid response rate of 60% in patients with AIED treated at their institution, though it is unclear over what length of time this was measured (Rauch 1997). In a cohort of patients with AIED reviewed by Broughton et al., 70% showed an initial response to steroids however the response was often not sustained over the mean follow-up period of 34.4 months (Broughton et al. 2004). Patients often required a repeat course of high-dose steroids and 71% of initial steroid responders ultimately required treatment with alternative immunosuppressive therapies at some point.

In a prospective study of 116 patients with AIED, Niparko et al. sought to describe with more precision the effect of prednisone treatment on the audiometric profile of patients after 4 weeks of therapy (Niparko et al. 2005). Most subjects experienced improvement in or stabilization of their hearing over that time period. Pure-tone averages (PTA) improved by 1 dB or more in 53.5% of subjects and remained stable in 29% of subjects while mean PTA improved from 52.4 dB to 48.3 dB in the better-hearing ear. Similarly, 59.5% of subjected experienced at least a 2% improvement in speech discrimination, with speech discrimination remaining stable in another 18.1% of subjects. Across all subjects, speech discrimination improved from 71.4 to 78.1% in the better-hearing ear. Loveman et al. reviewed 30 patients with a diagnosis of AIED (Loveman et al. 2004). In their series, the mean initial steroid dose was 35.1 mg and mean initial duration of treatment was 2.2 weeks, with a mean total duration of therapy of 7.3 weeks. Patients who did not respond to treatment were then given a 2- to 3-week course of prednisone at a higher dose of 1 mg/kg/day; steroids were discontinued if they did not respond to this dose, and patients were considered for treatment with methotrexate. Patients who were initial responders to steroids but relapsed after steroids were discontinued were given a second course of therapy at the previously successful dose. Fifty percent of patients in this cohort met criteria for audiometric improvement with steroids, while 12% experienced stable hearing. The authors found that this management strategy, while achieving hearing outcomes consistent with previous reports, resulted in a lower average dose and duration of steroid therapy. They suggest that the commonly recommended initial prednisone dose of 1 mg/kg and treatment duration of 4 weeks may be unnecessary for satisfactory outcomes, though other studies suggest a shorter duration of treatment may place patients at a higher risk of relapse (Rauch 1997).

Corticosteroids can have serious long-term side effects including osteoporosis, hypertension, glaucoma, weight gain, hyperglycemia, and adverse psychological effects, however one prospective, long-term study suggests that they are safe and generally well-tolerated in patients being treated for AIED (Alexander et al. 2009). Alexander et al. analyzed adverse events in 116 patients with AIED who were given high-dose prednisone as part of a prospective trial comparing methotrexate to prednisone treatment (Alexander et al. 2009). Study subjects received prednisone 60 mg/day as part of a 1-month challenge, and those whose hearing improved underwent an 18-week prednisone taper. Subjects were followed for a mean of 66 weeks, with few serious adverse events occurring during that period. A total of 16 patients (14%) experienced adverse events during the initial 1-month prednisone challenge, and 7 patients (6%) were unable to complete 1 month of treatment due to an adverse event. Of patients who completed the full 22-week prednisone course, the most common adverse events were hyperglycemia (17.6%), abdominal pain, shortness of breath, elevated liver function tests, and joint pains (5.9% each). Weight gain was also common. No incidences of osteonecrosis or fractures were reported.

In order to avoid the toxicities of systemic steroids as well as potentially benefit from higher inner ear drug levels, the use of intratympanic (IT) steroids has been investigated (Parnes et al. 1999). One animal study suggested IT steroids were not effective in improving hearing or reducing inner ear inflammatory infiltrates in a guinea pig model of immune-mediated labyrinthitis, and human studies have been limited to small case series (Parnes et al. 1999; Yang et al. 2000; Harris et al. 2013; García-Berrocal et al. 2006). Harris et al. described a series of 4 patients with AIED, of which 3 demonstrated improved hearing after IT steroid injections; however the patients were also receiving other immunosuppressive medications at the same time, including systemic steroids, so it is difficult to determine which, if any, intervention was effective (Harris et al. 2013). A retrospective case series of patients with AIED who were either refractory to or unable to wean from steroids included 11 patients who additionally failed or refused methotrexate therapy and were treated with IT methylprednisolone (García-Berrocal et al. 2006). Patients were given 6-methylprednisolone (0.3–0.5 mL of 40 mg/mL solution), weekly over a period of at least 2 months. Hearing was improved in 6 patients (54.5%) stable in 3 (27.3%), and worse in 2 (18.2%) and vestibular symptoms improved in all affected patients.

Identification of markers of steroid-responsiveness has been an active area of investigation. In 1990, Harris and Sharp detected antibodies to a 68-kD inner ear antigen in patients with suspected immune-mediated hearing loss using Western blot analysis of patient serum (Harris and Sharp 1990). In one series, 89% of patients with idiopathic, bilateral, progressive SNHL had antibodies to this protein and, moreover, 75% of those who were seropositive responded to treatment with prednisone while only 18% of seronegative patients responded (Moscicki et al. 1994). The authors suggested that the presence of these antibodies could therefore be used to predict which patients will have favorable responses to steroids in order to guide treatment decision-making. However, more recent studies have failed to find this correlation between anti-68-kD antibody status and steroid responsiveness (Broughton et al. 2004; Zeitoun et al. 2005). This discrepancy could be explained by the test's high specificity (90%), but low sensitivity (42%) in predicting steroid-responsiveness in a series of patients with suspected AIED; many patients who are antibody-negative will therefore also respond to steroid therapy (Hirose et al. 1999). Zeitoun and colleagues used an immunofluorescence-based assay to detect antibodies against an inner-ear supporting cell antigen in patients with suspected AIED (Zeitoun et al. 2005). Though they found no correlation between steroid-responsiveness and presence of the 68-kD protein based on the Western blot serum analysis as described by Harris and Sharp, they found that the presence of antibodies using the immunofluorescence was significantly associated with steroid-responsiveness. Immunofluorescence-positive patients were almost three times as likely to respond to treatment as those who were negative. The authors suggest that the Western blot test could be detecting other clinically-irrelevant proteins of a similar weight while the immunofluorescence test more specifically targets antibodies with a specific binding pattern on inner ear supporting cells. Therefore this assay may hold value in the future in guiding the use of steroids in patients with AIED.

In addition to specific antibodies, alterations in expression of the cytokine IL-1β and its receptor may be markers for steroid-responsiveness in AIED (Pathak et al. 2011; Vambutas et al. 2009). Recent studies have pointed to a potential role of cytokines in the pathogenesis of AIED, in particular, those in the interleukin-1 (IL-1) family (Pathak et al. 2011; Vambutas et al. 2009). For instance, IL-1β, a proinflammatory cytokine, is expressed by fibrocytes in the spiral ligament and spiral limbus in response to cochlear injury, and expressed by infiltrating inflammatory cells after the introduction of antigen into the cochlea of a systemically sensitized mouse (Satoh et al. 2002). Vambutas et al. examined interleukin 1 Receptor Type II (IL1R2), a protein expressed on the surface of B cells, macrophages, and neutrophils that sequesters IL-1β and thereby inhibits its proinflammatory effects (Vambutas et al. 2009). Expression of IL1R2 is induced by steroids. The authors found that patients with AIED who responded to corticosteroids had peripheral blood mononuclear cells (PBMC) that showed a robust increase in IL1R2 expression in vitro in response to dexamethasone, while PBMCs of steroid non-responders showed minimal increase. In a follow-up study, corticosteroid-responders also had lower circulating plasma levels of IL-1β and their PBMCs showed suppressed transcription of IL-1β in response to dexamethasone in vitro compared to non-responders (Pathak et al. 2011). These studies together suggest a potential method of predicting steroid-responsiveness as well as a mechanism through which steroids may exert an effect by altering IL-1β signaling pathways. However, the methodology incorporated in these studies has been questioned as their entry criteria and the audiometric criteria used do not conform to accepted norms.

Non-steroid Immunosuppressive Therapies for Autoimmune Inner Ear Disease

Given the undesirable side effects of long-term steroid treatment, significant proportion of patients with AIED who fail to respond to steroids, and frequent lack of sustained response to steroids over time, many have sought to identify alternative therapies (see Table 10.1 for summary). In McCabe's description of immune-mediated SNHL, he advocated the use of cyclophosphamide in addition to steroids (McCabe 1979). Other treatments described have included therapies such as plasmapharesis; immunosuppressive agents such as mycophenolate mofetil, methotrexate, and azathioprine; and biologic agents such as etanercept, adalimumab, infliximab, anakinra, and rituximab. Nonetheless, the relative rarity of AIED and often challenging diagnosis have resulted in a paucity of rigorous studies evaluating the efficacy of various treatment options relative to steroids. Indeed, a recent systematic review of non-steroid therapies for AIED concluded that "clear evidence of an effective treatment for AIED from high-quality prospective trials remains lacking" (Brant et al. 2015).

Table 10.1 Summary of immunosuppressive therapies investigated in AIED and their mechanisms

Therapy	Mechanism	Role in AIED
Corticosteroids (prednisone, prednisolone, dexamethasone, methylprednisolone)	Act on glucocorticoid receptor to influence transcription of a wide array of corticosteroid-responsive genes; results in up-regulation of anti-inflammatory cytokines and suppression of proinflammatory cytokines May also interact with mineralocorticoid receptor to control sodium reabsorption	Mainstay of treatment, showing benefit in both animal and human studies
Methotrexate	Inhibitor of dihydrofolate reductase, thereby interfering with DNA synthesis	Small retrospective and prospective series suggest benefit (Matteson et al. 2000; Sismanis et al. 2016; Lasak et al. 2001; Salley et al. 2001), however 1 RCT showed no effect compared to placebo (Harris et al. 2003)
Cyclophosphamide	Alkylating agent that cross-links DNA strands and disrupts cell growth and division; may act through other immunomodulatory mechanisms	Early case reports and small case series suggested benefit when used with steroids (McCabe 1979; Clements et al. 1989; Berrettini et al. 1998; Plester and Soliman 1989); equivocal results reported in retrospective studies (Broughton et al. 2004; Lasak et al. 2001; Veldman et al. 1993) Use limited by significant systemic toxicities
Azathioprine	Purine analog that interferes with nucleic acid metabolism	One uncontrolled prospective study suggested benefit when used with steroids (Saraçaydin et al. 2016); equivocal results in retrospective studies (Broughton et al. 2004; Lasak et al. 2001) Use limited by significant systemic toxicities

(continued)

Table 10.1 (continued)

Therapy	Mechanism	Role in AIED
TNF-α inhibitors (etanercept, infliximab, adalimumab, and golimumab)	Antagonists to the activity of TNF-α, a proinflammatory cytokine	Etanercept showed benefit in 1 uncontrolled prospective study (Rahman et al. 2001) while another was less favorable (Matteson et al. 2005); one RCT showed no benefit compared to placebo (Cohen et al. 2005) Infliximab improved hearing in case reports after failure of conventional therapies (Heywood et al. 2013; André et al. 2015); showed no benefit in 1 retrospective study (Liu et al. 2011); 1 small uncontrolled prospective study showed benefit of local infliximab infusion in patients who relapsed or could not wean from steroids (Van Wijk et al. 2006) IT gomalimumab did not clearly show benefit in a small prospective study (Derebery et al. 2014)
Anakinra	Competitive inhibitor of the IL-1 receptor type, part of IL-1 proinflammatory signaling pathway	Showed benefit in steroid non-responders in a small open-label, uncontrolled prospective study (Vambutas et al. 2014)
Rituximab	Monoclonal antibody against the CD20 antigen found on the surface of lymphocytes that results in elimination of B cells	Showed hearing benefit in a case report in a patient with Cogan's syndrome (Orsoni et al. 2010); no clear benefit in a retrospective study (Matsuoka and Harris 2013) 1 prospective study suggested patients able to maintain hearing gains from steroids with rituximab (Cohen et al. 2011)
Plasmapharesis	May remove autoantibodies, immune complexes, and other disease mediators from systemic circulation	Case reports show benefit in systemic autoimmune disease (Alpa et al. 2011; Hamblin et al. 1982; Kobayashi et al. 1992; Brookes and Newland 1986); one small prospective series did not show statistically significant hearing improvement with treatment (Luetje and Berliner 1997)

Abbreviations: *TNF-α* tumor necrosis factor-α, *IL-1* interleukin-1, *IT* intratympanic, *RCT* randomized controlled trial

Methotrexate

Methotrexate, an inhibitor of dihydrofolate reductase that is commonly used in the treatment of rheumatoid arthritis and other autoimmune and neoplastic disorders, is among the most studied alternative therapies to steroids for the treatment of AIED. In 1994, Sismanis et al. suggested methotrexate may hold benefit when they reported on a series of five patients with AIED who were treated with methotrexate, most of whom had discontinued steroid therapy due to adverse effects (Sismanis et al. 2016). Patients were given oral methotrexate 7.5 mg weekly, which was then increased to 15 mg weekly in most cases. The authors observed a significant improvement in speech discrimination after treatment as well as patient-reported symptoms of tinnitus and vertigo, though no significant change in PTAs. One patient experienced mild hair thinning, however methotrexate was otherwise tolerated well. Since then, other retrospective studies have indicated possible therapeutic benefit of methotrexate in stabilizing or improving auditory or vestibular symptoms associated with AIED, including one study that examined patients with bilateral Meniere's disease (Lasak et al. 2001). In particular, because methotrexate is well-tolerated over the long term, it was felt to be a promising substitute for prednisone in those patients whose disease was steroid-dependent.

Nonetheless, evidence from prospective studies is mixed. Several open-label, prospective studies of patients with AIED (two of which included patients with Meniere's disease and Cogan's syndrome) found hearing improved with methotrexate treatment in 53–82% of patients who had initially been treated with steroids (Matteson et al. 2000; Salley et al. 2001). However, the only randomized, placebo-controlled, double-blind trial aimed at assessing the efficacy of methotrexate in maintaining hearing improvements in patients with AIED after prednisone treatment did not show any benefit (Harris et al. 2003). One hundred and sixteen patients with AIED underwent a 1-month prednisone challenge and those patients deemed steroid-responders were then randomized to receive either methotrexate or placebo while being tapered from prednisone. Serial audiograms were obtained at defined time points during the study. The authors found no statistically significant difference in the rates of continued hearing loss in the methotrexate group compared to those in the placebo group, and concluded that methotrexate was no more effective than placebo in maintaining hearing improvement in patients with AIED who showed initial response to prednisone.

TNF-α Antagonists

Tumor necrosis factor-α (TNF-α), a cytokine that drives inflammation in many immune-mediated diseases, has been studied as a target for therapies treating AIED (Keithley et al. 2008). TNF-α has many actions in the inflammatory cascade, including attracting leukocytes to tissues and inducing apoptosis via the TNF receptor 1.

Moreover, it can induce recruitment of inflammatory cells from the systemic circulation into the cochlea (Keithley et al. 2008). Inflammatory cells infiltrating the inner ear expressed TNF-α and, to a lesser degree, IL-1β, in an animal model of immune-mediated labyrinthitis induced by systemic exposure to keyhole limpet hemocyanin (KLH) followed by injection of KLH into the cochlea (Satoh et al. 2002). Etanercept, a TNF receptor blocker, reduces hearing loss and the degree of inner ear inflammation in the KLH guinea pig labyrinthitis model when given systemically or when infused into the cochlea (Wang et al. 2003). Several biologic agents have been developed that act as antagonists to TNF-α activity and which are used in the treatment of various autoimmune conditions, including etanercept, infliximab, adalimumab, certolizumab, and golimumab.

Lobo et al. found etanercept to be as effective as corticosteroids in reducing hearing loss in the KLH guinea pig model, though human studies have shown mixed results (Lobo et al. 2006). A pilot study in human subjects with AIED who either failed or were intolerant of conventional therapies showed audiologic improvement in 7 of 12 subjects (58%) and stabilization in 4 of 12 subjects (33%) treated with subcutaneous injections of etanercept (Rahman et al. 2001). However, a subsequent open-label, prospective study of etanercept in 23 patients with AIED showed less favorable results, with improvement in only 30% of patients and stabilization in 57% of patients after 24 weeks of treatment (Matteson et al. 2005). Of note, both of these studies also included patients considered to have bilateral Meniere's disease. Cohen et al. conducted a blinded, randomized, placebo-controlled trial of 8 weeks of treatment with etanercept in 20 AIED patients and concluded that etanercept was no more effective than placebo in improving PTA or speech discrimination in this population (Cohen et al. 2005).

Infliximab, a monoclonal antibody that binds and inhibits TNF-α has also been investigated. A retrospective study of 8 patients with AIED refractory to steroids and cytotoxic therapy did not show any benefit of infliximab treatment in producing audiometric improvement, though one patient reported a subjective improvement in hearing (Liu et al. 2011). Case reports have described hearing improvement after infliximab treatment in a patient diagnosed with AIED who initially responded to steroids and azathioprine but continued to experience fluctuations and progressive decline in hearing, and in a congenitally blind woman with steroid-dependent episodes of SNHL and vertigo who was able to wean from steroids after starting infliximab (Heywood et al. 2013; André et al. 2015). Van Wijk et al. treated 9 AIED patients who relapsed from or could not be weaned from steroids with weekly infusions of infliximab delivered locally to the round window niche over a 4 week period (Van Wijk et al. 2006). The authors reported favorable results, with 4 of 5 patients who were steroid-dependent ultimately able to taper from steroids and 3 of 4 patients who had relapsed after the discontinuation of steroids showing improvements in their PTA.

In one case report, a patient with rheumatoid arthritis and SNHL recovered hearing after starting adalimumab, for treatment of her systemic autoimmune disease (Vergles et al. 2010). Intratympanic gomalimumab was studied as a treatment option in 10 patients with steroid-dependent AIED (Derebery et al. 2014). Gomalimumab was injected into the tympanic membrane of a single ear over a 6-week period. At the study end, 6 of 10 subjects experienced stable PTAs in the injected ear while the remainder showed progression. Word recognition improved in 3 patients, worsened in 3 and improved in 4 injected ears. However, the non-injected ears also showed stable thresholds in 7 patients and stable word recognition in 7 patients. Seven of the ten patients enrolled were able to taper from steroids with 3 of those patients maintaining stable hearing overall.

Anti-TNF-α agents were generally well-tolerated in each of these studies, with no adverse effects reports with the exception of minor injection-site reactions in the case of etanercept (Rahman et al. 2001). The safety profile of these therapies in other autoimmune diseases has been generally been very favorable, though rare adverse effects that have been reported include lymphoma, tuberculosis reactivation, congestive heart failure, a lupus-like syndrome, infections, and skin eruptions (Scheinfeld 2009).

Anakinra

Anakinra, a competitive inhibitor of the IL-1 receptor type, has been investigated as a potential therapeutic option in select patients who fail steroid treatment given preliminary data reviewed above suggesting a role for IL-1β signaling in the pathogenesis of AIED (Vambutas et al. 2014). Case reports have linked anakinra to improved hearing in patients with Muckle-Wells syndrome, a hereditary autoinflammatory disorder characterized by urticaria, rash, fever, and progressive SNHL and whose pathogenesis is felt to involve IL-1β dysregulation (Mirault et al. 2006; Yamazaki et al. 2008). Vambutas et al. conducted a phase I/II, open-label, prospective trial of subcutaneous injections of anakinra in 13 subjects with AIED whose hearing failed to respond to corticosteroid treatment (Vambutas et al. 2014). Seven of the ten subjects who completed the treatment protocol showed improvement on audiometric assessment as well as a reduction in IL-1β plasma levels. The most common adverse event was an injection site reaction, which occurred in 70% of patients and led 2 subjects to drop out of the study.

The incorporation of anakinra into the therapeutic armamentarium for AIED would be attractive both because of its excellent tolerance and because it could also address possible 'autoinflammatory' etiologies. However, the currently available studies pertaining to the use of anakinra in this patient population have questionable validity. It is not clear that the patients included in these studies were steroid resistant or simply patients whose hearing initially fluctuated spontaneously, with some of these fluctuations occurring at the same time as the administration of steroids. Furthermore, the audiometric criteria used to define a positive therapeutic response do not meet currently accepted standards.

Azathioprine

Azathioprine, an immunosuppressive agent that interferes with nucleic acid metabolism and affects the rapidly dividing cells of the immune system, has been investigated as a potential therapeutic agent for patients with AIED in several small studies. Case reports have described improvement in SNHL linked with systemic autoimmune disease after treatment with azathioprine (Dowd and Rees 1987; Khalidi and Rebello 2008). One prospective, open-label study investigated the effects of azathioprine in addition to prednisolone on hearing outcomes in 12 patients with AIED (Saraçaydin et al. 2016). Ten of the twelve patients experienced significantly improved PTAs and speech discrimination after 4 weeks of treatment; the other two patients experienced no change in their hearing. No adverse events were observed related to azathioprine use. The authors did not test the effects of azathioprine alone or report long-term outcomes, so it is unclear if these results are due to azathioprine rather than the steroids, or if the hearing improvements observed are sustained.

One retrospective study of patients with AIED treated with steroids alone or steroids with a cytotoxic medication suggested that azathioprine may be beneficial in some of these patients (Lasak et al. 2001). Seven of the thirty-nine patients in the study were treated with azathioprine as a second- or third-line cytotoxic agent and after failing steroid therapy. Five of seven patients experienced audiometric improvement after a mean treatment period of 7 months. However, significant toxicities were observed: one patient experienced lymphoblastic vasculitis and another was diagnosed with pancytopenia/sepsis. Another retrospective study by Broughton et al. failed to demonstrate benefit of azathioprine treatment. The authors reviewed a series of 42 patients with AIED treated at their institution, of whom five received azathioprine at some point during their treatment (Broughton et al. 2004). One patient was treated for 30 months with stabilization of hearing. Three improved with azathioprine and steroids but experienced a relapse after tapering of steroids. One patient experienced subjective improvement in hearing and vertigo however discontinued the medication due to gastrointestinal upset. Given the small numbers of patients treated with azathioprine, lack of standardized treatment protocols, and retrospective nature of these studies, it is difficult to draw any conclusions about which, if any, patients with AIED may benefit from this treatment.

Cyclophosphamide

Cyclophosphamide is an alkylating agent used in various neoplastic and inflammatory diseases which acts by cross-linking DNA strands and disrupting cell growth and division in rapidly proliferating cells (Langford 1997). Though its use was described in the treatment of AIED in McCabe's first characterization of the disease, no rigorous, prospective studies have demonstrated its efficacy (McCabe 1979). Case reports and small case series have shown improvement in SNHL linked to systemic vasculitis and Cogan's syndrome with cyclophosphamide, generally in conjunction with steroids (Clements et al. 1989; Berrettini et al. 1998; Plester and Soliman 1989). Veldman et al. found treatment with prednisone plus cyclophosphamide was no more effective in treating hearing loss than prednisone alone in a small series of patients with rapidly progressing SNHL (Veldman et al. 1993). In the retrospective review of 42 patients with AIED by Broughton et al., 6 patients received cyclophosphamide treatment (Broughton et al. 2004). Two of the six patients derived benefit, with hearing improving in one and stabilizing in another; 2 patients experienced continued hearing decline and 2 discontinued the medication due to adverse effects. In another retrospective review, 10 patients with AIED were given cyclophosphamide either as a first-line cytotoxic therapy after steroid failure or as a second-line therapy after methotrexate (Lasak et al. 2001). The authors reported a positive response in half of the patients over a mean treatment duration of 2.7 months.

Adverse effects related to cyclophosphamide use are frequent and severe. These include nausea and vomiting, serious infections, bone marrow suppression, hemorrhagic cystitis, infertility, teratogenicity, alopecia, pulmonary toxicity, and the malignancies including transitional cell carcinoma (Langford 1997). Thus, caution should be exerted with regard to its use in AIED given limited data showing efficacy.

Rituximab

Rituximab, a monoclonal antibody against the CD20 antigen found on the surface of lymphocytes, causes the elimination of B cells and is used in conditions felt to involve autoantibody production. Its use was reported to improve hearing in a woman with Cogan's syndrome with persistent hearing loss despite therapy with prednisone, cyclophosphamide, methotrexate, cyclosporine, and adalimumab (Orsoni et al. 2010). Matsuoka and Harris retrospectively reviewed treatment outcomes of 47 patients with AIED (including those considered to have immune-mediated Meniere's disease), of whom 5 had been treated with rituximab after failing steroids (Matsuoka and Harris 2013). Hearing improved in 2 patients, while all 5 experienced improvement in tinnitus, aural fullness, and vestibular symptoms and all reduced their dose of prednisone maintenance steroid. An open-label, pilot study of steroid-responsive AIED patients reported that 5 of 7 enrolled subjects were able to maintain the hearing improvement seen after steroids with rituximab infusions (Cohen et al. 2011). This effect persisted through 24 weeks of follow up, after steroids had been tapered off. No adverse events were reported.

Plasmapheresis

An alternative approach to the treatment of AIED that has been investigated is plasmapheresis, which may remove autoantibodies, immune complexes and other disease mediators from circulation. Luetje reported on eight patients with suspected

AIED, of whom four had been diagnosed with systemic autoimmune disease and who were treated with plasmapheresis (Luetje 1989). The patients in this small series had variable clinical courses with differing courses of treatment, including various combinations of steroids and cytotoxic medications. In some cases, plasmapheresis was used as an adjunct to steroids and/or cytotoxic medications, while in others, it was used as an alternative treatment while attempting to taper steroids. Thus, it is difficult to draw conclusions about the effect of plasmapheresis in each case. Overall, hearing was improved in 3 patients, declined in 2, and was essentially stable in the remaining 3 (though many patients experienced fluctuations throughout their clinical course). A follow up study, which included an additional 13 patients, reported longer-term results of plasmapheresis therapy (Luetje and Berliner 1997). Data was able to be collected on 28 ears from 16 patients who had at least 2 years of follow up (mean follow-up time: 6.7 years). Of those ears, 39.3% demonstrated audiometric improvement or stability during the follow-up period, though mean changes in speech reception thresholds and speech discrimination scores were not statistically significant. Only 4 of the 16 patients were using immunosuppressive medications at follow up. Other case reports have observed improvement after plasma exchange in a patient with sudden SNHL suspected to be of autoimmune origin, in patients with systemic lupus erythematous, and patients with SNHL associated with elevated serum immune complexes (Alpa et al. 2011; Hamblin et al. 1982; Kobayashi et al. 1992; Brookes and Newland 1986).

Additional Therapies

In the retrospective review by Broughton et al., intravenous gamma globulin was administered to one patient with AIED after failing treatment with methotrexate and discontinuing cyclophosphamide due to adverse effects (Broughton et al. 2004). The patient's hearing initially stabilized, but subsequently began to decline. However, the patient was able to taper to a lower dose of steroids and subjectively reported less severe fluctuations in hearing.

Mycophenolate mofetil, an immunosuppressive medication commonly used in solid organ transplantation, was successfully used to treat SNHL in a pediatric case of Cogan's syndrome, allowing the patient to taper off of steroids (Hautefort et al. 2009). Broughton et al. also report on one patient with AIED who failed azathioprine and methotrexate therapy and was treated with mycophenolate mofetil with good results (Broughton et al. 2004). The patient's hearing stabilized and steroids were able to be tapered to a low dose.

Cyclosporine, an immunosuppressive agent used in organ and bone marrow transplantation, was reported beneficial in a patient with steroid-dependent sudden SNHL, a patient with presumed AIED, and in a series of patients with SNHL associated with Behçet's disease (Elidan et al. 1991; McClelland et al. 2009; Di Leo et al. 2011). However, of note, this drug has also been associated with the development of hearing impairment in transplant patients (Gulleroglu et al. 2015; Rifai et al. 2005; Marioni et al. 2004).

3   Idiopathic Sudden Sensorineural Hearing Loss

Corticosteroids have long been the mainstay of treatment for idiopathic sudden sensorineural hearing loss (ISSHL), initially given empirically in early reports under
the hypothesis that their anti-inflammatory effects could be beneficial in presumed
cases of virally-mediated hearing loss (Glasscock et al. 1971; Whitaker 1980). The
first systematic, prospective, placebo-controlled, double-blinded study of the use of
steroids in patients with ISSHL appeared in 1980 when Wilson et al. showed recovery of hearing in patients with moderate hearing loss after treatment with steroids
(Wilson et al. 1980). Since then, an abundance of retrospective studies have appeared
which purport to show beneficial effects of systemic steroids in this population (Byl
1977; Moskowitz et al. 1984; Fetterman et al. 1996; Zadeh et al. 2003; Slattery et al.
2005; Chen et al. 2003). However, a Cochrane review first published in 2006 and
updated in 2013 found only 3 randomized controlled trials evaluating the used of
steroids in ISSHL that met the authors' inclusion criteria, including the study by
Wilson et al. (2006). Among the two other studies included in the review, neither
showed a statistically-significant difference in hearing recovery in steroid-treated
patients versus controls. Due to the small size of the included studies, inconsistent
treatment protocols, differing definitions of hearing recovery, and methodological
limitations of the included studies, the review authors write that "no conclusions can
be drawn about the effectiveness, or lack thereof, of steroids in the treatment of
idiopathic sudden sensorineural hearing loss." Two recent meta-analyses support
this finding in its failure to find a statistically-significant treatment effect of steroids
over placebo (Crane et al. 2015; Conlin and Parnes 2007).

Despite limited data to support the use of steroids in prospective studies, current
clinical practice guidelines recommend offering a short course of steroids in patients
without contraindications as steroids are one of the few treatment options with any
evidence to support its use (though it may only be retrospective in nature); moreover, the risk of serious adverse effects in short-term use of steroids is low while the
consequences of a major hearing loss can be quite significant (Stachler et al. 2012).
These guidelines suggest the use of oral prednisone 1 mg/kg/day (up to a maximum
of 60 mg daily) for 10–14 days, with therapy being initiated within the first 2 weeks
of symptom onset as recovery is greatest during this time window.

Intratympanic steroids represent a promising alternative to systemic steroids and
have been extensively studied recently both as a primary treatment for ISSHL or as
salvage therapy after failure of systemic steroids. A multi-center, randomized trial
demonstrated that IT methylprednisolone was not inferior to oral prednisone in the
treatment of ISSHL (Rauch et al. 2011). A meta-analysis examined 8 randomized
controlled trials evaluating the efficacy of IT dexamethasone in treating ISSHL
(Sabbagh El et al. 2016). The studies differed in the dosing regimen, technique of
drug administration, and whether or not dexamethasone was the first- or second-line
therapy. Hearing improvement was reported in 50–80% of subjects in the IT dexamethasone arms, though the meta-analysis did not find a statistically-significant difference between the steroid and control groups. However, two studies did show a

significant improvement in hearing in the treatment arm compared to controls, both of which used IT dexamethasone as a salvage therapy after failure of conventional treatment and both of which used a drug concentration of 4 mg/mL (Wu et al. 2011). Crane et al. examined randomized controlled trials involving any IT steroid, specifically the subset in which IT steroids were used as salvage therapy (Crane et al. 2015). In a meta-analysis of these studies, the authors did find a significant treatment effect of IT steroids with an odds ratio of 6.04, though they caution that poor quality of the studies comprising the analysis limit interpretation of these results. As IT steroids have been found beneficial, current practice guidelines recommend offering this therapy in patients with ISSHL who fail systemic steroids. In patients with diabetes, IT steroids may be attractive as an initial treatment option in order to avoid uncontrolled hyperglycemia (Han et al. 2009). Finally, side effects of IT steroids tend to be minor, and have included otalgia, aural fullness, headache, temporary dizziness/vertigo, and tympanic membrane perforation (Sabbagh El et al. 2016).

Steroids remain the only anti-inflammatory therapy whose use in ISSHL has been extensively investigated. Clarifying the role of these therapies in ISSHL remains challenging since, by definition, the pathogenesis of ISSHL is unknown and has been proposed to involve as varied mechanisms as viral infection, vascular occlusion, immune dysfunction, and membrane breaks within the inner ear. More recently, ISSHL was hypothesized to involve the abnormal activation of cellular stress pathways (Merchant et al. 2005). Likely, multiple etiologies may combine to result in a similar clinical presentation. If different pathogenic events are found to be mediated through common inflammatory or immunologic mechanisms, a new role for anti-inflammatory treatments may emerge. For example, a study in guinea pigs suggested that inhibitors of TNF signaling could reverse TNF-induced reductions in cochlear blood flow, suggesting a pathway through which modulation of inflammatory pathways could affect the microvascular disturbances which have been postulated to cause a subset of ISSHL (Sharaf et al. 2016). Further research into the complex interactions between inflammatory events and cochlear injury will lead to the identification of targets for future therapies. Many alternative, non-immunomodulating therapies have also been studied for use in ISSHL including antivirals, vasodilators, antioxidants, vitamins, fibrinogen or LDL apheresis, and hyperbaric oxygen; none of these interventions, however, with the exception of hyperbaric oxygen, are supported by enough evidence to merit recommendation in clinical practice guidelines (Conlin and Parnes 2007; Stachler et al. 2012; Suckfüll 2002; Agarwal and Pothier 2009; Angeli et al. 2012; Sano et al. 2010; Hatano et al. 2009; Westerlaken et al. 2016).

4   Conclusion

Anti-inflammatory therapies have played an important role in the treatment of AIED and ISSHL. Corticosteroids are the most commonly employed immunosuppressive medication in these disorders, though the exact mechanisms through which they act in the inner ear is unknown. Intratympanic steroids may also benefit patients with ISSHL who fail systemic steroid therapy, or may be a preferable first-line treatment in patients with diabetes. A variety of immunosuppressive therapies have been studied in the treatment of AIED on the basis of their effectiveness in other autoimmune and inflammatory conditions. These have included cyclophosphamide, methotrexate, azathioprine, rituximab, anakinra, anti- TNF-α agents, and plasmapharesis. While studies of these agents have suggested improvement or stabilization of hearing loss in some cases, these studies are generally limited by small sample sizes and are often retrospective or observational in nature, and lack adequate controls. Furthermore, they may have significant side effects, the risks of which may not be acceptable in an era in which cochlear implantation is a viable option. Further elucidation of potential immunologic or inflammatory mechanisms underlying different forms of SNHL may pave the way for the development of targeted therapies for inner ear disorders.

References

Agarwal L, Pothier DD. Vasodilators and vasoactive substances for idiopathic sudden sensorineural hearing loss. Cochrane Database Syst Rev. 2009;4:CD003422.

Alexander TH, Weisman MH, Derebery JM, Espeland MA, Gantz BJ, Gulya AJ, et al. Safety of high-dose corticosteroids for the treatment of autoimmune inner ear disease. Otol Neurotol. 2009;30(4):443–8.

Alpa M, Bucolo S, Beatrice F, Giachino O, Roccatello D. Apheresis as rescue therapy in a severe case of sudden hearing loss. Int J Artif Organs. 2011;34(7):589–92.

André R, Corlieu P, Crabol Y, Cohen P, Guillevin L. Infliximab reverses progressive deafness. Presse Med. 2015;44(6 Pt 1):675–7.

Angeli SI, Abi-Hachem RN, Vivero RJ, Telischi FT, Machado JJ. L-N-acetylcysteine treatment is associated with improved hearing outcome in sudden idiopathic sensorineural hearing loss. Acta Otolaryngol. 2012;132(4):369–76.

Berrettini S, Ferri C, Ravecca F, LaCivita L. Progressive sensorineural hearing impairment in systemic vasculitides. Semin Arthritis Rheum. 1998;27(5):301–18.

Brant JA, Eliades SJ, Ruckenstein MJ. Systematic review of treatments for autoimmune inner ear disease. Otol Neurotol. 2015;36(10):1585–92.

Brookes GB, Newland AC. Plasma exchange in the treatment of immune complex-associated sensorineural deafness. J Laryngol Otol. 1986;100(1):25–33.

Broughton SS, Meyerhoff WE, Cohen SB. Immune-mediated inner ear disease: 10-year experience. Semin Arthritis Rheum. 2004;34(2):544–8.

Byl FM. Seventy-six cases of presumed sudden hearing loss occurring in 1973: prognosis and incidence. Laryngoscope. 1977;87(5 Pt 1):817–25.

Chen C-Y, Halpin C, Rauch SD. Oral steroid treatment of sudden sensorineural hearing loss: a ten year retrospective analysis. Otol Neurotol. 2003;24(5):728.

Clements MR, Mistry CD, Keith AO, Ramsden RT. Recovery from sensorineural deafness in Wegener's granulomatosis. J Laryngol Otol. 1989;103(5):515–8.

Cohen S, Shoup A, Weisman MH, Harris J. Etanercept treatment for autoimmune inner ear disease: results of a pilot placebo-controlled study. Otol Neurotol. 2005;26(5):903–7.

Cohen S, Roland P, Shoup A, Lowenstein M, Silverstein H, Kavanaugh A, et al. A pilot study of rituximab in immune-mediated inner ear disease. Audiol Neurootol. 2011;16(4):214–21.

Conlin AE, Parnes LS. Treatment of sudden sensorineural hearing loss: II. A meta-analysis. Arch Otolaryngol Head Neck Surg. 2007;133(6):582–6.

Crane RA, Camilon M, Nguyen S, Meyer TA. Steroids for treatment of sudden sensorineural hearing loss: a meta-analysis of randomized controlled trials. Laryngoscope. 2015;125(1):209–17.

Derebery MJ, Fisher LM, Voelker CCJ, Calzada A. An open label study to evaluate the safety and efficacy of intratympanic golimumab therapy in patients with autoimmune inner ear disease. Otol Neurotol. 2014;35(9):1515–21.

Di Leo E, Coppola F, Nettis E, Vacca A, Quaranta N. Late recovery with cyclosporine-A of an autoimmune sudden sensorineural hearing loss. Acta Otorhinolaryngol Ital. 2011;31(6):399–401.

Dowd A, Rees WD. Treatment of sensorineural deafness associated with ulcerative colitis. Br Med J (Clin Res Ed). 1987;295(6589):26.

Elidan J, Cohen E, Levi H. Effect of cyclosporine A on the hearing loss in Behçet's disease. Ann Otol Rhinol Laryngol. 1991;100(6):464–8.

Fetterman BL, Saunders JE, Luxford WM. Prognosis and Treatment of Sudden Sensorineural Hearing Loss. Otol Neurotol. 1996;17(4):529.

Flammer JR, Rogatsky I. Minireview: glucocorticoids in autoimmunity: unexpected targets and mechanisms. Mol Endocrinol. 2011;25(7):1075–86.

García-Berrocal JR, Ramírez-Camacho R, Millán I, Górriz C, Trinidad A, Arellano B, et al. Sudden presentation of immune-mediated inner ear disease: characterization and acceptance of a cochleovestibular dysfunction. J Laryngol Otol. 2003;117(10):775–9.

García-Berrocal JR, Ibáñez A, Rodríguez A, González-García JÁ, Verdaguer JM, Trinidad A, et al. Alternatives to systemic steroid therapy for refractory immune-mediated inner ear disease: a physiopathologic approach. Eur Arch Otorhinolaryngol. 2006;263(11):977–82.

Glasscock ME, Nechtman C, Altenau M. Sudden loss of hearing: a medical emergency. South Med J. 1971;64(12):1485–9.

Gulleroglu K, Baskin E, Aydin E. Hearing status in pediatric renal transplant recipients. Exp Clin Transplant. 2015;13(4):324–8.

Hamblin TJ, Mufti GJ, Bracewell A. Severe deafness in systemic lupus erythematosus: its immediate relief by plasma exchange. Br Med J (Clin Res Ed). 1982;284(6326):1374.

Han CS, Park JR, Boo SH, Jo JM. Clinical efficacy of initial intratympanic steroid treatment on sudden sensorineural hearing loss with diabetes. Otolaryngol Head Neck Surg. 2009;141(5):572–8.

Harris JP, Sharp PA. Inner ear autoantibodies in patients with rapidly progressive sensorineural hearing loss. Laryngoscope. 1990;100(5):516–24.

Harris JP, Weisman MH, Derebery JM, Espeland MA, Gantz BJ, Gulya AJ, et al. Treatment of corticosteroid-responsive autoimmune inner ear disease with methotrexate: a randomized controlled trial. JAMA. 2003;290(14):1875–83.

Harris DA, Mikulec AA, Carls SL. Autoimmune inner ear disease preliminary case report: audiometric findings following steroid treatments. Am J Audiol. 2013;22(1):120–4.

Hatano M, Uramoto N, Okabe Y, Furukawa M, Ito DM. Vitamin E and vitamin C in the treatment of idiopathic sudden sensorineural hearing loss. Acta Otolaryngol. 2009;128(2):116–21.

Hautefort C, Loundon N, Montchilova M, Marlin S, Garabedian EN, Ulinski T. Mycophenolate mofetil as a treatment of steroid dependent Cogan's syndrome in childhood. Int J Pediatr Otorhinolaryngol. 2009;73(10):1477–9.

Heywood RL, Hadavi S, Donnelly S. Infliximab for autoimmune inner ear disease: case report and literature review. J Laryngol Otol. 2013;127(11):1145–7.

Hirose K, Wener MH, Duckert LG. Utility of laboratory testing in autoimmune inner ear disease. Laryngoscope. 1999;109(11):1749–54.

Hughes GB, Kinney SE, Barna BP, Calabrese LH. Practical versus theoretical management of autoimmune inner ear disease. Laryngoscope. 1984;94(6):758–67.

Keithley EM, Wang X, Barkdull GC. Tumor necrosis factor α can induce recruitment of inflammatory cells to the cochlea. Otol Neurotol. 2008;29(6):854–9.

Khalidi NA, Rebello R. Sensorineural hearing loss in systemic lupus erythematosus: case report and literature review. J Laryngol Otol. 2008;122(12):1371–6.

Kobayashi S, Fujishiro N, Sugiyama K. Systemic lupus erythematosus with sensorineural hearing loss and improvement after plasmapheresis using the double filtration method. Intern Med. 1992;31(6):778–81.

Langford CA. Complications of cyclophosphamide therapy. Eur Arch Otorhinolaryngol. 1997;254(2):65–72.

Lasak JM, Sataloff RT, Hawkshaw M. Autoimmune inner ear disease: steroid and cytotoxic drug therapy. Ear Nose Throat J. 2001;80(11):808–11.

Liu YC, Rubin R, Sataloff RT. Treatment-refractory autoimmune sensorineural hearing loss: response to infliximab. Ear Nose Throat J. 2011;90(1):23–8.

Lobo D, Trinidad A, García-Berrocal JR, Verdaguer JM, Ramírez-Camacho R. TNFα blockers do not improve the hearing recovery obtained with glucocorticoid therapy in an autoimmune experimental labyrinthitis. Eur Arch Otorhinolaryngol. 2006;263(7):622–6.

Loveman DM, de Comarmond C, Cepero R, Baldwin DM. Autoimmune sensorineural hearing loss: clinical course and treatment outcome. Semin Arthritis Rheum. 2004;34(2):538–43.

Luetje CM. Theoretical and practical implications for plasmapheresis in autoimmune inner ear disease. Laryngoscope. 1989;99(11):1137–46.

Luetje CM, Berliner KI. Plasmapheresis in autoimmune inner ear disease: long-term follow-up. Otol Neurotol. 1997;18(5):572.

Marioni G, Perin N, Tregnaghi A, Bellemo B. Progressive bilateral sensorineural hearing loss probably induced by chronic cyclosporin A treatment after renal transplantation for focal glomerulosclerosis. Acta Otolaryngol. 2004;124(5):603–7.

Matsuoka AJ, Harris JP. Autoimmune inner ear disease: a retrospective review of forty-seven patients. Audiol Neurootol. 2013;18(4):228–39.

Matteson EL, Tirzaman O, Kasperbauer J, Facer GW, Beatty CW, Fabry DA, et al. Use of methotrexate for autoimmune hearing loss. Ann Otol Rhinol Laryngol. 2000;109(8):710–4.

Matteson EL, Choi HK, Poe DS, Wise C, Lowe VJ, McDonald TJ, et al. Etanercept therapy for immune-mediated cochleovestibular disorders: a multi-center, open-label, pilot study. Arthritis Rheum. 2005;53(3):337–42.

McCabe BF. Autoimmune sensorineural hearing loss. Ann Otol. 1979;88(5):585–9.

McClelland L, Powell RJ, Birchall J. Role of ciclosporin in steroid-responsive sudden sensorineural hearing loss. Acta Otolaryngol. 2009;125(12):1356–60.

Merchant SN, Adams JC, Nadol JBJ. Pathology and pathophysiology of idiopathic sudden sensorineural hearing loss. Otol Neurotol. 2005;26(2):151.

Mirault T, Launay D, Cuisset L, Hachulla E, Lambert M, Queyrel V, et al. Recovery from deafness in a patient with Muckle-Wells syndrome treated with anakinra. Arthritis Rheum. 2006;54(5):1697–700.

Moscicki RA, Martin JES, Quintero CH, Rauch SD, Nadol JB, Bloch KJ. Serum antibody to inner ear proteins in patients with progressive hearing loss: correlation with disease activity and response to corticosteroid treatment. JAMA. 1994;272(8):611–6.

Moskowitz D, Lee KJ, Smith HW. Steroid use in idiopathic sudden sensorineural hearing loss. Laryngoscope. 1984;94(5 Pt 1):664–6.

Niparko JK, Wang N-Y, Rauch SD, Russell GB, Espeland MA, Pierce JJ, et al. Serial audiometry in a clinical trial of AIED treatment. Otol Neurotol. 2005;26(5):908–17.

Orsoni JG, Laganà B, Rubino P, Zavota L, Bacciu S, Mora P. Rituximab ameliorated severe hearing loss in Cogan's syndrome: a case report. Orphanet J Rare Dis. 2010;5(1):18.

Parnes LS, Sun AH, Freeman DJ. Corticosteroid pharmacokinetics in the inner ear fluids: an animal study followed by clinical application. Laryngoscope. 1999;109(S91):1–17.

Pathak S, Goldofsky E, Vivas EX, Bonagura VR, Vambutas A. IL-1β Is overexpressed and aberrantly regulated in corticosteroid nonresponders with autoimmune inner ear disease. J Immunol. 2011;186(3):1870–9.

Plester D, Soliman AM. Autoimmune hearing loss. Otol Neurotol. 1989;10(3):188.

Rahman MU, Poe DS, Choi HK. Etanercept therapy for immune-mediated cochleovestibular disorders: preliminary results in a pilot study. Otol Neurotol. 2001;22(5):619.

Rauch SD. Clinical Management of Immune-mediated Inner-ear Disease. Ann N Y Acad Sci. 1997;830(1):203–10.

Rauch SD, Halpin CF, Antonelli PJ, Babu S, Carey JP, Gantz BJ, et al. Oral vs intratympanic corticosteroid therapy for idiopathic sudden sensorineural hearing loss: a randomized trial. JAMA. 2011;305(20):2071–9.

Rifai K, Bahr MJ, Cantz T, Klempnauer J. Severe hearing loss after liver transplantation. Transplant Proc. 2005;37(4):1918–9.

Ruckenstein MJ. Autoimmune inner ear disease. Curr Opin Otolaryngol Head Neck Surg. 2004;12(5):426.

Ruckenstein MJ, Hu L. Antibody deposition in the stria vascularis of the MRL-Faslpr mouse. Hear Res. 1999;127(1–2):137–42.

Ruckenstein MJ, Sarwar A, Hu L, Shami H, Marion TN. Effects of immunosuppression on the development of cochlear disease in the MRL-Fas Mouse. Laryngoscope. 1999;109(4):626–30.

Ruckenstein MJ, Mount RJ, Harrison RV. The MRL-lpr/lpr mouse: a potential model of autoimmune inner ear disease. Acta Otolaryngol. 2009;113(2):160–5.

Ryan AF, Harris JP, Keithley EM. Immune-mediated hearing loss: basic mechanisms and options for therapy. Acta Otolaryngol. 2009;122(5):38–43.

Sabbagh El NG, Sewitch MJ, Bezdjian A, Daniel SJ. Intratympanic dexamethasone in sudden sensorineural hearing loss: a systematic review and meta-analysis. Laryngoscope. 2016;88(suppl):524.

Salley LH, Grimm M, Sismanis A, Spencer RF, Wise CM. Methotrexate in the management of immune mediated cochleovesitibular disorders: clinical experience with 53 patients. J Rheumatol. 2001;28(5):1037–40.

Sano H, Kamijo T, Ino T, Okamoto M. Edaravone, a free radical scavenger, in the treatment of idiopathic sudden sensorineural hearing loss with profound hearing loss. Auris Nasus Larynx. 2010;37(1):42–6.

Saraçaydin A, Katircioğlu S, Katircioğlu S, Karatay MC. Azathioprine in combination with steroids in the treatment of autoimmune inner-ear disease. J Int Med Res. 2016;21(4):192–6.

Satoh H, Firestein GS, Billings PB, Harris JP, Keithley EM. Tumor necrosis factor-α, an initiator, and etanercept, an inhibitor of cochlear inflammation. Laryngoscope. 2002;112(9):1627–34.

Scheinfeld N. A comprehensive review and evaluation of the side effects of the tumor necrosis factor alpha blockers etanercept, infliximab and adalimumab. J Dermatol Treat. 2009;15(5):280–94.

Sharaf K, Ihler F, Bertlich M, Reichel CA, Berghaus A, Canis M. Tumor necrosis factor-induced decrease of cochlear blood flow can be reversed by etanercept or JTE-013. Otol Neurotol. 2016;37(7):e203–8.

Sismanis A, Wise CM, Johnson GD. Methotrexate management of immune-mediated cochleovestibular disorders. Otolaryngol Head Neck Surg. 2016.

Slattery WH, Fisher LM, Iqbal Z. Oral steroid regimens for idiopathic sudden sensorineural hearing loss. Otolaryngol Head Neck Surg. 2005;132(1):5–10.

Stachler RJ, Chandrasekhar SS, Archer SM, Rosenfeld RM, Schwartz SR, Barrs DM, et al. Clinical practice guideline: sudden hearing loss. Otolaryngol Head Neck Surg. 2012;146(3 Suppl):S1–S35.

Suckfüll M. Fibrinogen and LDL apheresis in treatment of sudden hearing loss: a randomised multicentre trial. Lancet. 2002;360(9348):1811–7.

Trune DR, Canlon B. Corticosteroid therapy for hearing and balance disorders. Anat Rec. 2012;295(11):1928–43.

Trune DR, Wobig RJ, Kempton JB, Hefeneider SH. Steroid treatment in young MRL.MpJ-Faslpr autoimmune mice prevents cochlear dysfunction. Hear Res. 1999a;137(1–2):167–73.

Trune DR, Wobig RJ, Kempton JB, Hefeneider SH. Steroid treatment improves cochlear function in the MRL.MpJ-Faslpr autoimmune mouse. Hear Res. 1999b;137(1–2):160–6.

Trune DR, Kempton JB, Gross ND. Mineralocorticoid receptor mediates glucocorticoid treatment effects in the autoimmune mouse ear. Hear Res. 2006;212(1–2):22–32.

Vambutas A, DeVoti J, Goldofsky E, Gordon M, Lesser M, Bonagura V. Alternate splicing of interleukin-1 receptor type II (IL1R2) in vitro correlates with clinical glucocorticoid responsiveness in patients with AIED. PLoS One. 2009;4(4):e5293.

Vambutas A, Lesser M, Mullooly V, Pathak S, Zahtz G, Rosen L, et al. Early efficacy trial of anakinra in corticosteroid-resistant autoimmune inner ear disease. J Clin Invest. 2014;124(9):4115–22.

Van Wijk F, Staecker H, Keithley E, Lefebvre PP. Local perfusion of the tumor necrosis factor α blocker infliximab to the inner ear improves autoimmune neurosensory hearing loss. Audiol Neurootol. 2006;11(6):357–65.

Veldman JE, Hanada T, Meeuwsen F. Diagnostic and therapeutic dilemmas in rapidly progressive sensorineural hearing loss and sudden deafness a reappraisal of immune reactivity in inner ear disorders. Acta Otolaryngol. 1993;113(3):303–6.

Vergles JM, Radic M, Kovacic J, Salamon L. Successful use of adalimumab for treating rheumatoid arthritis with autoimmune sensorineural hearing loss: two birds with one stone. J Rheumatol. 2010;37(5):1080–1.

Wang X, Truong T, Billings PB, Harris JP, Keithley EM. Blockage of immune-mediated inner ear damage by etanercept. Otol Neurotol. 2003;24(1):52.

Westerlaken BO, Stokroos RJ, Wit HP, Dhooge IJM, Albers FWJ. Treatment of idiopathic sudden sensorineural hearing loss with antiviral therapy: a prospective, randomized, double-blind clinical trial. Ann Otol Rhinol Laryngol. 2016;112(11):993–1000.

Whitaker S. Idiopathic sudden hearing loss. Otol Neurotol. 1980;1(3):180.

Wilson WR, Byl FM, Laird N. The efficacy of steroids in the treatment of idiopathic sudden hearing loss: a double-blind clinical study. Arch Otolaryngol. 1980;106(12):772–6.

Wilson WR, Stathopoulos D, O'Leary S. Steroids for idiopathic sudden sensorineural hearing loss. Cochrane Database Syst Rev. 2006;7:CD003998.

Wu H-P, Chou Y-F, Yu S-H, Wang C-P, Hsu C-J, Chen P-R. Intratympanic steroid injections as a salvage treatment for sudden sensorineural hearing loss: a randomized, double-blind, placebo-controlled study. Otol Neurotol. 2011;32(5):774–9.

Yamazaki T, Masumoto J, Agematsu K, Sawai N, Kobayashi S, Shigemura T, et al. Anakinra improves sensory deafness in a Japanese patient with Muckle-Wells syndrome, possibly by inhibiting the cryopyrin inflammasome. Arthritis Rheum. 2008;58(3):864–8.

Yang GSY, Song H-T, Keithley EM, Harris JP. Intratympanic immunosuppressives for prevention of immune-mediated sensorineural hearing loss. Otol Neurotol. 2000;21(4):499.

Zadeh MH, Storper IS, Spitzer JB. Diagnosis and treatment of sudden-onset sensorineural hearing loss: a study of 51 patients. Otolaryngol Head Neck Surg. 2003;128(1):92–8.

Zeitoun H, Beckman JG, Arts HA. Corticosteroid response and supporting cell antibody in autoimmune hearing loss. Arch Otolaryngol Head Neck Surg. 2005;131(8):665–72.

Chapter 11
Implementation and Outcomes of Clinical Trials in Immune-Mediated Hearing Loss and Other Rare Diseases

Andrea Vambutas and Martin L. Lesser

Abstract Clinical trials for rare diseases can be challenging to design, meet targeted enrollment, and obtain sufficient evidence of efficacy for FDA labeling of new drugs to treat these orphan diseases. Autoimmune inner ear disease, and related diseases of immune mediated hearing loss are yet to be classified orphan diseases. In this chapter, we have addressed some of the unique challenges in designing clinical trials for Autoimmune Inner Ear Disease, Sudden Sensorineural Hearing Loss, Meniere's Disease and Autoinflammatory Diseases.

Keywords Autoimmune Inner Ear Disease (AIED) · Sudden Sensorineural Hearing Loss (SSNHL) and Meniere's Disease (MD) · Autoinflammatory Disease · Rare disease · Orphan disease

There are several clinical diseases of hearing loss that are potentially immune-mediated and may benefit from intervention with traditional or experimental immunomodulators. The way in which we rigorously test the safety and efficacy of these agents is through clinical trials. The purpose of this chapter is to both discuss some of the challenges in implementing clinical trials in rare diseases and to review outcomes of some of the trials completed to date for new therapies for these diseases. Immune-mediated hearing loss potentially includes Autoimmune Inner Ear Disease (AIED), Sudden Sensorineural Hearing Loss (SSNHL) and Meniere's Disease (MD). Additionally, a family of rare monogenic autoinflammatory diseases has emerged that has sensorineural hearing loss among their clinical features. Clinical trials for hearing restoration in immune-mediated hearing loss have been difficult to

A. Vambutas (✉)
Department of Otolaryngology and Molecular Medicine, Barbara and Donald Zucker School of Medicine at Hofstra/Northwell, Hempstead, NY, USA
e-mail: avambuta@northwell.edu

M. L. Lesser
Biostatistics Unit, Feinstein Institute for Medical Research, Manhasset, NY, USA

Department of Molecular Medicine and Population Health, Barbara and Donald Zucker School of Medicine at Hofstra/Northwell, Hempstead, NY, USA

© Springer International Publishing AG, part of Springer Nature 2018
V. Ramkumar, L. P. Rybak (eds.), *Inflammatory Mechanisms in Mediating Hearing Loss*, https://doi.org/10.1007/978-3-319-92507-3_11

execute for a variety of reasons. All of these diseases would be classified as rare diseases, and have the advantage of qualifying for orphan status for drug and biologic therapy development that provides incentives to pharmaceutical companies, as less than 200,000 individuals are afflicted for each disease in the population in any given year (Wellman-Labadie and Zhou 2010). Moreover, unequivocally establishing a clinical diagnosis of AIED or autoinflammatory disease can be challenging. The following chapter is devoted to describing and defining some of the components required for clinical trial implementation and results achieved to date as it relates to immune-mediated hearing loss.

1   Study Population

Adequately defining the population to be studied such that other investigators can recruit patients into clinical trials, and results can be generalizable to the population in the event of a trial that demonstrates efficacy is one of the most critical components in trial design. Inclusion and exclusion criteria must be carefully stated such that the population to be studied is unambiguous. Methodology to enhance the homogeneity of a cohort of potential participants to be studied include presence of a particular biomarker, or response (or lack of response) to another non-study drug for trial entry. Whereas this may enhance homogeneity, it also may result in diminished recruitment, especially in rare diseases. Thus one must consider both aspects in appropriately designing a trial.

AIED   Autoimmune Inner Ear Disease (AIED) can be a difficult disease to appropriately define. Typically, both ears are afflicted, and the patients experience a progressive decline in hearing in one or both ears. The disease affects fewer than 50,000 individuals annually (Vambutas and Pathak 2016), and although 70% of patients are initially steroid responsive, only 14% remain responsive after 3 years (Broughton et al. 2004), highlighting the need to develop alternate treatments. During the methotrexate trial, a phase 2 trial in which investigators hoped to prove that in corticosteroid responsive patients with AIED, that methotrexate was superior to placebo in maintaining hearing, the investigators defined audiometric enrollment criteria as hearing loss progressing in greater than 3 but less than 90 days (Harris et al. 2003). Some patients with AIED may have steroid dependent SNHL or fluctuating SNHL. Those with fluctuating SNHL are particularly problematic to study as it is difficult to distinguish clinical efficacy from natural fluctuations in this cohort. Similarly, steroid dependent SNHL is also problematic as steroids must be tapered as the new drug to be studied is added: this may result in an increased number of adverse events from drug interactions, and difficulties in ascertaining efficacy depending on the outcome to be measured.

SSNHL   Sudden sensorineural hearing loss (SSNHL) is defined as a loss of 30 dB or greater at 3 contiguous frequencies that evolves within 3 days (Wilson et al. 1980), with 15,000 new cases worldwide per year (Hughes et al. 1996). Although timely oral corticosteroid use has been the gold standard, a multicentered phase 3 clinical trial identified that intratympanic corticosteroid therapy was not inferior to the gold standard (Rauch et al. 2011). It became apparent, during a recent phase 2 clinical trial assessing the efficacy of a c-Jun N-terminal kinase (JNK) inhibitor, that a high rate of spontaneous improvement was observed in the placebo group (Suckfuell et al. 2014). This lead the investigators to determine the drug of interest was effective in severe-profound sudden hearing loss, however, no concrete conclusions could be reached about those participants with mild-moderate SSNHL.

MD Although more prevalent than AIED or SSNHL, the incidence of Meniere's Disease is still under 100,000 individuals (Stahle et al. 1978). For some, Meniere's Disease (MD) may be immune-mediated (Hietikko et al. 2014). Given that some patients experience control of vertigo with corticosteroid use in this disease, an immunologic role in the pathogenesis of this disease is possible (Barrs et al. 2001), although hearing improvement with intratympanic corticosteroids in Meniere's Disease has been disappointing (Arriaga and Goldman 1998; Silverstein et al. 1998).

Autoinflammatory Diseases Sensorineural hearing loss has been observed in the genetically inherited family of Cryopyrin-Associated Autoinflammatory Syndromes (CAPS) such as Neonatal-Onset Multisystem Inflammatory Disease (NOMID) and Muckle-Wells Syndrome (MWS) whose hallmarks are IL-1β dysregulation. This family of autoinflammatory diseases have a number of systemic signs including transient skin rashes, periodic fevers and sensorineural hearing loss (see www.auto-inflammatory.org for detailed clinical manifestations). Although case reports exist attesting to the hearing improvement with IL-1 inhibition (Mirault et al. 2006; Yamazaki et al. 2008), in general the improvement in auditory acuity is limited (Ombrello and Kastner 2011), with only 25% improvement at 4000 Hz and below (Kuemmerle-Deschner et al. 2015). Here, a diagnosis can be obtained through genetic testing for a mutation in NLRP3, in the case of Muckle-Wells Syndrome (Kuemmerle-Deschner et al. 2013).

2 Study Design/Trial Phase

What question are you answering? Clinical trials can be divided into phases. Each phase serves a different purpose. Typically, the phase of the clinical trial corresponds to what question is to be answered. If questions of safety of a new drug are to be addressed, this is typically through a phase I trial. If questions of efficacy are to be addressed, this may be initiated in a phase 2 trial and refined during a phase 3 trial. The purpose of each phase is outlined in the below table.

In immune-mediated hearing losses, for many of the diseases listed, current standard of care is to treat with corticosteroids. Development of new therapies therefore requires a comparison to this reference therapy. An ethical concern that exists is whether there is compelling evidence that the experimental therapy is comparable to the reference therapy, because delayed treatment with the reference therapy may render the patient refractory to responding whereas earlier treatment would have resulted in a more favorable outcome. One method that has been employed in AIED is to enroll patients that have either responded to or failed to respond to corticosteroids as a strategy to be able to compare a new drug against placebo. This approach, while increasing the homogeneity of the study group, may biologically alter the patients' ability to respond to the new therapy and therefore could potentially lead erroneously discarding the new therapy for lack of efficacy. Another method is to use the new drug in combination with corticosteroids as compared to corticosteroids alone, however this does not provide assurance that the new drug, if more effective in combination with corticosteroids, will act similarly as monotherapy.

Basic Trial Design Although there are many different designs for later phase comparative clinical trials, two of the more common designs are parallel arm and two period crossover. In a randomized parallel arm trial, participants receive one treatment or the other. In a randomized crossover design, patients receive both treatments in opposite sequence. Therefore, crossover trials in immune-mediated hearing loss, whereas gold standard for other diseases, may be difficult to execute or interpret here because the delay in treatment may preclude response. Furthermore in a crossover design, all patients must return to the original disease state for the second phase, which may not possible if there is a carryover effect of an antibody-mediated biologic therapy with a long half-life.

Trial phase	Purpose	Types of design
Preclinical	Animal studies to determine mechanism, potential efficacy and safety	
0	Pharmacokinetics, bioavailability	First in human, very small cohort, open label
1	Exploratory, assessment of toxicity, pharmacokinetics, determination safe dose	3 + 3 dose escalation; continual reassessment model (CRM); fixed multiple dose; open label studies for toxicity
2 (2a)	Evaluation of dosing requirements	
2 (2b)	Evaluation of efficacy in a select group, estimation of treatment effect	Single primary outcome; single arm-open label; single arm-blinded evaluator; Simon 2 stage design
3	Comparative trial of new therapy to commonly used treatments, hypothesis testing	Larger cohorts; randomized-placebo controlled; multicentered
4	Establish new indication, post-marketing surveillance for side effects from long term use	

Rare Disease Considerations In rare diseases, there is clear precedent for accelerated clinical trials and novel trial designs that may combine several phases. Clinical trials for autoinflammatory diseases have taken advantage of accelerated trial designs to bring new therapies to rapid FDA approval. Several IL-1b inhibitors that are currently FDA approved for the rare diseases of Muckle-Wells, CAPS and NOMID have been approved on the basis of a positive clinical result in small cohorts of patients.

Indication	Study number	Phase/ type of study	Study design and type of control	Test products; dosage regimen; route of administration	FDA action	Number subjects enrolled		
						Total drug	Total placebo	Total subjects
CAPS	NCT00685373 (Kuemmerle-Deschner et al. 2011)	III	Open label	Canakinumab, 150 mg SQ q8wk × 2 years	**FDA approval for FCAS and MWS**	166* rollover trial, actual patient recruitment is 78	0	166
NOMID	NCT00069329 (Goldbach-Mansky et al. 2006)	Orphan	Single center Open label	Anakinra 1-5 mg/kg/ day × 36 months	**FDA approved for NOMID**	26	0	26

* indicated that the 166 was a total number but only 78 patients were enrolled in the first phase of this trial

Packaging the design: components of the protocol and Manual of Procedures (MOP) Once a design has been identified, a detailed protocol must be drafted. Requisite components include: background/rationale for the study; study objectives; study design; subject selection and withdrawal; study drug; study procedures; statistical plan; safety and adverse events including Data Safety Monitoring Board (DSMB); data handling and record keeping; data handling and record keeping; study monitoring, auditing and inspecting; ethical considerations; study finances and publication plan. The protocol is submitted to both the FDA to receive an Investigational New Drug application (IND) and to the local IRBs. Some studies may use centralized IRBs, but ultimately it is at the discretion of the local IRB to accept the central IRBs' approval of a study. The Manual of Procedures (MOP) provides an even greater amount of detail regarding operationalizing of study procedures.

IND Submissions/Trial Registry: All clinical trials require trial submission to the FDA as an IND, or Investigational New Drug, regardless of the phase of study. The trial may be sponsored by either the Principal Investigator or by the pharmaceutical company. In the event the trial is successful in achieving its efficacy target, all subsequent phases are submitted as "protocol amendments", but are covered under the original IND. The ultimate goal is if efficacy is determined, the FDA will approve this new drug for the indication for which it was tested.

All phases of clinical trials, in addition to obtaining local IRB approval, require registry on clinicaltrials.gov in the United States prior to trial inception (there are similar European and Japanese trial registries).

3   Special Considerations

Especially in rare diseases, trial design can be complicated, and potential bias can arise. Methods to mitigate bias include randomization, and blinding. Determination of sample size, and recruitment pose special challenges in rare diseases. Moreover, comparison of new therapies to placebo in lieu of steroids is controversial, as patients may miss the timely opportunity to recover hearing. Therefore, trials in this area have been designed to test the new drug either compared to oral corticosteroids, following corticosteroid therapy, in conjunction with corticosteroids or in patients that have not responded to corticosteroids.

Randomization Clearly, if all patients with mild hearing loss received standard treatment and all patients with profound hearing loss received the experimental treatment, it is possible that the standard treatment may be deemed superior because the milder hearing loss was more responsive to treatment rather than a true drug effect. In order to avoid this type of "selection bias" random allocation of patients to the treatment groups is preferable. In many situations, it is advisable to classify participants with respect to prognostic factors and may refer to clinical site, degree of hearing loss, gender, or other variables. This process is known as stratification whereby participants are randomized separately within each of the strata combinations. In a large clinical trial confounding variables often will even out and bear minimal effect on the results in each arm. However, in small clinical trials, strong confounders may exert large effects.

Blinding Early stage clinical trials, especially in rare diseases, may involve open-label assessment of efficacy, where all participants are given the active drug to be studied. Whereas this certainly assists recruitment efforts (who would not want to receive the active drug rather than placebo), bias may be incurred, as participants may experience a "placebo effect" by virtue of feeling they should achieve benefit with the active drug, especially in clinical trials that are more subjective in nature such as mood improvement, reduction of joint pain, or tinnitus. As the stage of clinical trial advances, there is a clear need to exclude a placebo effect. Therefore, in phase 2 and 3 clinical trials, the drug of interest is compared to either placebo (when there is no acceptable standard therapy) or an active reference therapy (one that is known to be clinically effective). Issues that may arise for immune-mediated hearing loss is the timing of the placebo period: for instance, a delay in treatment, especially in SSNHL, may render the patient incapable of clinically responding. Thus for AIED or SSNHL, crossover trials that involve use of a placebo may be difficult to execute. Ideal blinding for clinical trials is double-blind trial, where both the investigator and the participant is unaware of which treatment are they have been randomized to. In these studies, one study team member, usually from the biostatistics and/or pharmacy group remains unblinded, and in the event of serious adverse events, can provide data to the Data Safety Monitoring Board (DSMB) of a clinical trial.

Sample Size Although estimated numbers of patients for each phase of clinical trials are readily available, these numbers are generally applicable to common diseases. Rare diseases such as AIED, SSNHL, Meniere's Disease and the autoinflammatory diseases may, based on statistical parameters, require larger samples than may be available. In such cases, the trial may only be able to detect large differences between treatment groups as opposed to more modest effect sizes if the sample size were larger. Given the rare nature of the diseases studied in immune-mediated hearing loss, sample size is of paramount consideration. The sample size needs to be low enough for realistic recruitment, but robust enough to detect differences between arms of the drug to be studied. The trap the investigative team may fall into is proposing an unrealistically large treatment effect between arms (either between active drug and placebo or between new drug and standard reference therapy (i.e. oral corticosteroids) in order to reduce sample size requirements based on the power calculation. Many of the studies performed to test new drugs in immune-mediated hearing loss have fallen just below primary efficacy targets rendering the overall trial as "unsuccessful".

Recruitment Recruitment for clinical trials in immune-mediated hearing loss can be exceedingly challenging, and requires multiple centers to participate. One important potential challenge to conducting such trials is to gain interest and commitment from many geographically separate sites in order to maximize recruitment. Study design and perceived benefit to the patient is critically important to recruitment. Here, investigators should view the study from both the participant's and referring physician's perspective. If the clinical trial is comparing a new drug to standard therapy without compelling evidence of efficacy of the new drug, it is unlikely recruitment would be successful. Furthermore, in the case of the sudden sensorineural hearing loss trial (Rauch et al. 2011), comparing oral to intratympanic therapy, recruitment challenges existed from referring physicians, as the referring physician had access to and could offer the same therapy in their office and receive reimbursement, or they could refer the patient to a participating investigator and potentially lose that patient and that revenue.

4   Data Collection/Quality

Subject Compliance In studies of intratympanic therapies, subject compliance is not an issue as the investigator at each site is administering the study medication. In studies of oral or daily injectable medications, subject compliance is more difficult to monitor. Some medications can be monitored by measuring blood levels of the study medication. For many biologic therapies, since the drug is intended to modulate the immune system, the drug and the endogenous proteins in the circulation may be indistinguishable making compliance monitoring more difficult. Counting remaining pills or syringes is routinely performed, but does not exclude the possibility that the participant is discarding the study medication. Ultimately, the study team must rely on the integrity of the participant for adequate data.

Response Variables Some trials may have a numerical value for response such as decibel, pure tone average (PTA) or Word Recognition Score (WRS). Some outcomes may be simpler dichotomous binary outcomes. However, some of the measures used in hearing loss research are numerical and might be consider "continuous" variables. However, it is not uncommon to define a binary outcome (i.e. response yes/no) which is a function of one or more variables (i.e. PTA of greater than x and WRS >y%). The investigators should make very attempt to understand and control for these variables *apriori* in the randomization process. Furthermore, restriction of certain activities during the course of the trial may also reduce response variables. Recording and restriction of certain concomitant medications known to interact with the study drug should be performed. Finally, response variables may be determined at the end of the trial, once the data is unblinded to the investigative team.

5   Reporting Results

Consort Diagrams: In order to increase transparency and reduce inadequate reporting of clinical trials, clinical trials should be reported according to a minimum set of recommended criteria as set forth in the CONSORT Statement. The acronym CONSORT stands for Consolidated Standards of Reporting Trials, and includes a 25 item checklist and flow diagram recommended for reporting clinical trial results (www.consort-statement.org).

Adverse Events During any trial, adverse events may occur. These adverse events may be from mild, moderate, severe or life-threatening. Severity for any condition can be graded using the NCI Common Terminology Criteria for Adverse Events (CTCAE) manual which is regularly updated, but allows for consistency of grading. Serious adverse events require immediate reporting to the local IRB and the study sponsor who, in turn, is responsible for reporting the event to the FDA. Furthermore, the adverse event must be designated whether it is attributed to the drug undergoing study, where the attributions to be considered are: definite, probable, possible, or unlikely. Ultimately it is up to the local site investigator to attribute whether the adverse event is the result of the study medication. All trials should have pausing and stopping rules based on the number of adverse events incurred. If this threshold is reached, the study would be unblinded to the DSMB for a decision whether the trial should be suspended for safety concerns. The following table is standardly used to define whether the adverse event is related to the study drug.

Adverse event severity

The intensity or severity of Adverse Events are as follows:

- *Mild*—awareness of sign or symptom, but easily tolerated. Not expected to have a clinically significant effect on the subject's overall health and well-being. Not likely to require medical attention
- *Moderate*—discomfort enough to cause interference with usual activity or affects clinical status. May require medical intervention
- *Severe*—incapacitating or significantly affecting clinical status. Likely requires medical intervention and/or close follow-up
- *Life Threatening*—patient is in imminent danger of death
- *Fatal*—results in death of patient

Adverse event attribution

Adverse events may be attributed to use of the experimental drug/investigational agent as follows:

- *Definite*—a clinical event, including a laboratory test abnormality, that is a known effects of the drug and/or procedure, there is a clear temporal association with the use of the drug and/or procedure, and there is improvement upon withdrawal of the drug if induced by drug, not procedure
- *Probable*—a clinical event, including a laboratory test abnormality, in which a relationship to the study drug and/or procedure seems probable because of such factors and consistency with known effects of the drug and/or procedure, a clear temporal association with the use of the drug and/or procedure, improvement upon withdrawal of the drug, lack of alternative explanations for the experience, or other factors
- *Possible*—a clinical event, including a laboratory test abnormality, with a reasonable time sequence to administration of the study drug and/or procedure, but which concurrent disease, procedure or other drugs or chemicals could not explain
- *Unlikely*—a clinical event, including a laboratory test abnormality, with a temporal relationship to administration of the study drug and/or procedure, which makes a causal relationship highly improbable and in which other factors suggesting an alternative etiology exist. Such factors include a known relationship of the adverse experience to concomitant medication, the subject's disease state, or environmental factors including common infectious diseases

Intention-to-treat (ITT) vs. per protocol Intention-to-Treat principle in clinical trials analyzes each subject in the group they were randomized to rather than how they were actually treated. For instance, in our early efficacy trial of anakinra for corticosteroid resistant AIED, seven out of ten participants improved by pure tone average, as these ten participants all received the requisite 84 days of anakinra, thereby resulting in a Per Protocol response rate of 7/10 or 70%. Twelve patients actually were recruited to, and commenced anakinra therapy, however two dropped out prior to day 28 because of intolerable injection site reactions, thus, in the ITT analysis 7 out of 12 patients responded to anakinra as measured by pure tone average, corresponding to a 7/12 or 56% response rate (Vambutas et al. 2014).

6   Completed Trials in IMED

In order to understand the rationale for new drug therapies in immune-mediated hearing loss, the potential molecular targets for these drugs is shown in the following diagram (Fig. 11.1). Notably, many of these agents act on different targets in the same signaling pathway.

AIED Several clinical trials have been conducted in AIED. Perhaps the earliest observational study, which would loosely be considered a retrospective analysis of an open label trial is the description of clinical benefit of the combination of steroids and cyclophosphamide put forth by McCabe in the late 1970s (McCabe 1979). Methotrexate has been shown to be no better than placebo in patients that responded to corticosteroids in a large clinical trial (Harris et al. 2003). Similarly, clinical trials of anti-TNF-α therapy have also shown limited benefit (Cohen et al. 2005; Matteson et al. 2005). In both the methotrexate and TNF trials, these agents were used following successful corticosteroid therapy. We have previously shown in corticosteroid responsive patients, that plasma TNF levels drop in those patients that respond to corticosteroids, thereby potentially removing the antigenic target needed for TNF therapies to be effective. Of the 70% of AIED patients that initially respond to corticosteroids, that response is lost over time with only 14% demonstrated improvement after 34 months follow-up (Broughton et al. 2004). We performed an early

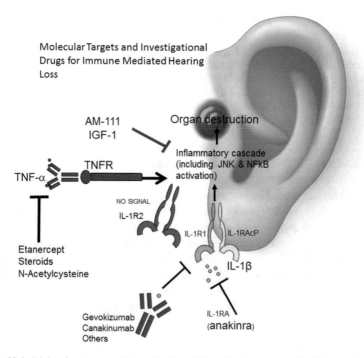

Fig. 11.1 Molecular targets and investigational drugs for immune mediated hearing loss

phase-open label clinical trial in corticosteroid resistant patients using anakinra, an interleukin-1 receptor antagonist. The trial was prematurely closed as the efficacy target was achieved before total anticipated subject recruitment occurred, and hearing improvement correlated with a reduction of plasma IL-1 (Vambutas et al. 2014).

SSNHL Because SSNHL is readily defined, and more common than AIED, more new drugs or new treatment modalities have been identified through clinical trials. Notably, given the proven efficacy of corticosteroids for SSNHL, placebo controlled trials are virtually non-existent. Comparators in clinical trials has been again the reference therapy corticosteroids. In a non-inferiority, multicentered phase 3 clinical trial comparing oral and intratympanic steroids for sudden deafness (NCT00097448), intratympanic therapy was deemed "non-inferior to traditional oral corticosteroids (Rauch et al. 2011). Other intratympanic therapies have also been studied for SSNHL. An inhibitor of JNK has been noted to be effective in a phase 2 clinical trial of SSNHL patients (Suckfuell et al. 2014). Interestingly, this study identified a higher than expected rate of spontaneous improvement in mild-moderate SSNHL, leading the investigators to determine efficacy in the severe-profound cohort of SSNHL participants. Intratympanic Insulin-like growth factor (IGF-1) was compared to intratympanic dexamethasone in patients with SSNHL that failed oral corticosteroid therapy in a multicentered clinical trial of 120 subjects, and IGF-1 was found to be superior to dexamethasone (66.7% vs. 53.6%, Japanese clinical trial registry number UMIN000004366) (Nakagawa et al. 2014). The IGF-1 signaling pathway is antagonistic to JNK signaling, and therefore may represent another method to inhibit JNK signaling (Yin et al. 2013). Furthermore, Presence of inflammatory mediators TNF and IL-1 may repress IGF signaling and induce resistance to IGF-1 (O'Connor et al. 2008). Recently, N-acetylcysteine was shown to be potentially more effective in SSNHL than corticosteroids (Chen and Young 2016). This is consistent with our observations that N-acetylcysteine lowers TNF levels in corticosteroid responsive patients (Pathak et al. 2015), and that when corticosteroids are used in combination with N-acetylcysteine, the clinical response was greater than with steroids alone (Angeli et al. 2012). Of note, these observations were made in studies that were retrospective observational studies rather than prospective clinical trials.

Meniere's Disease Corticosteroids have been studied in the treatment of Meniere's disease. Here, corticosteroids have been used for the control of vertigo, administered through an intratympanic route. In a phase 2 study of OTO-104, a proprietary injectable steroid formulation that, upon intratympanic administration, transforms from a liquid to a gel that provides sustained release of corticosteroids over the round window was tested to see if control of vertigo was superior to placebo. Efficacy for control of vertigo exceeded placebo for several secondary endpoints, but missed significance for improvement at 90 days (primary endpoint), with the drug achieving 61% compared to placebo at 43% (Lambert et al. 2016). These results are in contrast to an earlier study of dexamethasone perfusion of the inner ear for Meniere's that failed to demonstrate benefit (Silverstein et al. 1998).

Autoinflammatory Diseases Notably, no prospective clinical trials have been performed specifically to determine the effect of IL-1 inhibition on hearing restoration in this family of diseases. Perhaps, the most well characterized, genetically inherited, autoinflammatory disease associate with sensorineural hearing loss has been Muckle-Wells Syndrome (Vambutas and Pathak 2016). Observational studies in Muckle Wells Syndrome patients demonstrate 25% improve hearing at 4000 Hz or below, however this has not been rigorously tested (Kuemmerle-Deschner et al. 2015).

7 Conclusions

In summary, there are a number of considerations that should be addressed in designing effective clinical trials for rare diseases. As seen in the trials completed to date, a number of well-designed trials missed their efficacy targets, possibly due to lack of efficacy, but potentially because of issues of a requisite large effect size for studies with limited numbers of available patients.

References

Angeli SI, Abi-Hachem RN, Vivero RJ, Telischi FT, Machado JJ. L-N-Acetylcysteine treatment is associated with improved hearing outcome in sudden idiopathic sensorineural hearing loss. Acta Otolaryngol. 2012;132:369–76.

Arriaga MA, Goldman S. Hearing results of intratympanic steroid treatment of endolymphatic hydrops. Laryngoscope. 1998;108:1682–5.

Barrs DM, Keyser JS, Stallworth C, McElveen JT Jr. Intratympanic steroid injections for intractable Meniere's disease. Laryngoscope. 2001;111:2100–4.

Broughton SS, Meyerhoff WE, Cohen SB. Immune mediated inner ear disease: 10-year experience. Semin Arthritis Rheum. 2004;34:544–8.

Chen CH, Young YH. N-acetylcysteine as a single therapy for sudden deafness. Acta Otolaryngol. 2016;137(1):58–62.

Cohen S, Shoup A, Weisman MH, Harris J. Etanercept treatment for autoimmune inner ear disease: results of a pilot placebo-controlled study. Otol Neurotol. 2005;26:903–7.

Goldbach-Mansky R, et al. Neonatal-onset multisystem inflammatory disease responsive to interleukin-1beta inhibition. N Engl J Med. 2006;355:581–92.

Harris JP, Weissman MH, Derebery JM, et al. Treatment of corticosteroid-responsive autoimmune inner ear disease with methotrexate: a randomized controlled trial. JAMA. 2003;290:1875–83.

Hietikko E, Sorri M, Mannikko M, Kotimaki J. Higher prevalence of autoimmune diseases and longer spells of vertigo in patients affected with familial Meniere's disease: a clinical comparison of familial and sporadic Meniere's disease. Am J Audiol. 2014;23:232–7.

Hughes GB, Freedman MA, Haberkamp TJ, Guay ME. Sudden sensorineural hearing loss. Otolaryngol Clin N Am. 1996;29:393–405.

Kuemmerle-Deschner JB, et al. Two-year results from an open-label, multicentre, phase III study evaluating the safety and efficacy of canakinumab in patients with cryopyrin-associated periodic syndrome across different severity phenotypes. Ann Rheum Dis. 2011;70:2095–102.

Kuemmerle-Deschner JB, et al. Hearing loss in Muckle-Wells syndrome. Arthritis Rheum. 2013;65:824–31.

Kuemmerle-Deschner JB, Koitschev A, Tyrrell PN, et al. Early detection of sensorineural hearing loss in Muckle-Wells-syndrome. Pediatr Rheumatol Online J. 2015;13:43.

Lambert PR, Carey J, Mikulec AA, LeBel C, Otonomy Meniere's Study, G. Intratympanic sustained-exposure dexamethasone thermosensitive gel for symptoms of Meniere's disease: randomized phase 2b safety and efficacy trial. Otol Neurotol. 2016;37(10):1669–76.

Matteson EL, Choi HK, Poe DS, Wise C, Lowe VJ, McDonald TJ, Rahman MU. Etanercept therapy for immune-mediated cochleovestibular disorders: a multicenter, open-label, pilot study. Arthritis Rheum. 2005;53:337–42.

McCabe BF. Autoimmune sensorineural hearing loss. Ann Otol Rhinol Laryngol. 1979;88:585–9.

Mirault T, Launay D, Cuisset L, Hachulla E, Lambert M, Queyrel V, Quemeneur T, Morell-Dubois S, Hatron PY. Recopvery from deafness in a patient with Muckle-wells syndrome treated with anakinra. Arthritis Rheum. 2006;54:1697–700.

Nakagawa T, et al. A randomized controlled clinical trial of topical insulin-like growth factor-1 therapy for sudden deafness refractory to systemic corticosteroid treatment. BMC Med. 2014;12:219.

O'Connor JC, et al. Regulation of IGF-I function by proinflammatory cytokines: at the interface of immunology and endocrinology. Cell Immunol. 2008;252:91–110.

Ombrello MJ, Kastner DL. Autoinflammation in 2010: expanding clinical spectrum and broadening therapeutic horizons. Nat Rev Rheumatol. 2011;7:82–4.

Pathak S, Stern C, Vambutas A. N-Acetylcysteine attenuates tumor necrosis factor alpha levels in autoimmune inner ear disease patients. Immunol Res. 2015;63:236–45.

Rauch SD, et al. Oral vs intratympanic corticosteroid therapy for idiopathic sudden sensorineural hearing loss: a randomized trial. JAMA. 2011;305:2071–9.

Silverstein H, Isaacson JE, Olds MJ, Rowan PT, Rosenberg S. Dexamethasone inner ear perfusion for the treatment of Meniere's disease: a prospective, randomized, double-blind, crossover trial. Am J Otol. 1998;19:196–201.

Stahle J, Stahle C, Arenberg IK. Incidence of Meniere's disease. Arch Otolaryngol. 1978;104:99–102.

Suckfuell M, Lisowska G, Domka W, et al. Efficacy and safety of AM-111 in the treatment of acute sensorineural hearing loss: a double-blind, randomized, placebo-controlled phase II study. Otol Neurotol. 2014;35:1317–26.

Vambutas A, Pathak S. AAO: Autoimmune and Autoinflammatory (Disease) in Otology: what is new in immune-mediated hearing loss. Laryngoscope Investig Otolaryngol. 2016;1(5):110–5.

Vambutas A, et al. Early efficacy trial of anakinra in corticosteroid-resistant autoimmune inner ear disease. J Clin Invest. 2014;124:4115–22.

Wellman-Labadie O, Zhou Y. The US orphan drug act: rare disease research stimulator or commercial opportunity? Health Policy. 2010;95:216–28.

Wilson WR, Byl FM, Laird N. The efficacy of steroids in the treatment of idiopathic sudden hearing loss. A double-blind clinical study. Arch Otolaryngol. 1980;106:772–6.

Yamazaki T, Masumoto J, Agematsu K, Sawai N, Kobayashi S, Shigemura T, Yasui K, Koike K. Anakinra improves sensory deafness in a japanese patient with Muckle-Wells syndrome, possibly inhibiting the cryopyrin inflammasome. Arthritis Rheum. 2008;58:864–8.

Yin F, Jiang T, Cadenas E. Metabolic triad in brain aging: mitochondria, insulin/IGF-1 signalling and JNK signalling. Biochem Soc Trans. 2013;41:101–5.

Index

© Springer International Publishing AG, part of Springer Nature 2018
V. Ramkumar, L. P. Rybak (eds.), *Inflammatory Mechanisms in Mediating Hearing Loss*, https://doi.org/10.1007/978-3-319-92507-3

Printed in the United States
By Bookmasters